# Gravitomagnetism

Gravity's Secret

# Gravitomagnetism

## Gravity's Secret

DR RONALD A. EVANS

Copyright © 2021 Dr Ronald A. Evans

The moral right of the author has been asserted.

Apart from any fair dealing for the purposes of research or private study, or criticism or review, as permitted under the Copyright, Designs and Patents Act 1988, this publication may only be reproduced, stored or transmitted, in any form or by any means, with the prior permission in writing of the publishers, or in the case of reprographic reproduction in accordance with the terms of licences issued by the Copyright Licensing Agency. Enquiries concerning reproduction outside those terms should be sent to the publishers.

Matador
9 Priory Business Park,
Wistow Road, Kibworth Beauchamp,
Leicestershire. LE8 0RX
Tel: 0116 279 2299
Email: books@troubador.co.uk
Web: www.troubador.co.uk/matador
Twitter: @matadorbooks

ISBN 978 1 80046 217 5

British Library Cataloguing in Publication Data.
A catalogue record for this book is available from the British Library.

Printed and bound in the UK by TJ Books LTD, Padstow, Cornwall
Typeset in 10.5pt Aldine by Troubador Publishing Ltd, Leicester, UK

Matador is an imprint of Troubador Publishing Ltd

To the university research scientists
who worked on Project Greenglow
and to the BAE Systems engineers
who made it happen.

# CONTENTS

| | | |
|---|---|---|
| Acknowledgements and Introduction | | ix |
| 1. | Mysterious gravity and its secret | 1 |
| 2. | Overcoming gravity | 5 |
| 3. | Galileo updates Aristotle | 9 |
| 4. | Kepler's elliptical planetary orbits | 13 |
| 5. | Newton's masterpiece | 18 |
| 6. | Measuring vast distances and the speed of light | 32 |
| 7. | Gravity is a universal force | 39 |
| 8. | Relativity – An important diversion | 44 |
| 9. | The Lorentz transform and special relativity | 52 |
| 10. | Gravity and acceleration are equivalent | 63 |
| 11. | Space and Minkowski's space-time | 70 |
| 12. | Einstein's early triumphs | 79 |
| 13. | Gravity, space and space-time curvature | 89 |
| 14. | An outline of Einstein's general relativity | 96 |
| 15. | The means of gravity control remains unknown | 113 |
| 16. | Fluid dynamics with sources and sinks | 120 |
| 17. | The source–sink combination | 126 |
| 18. | The line vortex and rotating flows | 129 |
| 19. | Vortex rings | 134 |
| 20. | Fixed line vortex in a uniform stream | 138 |
| 21. | Aerodynamic flight requires vortex control | 142 |

| | |
|---|---:|
| 22. Magnetism and Electricity | 146 |
| 23. The electric vortex and electromagnetism | 155 |
| 24. The electric vortex in a magnetic field | 161 |
| 25. Faraday and the control of electromagnetism | 165 |
| 26. Maxwell, Hertz and the revolution in communications | 172 |
| 27. Heat and Thermodynamics | 181 |
| 28. The steam-driven Industrial Revolution | 194 |
| 29. The Thermodynamic laws | 198 |
| 30. Acoustics, diffusion and extended thermodynamics | 212 |
| 31. Stars, black holes and thermodynamics | 225 |
| 32. Faraday's gravity experiments | 232 |
| 33. Extending Newton's gravity and the gravity vortex | 249 |
| 34. Detecting magnetism and gravitomagnetism | 270 |
| 35. NASA Gravity Probe-B | 282 |
| 36. Gravity waves and their detection | 287 |
| 37. Some speculations concerning gravity | 295 |
| 38. Project Greenglow | 319 |
| 39. The book and the BBC *Horizon* programme | 335 |
| 40. The search for gravitomagnetism | 343 |
| | |
| Appendix: Extended gravity | 366 |
| Names index | 375 |
| Subject index | 380 |

# ACKNOWLEDGEMENTS AND INTRODUCTION

In 2015, I published my book, *Greenglow & The Search for Gravity Control*. Afterwards, I took part in a BBC *Horizon* documentary which was largely based on the contents of the book. The programme was transmitted in March 2016. That, I thought, was that. I had no intention of writing another book.

In early 2017, a friend of my family, Dr Peter Roach, told us that he had given a chemistry presentation at the Savile Club in Mayfair, London, which had gone extremely well. Peter suggested that I might like to contact the club secretary, Ken Allen, to see whether they might be interested in a talk by me about gravity. I contacted the secretary, who told me that they ran a series of talks at the club under the theme, *Science at the Savile Club*. I learnt that the club had several distinguished scientists as members and that Professor Ernest Rutherford, of atomic physics fame, had been a member. However, most club members were not scientists, although they had a keen interest in the various fields of science. So, a presentation should not be too deep. Ken suggested that I might like to prepare a draft slide presentation and let him see it, so that he could assess whether it was suitable for their series of talks. And so, I set to work, little realising what my effort would lead to.

After several months, I had created a lot of PowerPoint slides, some of which were far too busy for a presentation. I began writing notes for each slide. I was quite happy, bumbling along, getting my ideas in order. Gradually, it dawned on me that what I was preparing was not a presentation, but another book. It was different from my first book, in which I had described the work undertaken by university academics for BAE Systems' *Project Greenglow*. The new book was my own view (although greatly influenced by others) on the search for gravity control, which I believed should be focussed on the newly discovered force field of gravitomagnetism. Gravitomagnetism arises when masses and their associated gravity fields move. Although the slides contain far too much detail to form the presentation for a talk, each chapter of the new book is based on a single slide. So, I acknowledge that Peter Roach and Ken Allen were responsible for me starting on my new book.

The book is separated into two parts. The first part (Chapters 1–14) describes the history of gravity, from the ancient Greek view that bodies naturally fall down towards the centre of the Earth, that being the centre of the Universe; via the important developments during the Renaissance when gravity's role in the Solar System was uncovered, up to Einstein's explanation that gravity is due to the curvature of space-time. Einstein published his paper on general relativity more than one hundred years ago and we are still waiting for the breakthrough that will lead to gravity control. We are at an impasse. We are unable to control gravity and no one in the world can see a way forward. Well, no one in the white world, at least. There are rumours that in the black world progress has been made. The second part of the story (Chapters 15–40) investigates ways around the impasse using mathematical patterns, or analogues, as a guide. The watchword for this part of the book is "Look for the vortex". This approach offers movement in the search for gravity control. Faraday's gravity experiments are examined in detail, and the results of more recent gravity experiments are also explored. Finally, some new experimental ideas involving gravitomagnetism are proposed. If they are carried out and are successful, they should eventually open up the way leading to gravity control. Only then can the last part of the story of gravity be written.

Thus, the important end goal for the book was to devise some ground-based experiments which might reveal the role played by gravitomagnetism in gravity control. I contacted George Seyfang, my former work colleague and retired engineer from BAE Systems, and we talked about my idealised experiments, some involving fibre optics. George is a dab hand at simple experiments and he conducted some initial tests at home. But my ideas for gravity experiments required more technical expertise. So, through Tony Cuthbert (an inventor based in Wales), I contacted Dr Frank Kvasnik, a retired Senior Lecturer from the University of Manchester Institute of Science and Technology. Frank has experience working with fibre optics. George, Frank and I had an interesting meeting in August 2017, where we sorted out some of the experimental ideas. The main difficulty was getting access to a laboratory and funding. Time passed.

In October 2017, George and I attended a *Moonclub* meeting initiated by the late Professor John Allen. John was the technical consultant for *Project Greenglow* and was formerly the Chief of Future Projects at British Aerospace (BAE Systems) Kingston, the original home of the Harrier aircraft. John had a life-long interest in the possibility of controlling gravity. Other attendees were Dr Mike Provost (ex-Rolls-Royce and an early *Greenglow* supporter), Professor Alan Wickens (formerly Head of British Rail Research) and Mike Rockall (ex-Barclay's Bank Director and entrepreneur). During the meeting, I used some of the PowerPoint slides that I had started to prepare for the Savile Club, which now formed the nucleus for my new book. It was an interesting discussion meeting but, in my view, the outcome of the meeting was inconclusive. Unfortunately, the subject of gravity control is tainted by *anti-gravity*, which makes it very difficult for academics to show any interest in the subject, and it puts off possible funding agencies. More time passed.

I have remained in occasional correspondence with Dr H. Ron Harrison, who wrote the Foreword to my earlier book. Ron was involved in the arguments with Professor Eric Laithwaite about the inertial effects associated with gyroscopes. Ron has his own view of gravity, developed in his book entitled *Gravity; Galileo to Einstein and Back*. Ron is keen that I should mention his paper, *Post Newtonian Gravity, a new simpler approach* (Int. J. *Space Science and Engineering*, Vol. 4, No. 2, 2016).

I have also kept in occasional contact with Rob Chambers (ex-British Aerospace Plymouth, formerly Sperry Gyroscopes), a supporter of the *Greenglow* Programme from its inception, who keeps me updated on stories related to gravity.

From time to time, I have sought advice on various matters involving gravity from Professor Robin Tucker of Lancaster University, who was the *Project Greenglow* Academic Adviser. Robin is always helpful, although he may not agree with some of the views expressed in my book.

Only a handful of MoD scientists took an overt interest in BAE Systems' *Project Greenglow*. Among these were Dr Gari Owen and Dr Andrew May. Both have expressed support for my latest book effort. Andrew read through the chapter giving an outline of general relativity and suggested changes, which I made.

Darren Moss, a family friend with a technical background, read through an early draft of my book. He pointed out errors and some passages in the text which he found difficult to read and suggested that they needed improving. This was not the mathematical content, which he felt he could just skip over, without losing too much of the flow. I made use of his comments.

On encountering a mathematical equation in this book, some people's eyes will glaze over. Although I'm a mathematician, this happens to me, too, when I look in some technical books. My advice is the same as Darren's; skip over the maths and keep reading, as you must be interested in the subject of gravity control. Others, who can interpret the mathematical hieroglyphs, will probably only give the equations a passing glance and carry on reading, too. They may come back to them later. In my view, the equations are part of the structure holding up our understanding of nature. They are there, if you want them.

Trying to get permission to use copyrighted photos was time-consuming and I thank those people who helped me, including Andrea Kay (BAE Systems), Steve Crabtree (BBC), Noah McMahon (Zero G Corporation), Geoff Russell (Leonardo Helicopters) and Catalina George (Virgin Balloon Flights).

The drawing on the front cover of this book and the illustrations of the famous scientists are by Dave Windett (www.davewindett.com). His work adds a bit of colour and lightness (dare I say levity) to my book, which can be heavy (an analogue suggesting that mathematics has gravity) in a few places. As the 19th century Oxford mathematician Charles Dodgson (Lewis Carroll) wrote in his children's book, *Alice in Wonderland*:

"*and what is the use of a book,*" *thought Alice,* "*without pictures or conversations?*"

Of course, my children (Nicholas, Claire, Richard and Emma) have got used to my obsession with gravity by now. However, they always show an interest in how I am doing and help me if they can, especially Nicholas with computing.

In trying to get my earlier semi-technical book published, I spent ages writing to UK agents and publishers, nearly fifty in all, before I gave up and self-published with Troubador Publishing Ltd. This time, I only half-heartedly tried a few agents and publishers. As expected, I got no interest, so I turned once again to my trusted self-publishing company, Troubador.

There is a noticeable dearth of scientists studying the possibility of gravity control. It may be that following on from Einstein's theory of general relativity is too daunting a prospect. Or the fact that the force of gravity is so weak that only astronomical experiments, at great expense, seem worth pursuing. Or it may be that fundamental gravity research is blighted by its association with the notion of anti-gravity. For whatever reason, few scientists are willing to risk their careers in the search for a means of gravity control. Conjuring up new ground-based experiments to try in the search for a breakthrough in understanding is very difficult. It needs imagination and risks absurdity. Those few who have dared to carry out investigations linked with fundamental gravity control research have not been successful, and they have received little encouragement from the rest of the scientific community for their efforts. That is not to say that most scientists are not interested; they are.

So, lastly, I acknowledge the work carried out by those scientists who dared to take part in BAE Systems' *Project Greenglow* and the support from the BAE engineers who made it happen.

As soon as there is a breakthrough in understanding, there will be a great rush by many scientists and engineers eager to enter the new field of study. If history is anything to go by, once the breakthrough occurs there will be rapid advances made, carried out by bright-minded academics, inventors and entrepreneurs. I'm convinced that gravity control is on the way. For the moment, I'm lucky to be in at the beginning of the study to control gravity, before being left behind in the rush.

**Ron Evans**
St Anne's-on-the-sea
Lancashire
March 2020

CHAPTER 1

# MYSTERIOUS GRAVITY AND ITS SECRET

Most of us have played with a magnet. It has two ends, or poles, often labelled north and south. The north pole of one magnet will repel the north pole of another across empty space. We have felt that curious repulsive effect as the two north poles avoid coming together. Magnetism is a mysterious invisible force. Similarly, two south poles repel each other, but a north pole will attract a south pole. Either end, or pole, of a magnet will attract certain metal objects, especially those made of iron, across empty space. If the objects are free to move, they will stick to the magnet. The magnetic influence in space around a magnet can be made visible if we place the magnet on a sheet of paper, sprinkle iron filings around the magnet and give the paper a jerk. We say that the pattern exhibited by the iron filings shows the existence of a magnetic field.

Likewise, or analogously, a mass has a gravity field, although it only has one pole. Gravity is another mysterious invisible force which extends across empty space, attracting all objects with mass. Mass size is important. Large astronomical bodies have large gravity fields and attract masses of any size. Small bodies, say, people-sized, have small gravity fields and do not noticeably attract other small bodies. Young babies placed on a blanket on the floor lie stuck to the Earth's surface, like an iron object stuck to a magnet. At a very early age, we are all aware of something pulling us down to the ground and we spend our first years struggling to overcome its pull and stand upright. Gravity has an effect on us from birth until death. It affects the way we grow, particularly our muscle development. The absence of gravity, as astronauts are aware, causes medical problems.

Young children quickly learn to respect gravity. Falling over can hurt and the further the fall, the more dangerous the consequences. Children also learn that they can defeat gravity by jumping, but it's only a very brief victory. At present, we are slaves to gravity, although we know ways to overcome it. Sometime in the future, we might become gravity's master and make it do our bidding. Instead of unchangeable gravity, we may learn how to change gravity. But we must be careful, as it is gravity that holds the Solar System together and it would be very dangerous if our meddling on a large scale disturbed the natural balance of the planetary motions.

There are benefits to living in an environment dominated by gravity. Lifting a mass up requires us to do some work against gravity, but the mass then has some potential energy. Although we might not realise it, the energy is stored in the Earth's gravity field. Knocking an object off its perch and causing it to fall causes gravity to do some work. We can exploit this result by channelling water to fall over the blades of a water wheel, causing it to turn and using the rotational motion to work machines for us. Fortunately, the Sun's radiant energy levitates the water up for us in the first place, via ocean evaporation. Even here, gravity plays a part, through buoyancy and water vapour being less dense than air at ground level. Rainfall completes the circuit, filling lakes and rivers and allowing water to be channelled. Nowadays, water wheels have largely been replaced by hydroelectric schemes, where the energy stored in the Earth's gravity field can be extracted from the falling water and converted into electrical power. So, at the moment, we know how to make use of gravity, even though we can't control it.

As a retired engineer from the Aerospace business, let me quote from Arthur C. Clarke's book, *Profiles of the Future*, to partly set the scene for this book. From Chapter 5 of his book, entitled *Beyond Gravity*, he considers the term "Space Drive":

> *It is an act of faith among science-fiction writers, and an increasing number of people in the astronautics business, that there must be some safer, quieter, cheaper and generally less messy way of getting to the planets than the rocket.*
>
> *It may seem a little premature to speculate about the uses of a device which may not even be possible, and is certainly beyond the present horizon of science. But it is a general rule that, whenever there is a technical need, something always comes along to satisfy it – or by-pass it. For this reason, I feel sure that eventually we will have some means of either neutralizing gravity or overpowering it by brute force. In any event, it will give us both*

*levitation and propulsion, in amounts determined only by the available power.*

The Space Drive is only one aspect of the technology that will be opened up to us once we learn how to control gravity. There will be many other developments, some of which haven't occurred to us yet.

It may seem strange, but gravity is actually an extremely weak force, especially when compared to the electromagnetic force. It needs a massive object, such as a star or planet, to create a noticeable gravity field. The Earth's gravity field itself has minor surface variations, due to changes in surface geology, which can be measured using very sensitive gravimeters and gradiometers. One important effect is the variation of surface gravity with latitude, due to the Earth's rotation and radius, which has nothing to do with mass. This is a dynamic effect and this book is about gravity dynamics. A recent NASA satellite experiment has shown that moving mass is linked with the little-known force field called gravitomagnetism. It is gravity's secret dynamic companion. It is speculated that control of gravity lies with the control of gravitomagnetism.

I am very aware that weak electromagnetic radiation can be greatly amplified using quantum mechanical means. It seems likely, to me, that weak gravity radiation may be amplified in a similar way, once we have a quantum gravity theory in place. This may lead to gravity beams like lasers. But, to start, we need to get to grips with gravity field dynamics.

All research has to start somewhere, so we will begin our investigation of gravitomagnetism by collecting together our knowledge about gravity. Then we will investigate a speculative theory for gravitomagnetism and we will propose and consider ideas for various experiments, based on the idea of gravitomagnetic fields, to test the theory. This book goes no further than that.

So, the major purpose of this book is, firstly, to stimulate interest in the idea of gravity control, secondly, to inspire others to set in motion a campaign to secure funding for those experiments deemed worthy of pursuing and, thirdly, to encourage scientists to carry out those experiments. All this will have to be done by younger people interested in this exciting, not yet ripe, futuristic area of science. Successful experiments will lead to further experimentation, which will eventually lead to the technology needed to create gravity beams and to build Space Drives. But these and other developments are for the future and are far outside the remit of this book.

## Mysterious gravity and its secret

- Awareness of gravity from birth
- All masses have a gravity field
- Masses attract other masses
- Gravity holds us pinned to the ground
- Lack of gravity affects health
- Energy can be extracted from a gravity field
- Gravity can be overcome but not controlled
- At the moment gravity seems to be unchangeable
- Gravitomagnetism may lead the way to gravity control

CHAPTER 2

# OVERCOMING GRAVITY

We now have vehicles that employ means to overcome gravity. Most of them rely on the presence of an atmosphere. For example, a balloon can levitate in the air, defeating gravity. The reasoning behind this effect was first explained by Archimedes, an ancient Greek scientist, in the 3$^{rd}$ century BC. However, Archimedes was more concerned with fluids, and his buoyancy principle explained why some bodies float on the surface of water and are not pulled under by gravity. At the time, no thought was given to floating a body in the air and, yet, people must have been aware that hot air rises. It took another 2,000 years before the idea of floating a balloon of hot air in the colder atmospheric air was realised. The Montgolfier brothers, Joseph and Jacques, first flew their hot air balloon in 1782. Since then, ballooning has undergone various developments, leading to airships, filled with lighter than air gas, for military and commercial applications. Hot air balloon rides have become a popular form of entertainment for those seeking anti-gravity thrills. These are offered by a number of concerns in the UK, including Virgin Balloon Flights.

Another means of defying gravity is with a kite. The Chinese suspended look-outs from man-carrying kites deployed from ships at sea in the 13$^{th}$ century; the purpose being to spot the enemy first and take advantage of the knowledge. The simple one-piece kite is really a fixed wing which develops lift in a strong wind. Kites led on to gliders with a main wing and other smaller wings for stability and control. To get a glider to lift in stationary air requires some forward motion to create a wind over the main wing; hence the catapult launch. But this does not provide a continuous means of propulsion. A powered rotating wing, or airscrew, develops a thrust roughly at right angles to its plane of rotation. In effect, the airscrew sucks air in from the front and pushes it out at the back, the reaction being to thrust the airscrew forward. It's only a variation of the Archimedean screw, which has been around for several thousand years, which is used to lift water up from rivers. The airscrew was first tried out for balloon propulsion, using hand-cranking. Later, a steam-powered airscrew was fitted to a glider, propelling it forward and causing it to lift. It wasn't a great success; the engine being too heavy and the airscrew thrust too weak, but it gave a hint of the future. The development of the internal combustion engine in 1885, by the German engineers Gottlieb Daimler and Wilhelm Maybach, provided the rotary power needed for the airscrew. The Wright brothers, Orville and Wilbur, built a lightweight combustion engine to turn a pair of airscrews (now called propellers), which they fitted to their Flyer I glider, and the history of manned powered flight began in the United States in December 1903.

The helicopter is a body with no fixed wings for lift, that relies on a large powered rotating wing, or rotor, in the horizontal plane for vertical lift to overcome gravity. A small rotary wing, with its plane perpendicular to the main rotor, counters any rotation of the body of the vehicle. In the UK, helicopters were built by Westland, which is now part of the global company Leonardo Helicopters. A Second World War variation of the 13th century Chinese kite-borne observer was the observer sat in an auto-gyro, or unpowered helicopter, towed behind some German submarines. As the wind passed through the auto-gyro rotor, it rotated and developed lift, raising the observer aloft.

The development of the jet engine led to an increase in thrust. In one form of engine, a number of small turbine blades, are mounted on a common axle within a duct. As the blades rotate, air is sucked in at the front of the duct which is combined with other chemical vapours and the mixture ignited, creating a powerful gas flow which is ejected from the rear of the duct. The reaction is a powerful forward thrust on the engine. The BAE Systems Harrier aircraft has a single jet engine with a pair of intake ducts. It also has a pair of swivelling exhaust ducts which enables the jets to be aimed downwards. The upward reaction is enough to lift the aircraft vertically against the force of gravity, allowing it to hover.

Finally, we have the rocket. This is a self-contained reaction engine which can operate outside of the Earth's atmosphere. In July 1969, NASA's

Apollo 11 took men to the Moon. Since then, rockets have been used to send probes to explore the planets and their moons in the Solar System.

None of today's devices used to overcome gravity are able to control gravity and make use of the phenomenon. During the 18th century, the famous American scientist Benjamin Franklin commented that he was sorry that he had been born so soon when so much rapid progress was being made in the sciences, because he wouldn't live to see the results. He mused: *Imagine the power that man will have over matter a few hundred years from now. We may learn how to remove gravity from large masses and float them over great distances.* We are not there, yet. The means of gravity control is still a mystery.

# Overcoming gravity

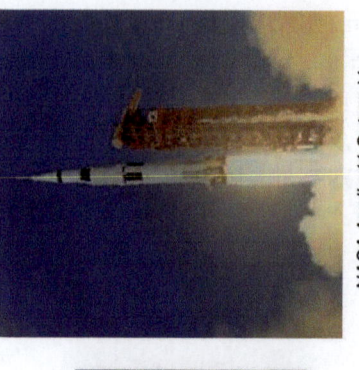

Moon Rocket

NASA Apollo 11 Saturn V

Benjamin Franklin 1776

### Helicopter

Leonardo AW169 air ambulance
Copyright Leonardo Helicopters

### Hot air balloon

Photo courtesy of ©Virgin Balloon Flights

### Vertical take-off and landing aircraft

Harrier (copyright BAE Systems) / Ray Troll

CHAPTER 3

# GALILEO UPDATES ARISTOTLE

Aristotle, the famous Greek philosopher, lived in the 4th century BC. He was a student of Plato and later the tutor of Alexander the Great. Aristotle was held in such high regard for his vast encyclopaedic knowledge about natural phenomena that his recorded views went unchallenged in Europe and the Middle East for nearly 2,000 years after his death. It was Aristotle's view that the Universe was a perfect sphere, with the Earth at its centre. Matter was thought to be composed of four elements; namely earth, air, fire and water. The Universe itself was filled with a fifth element called the ether. Aristotle argued that it was a natural property of earth and water to fall in straight lines towards their proper place at the centre of the Universe. They possessed gravity, or heaviness, whereas air and fire were displaced straight upwards as they possessed levity, or lightness. Aristotle was also of the opinion that heavier bodies fell faster than lighter ones.

With the Renaissance in Europe, scientists began to question some of Aristotle's assertions. In Italy, the mathematical physicist Galileo Galilei was dubious about the idea that heavier objects fell more quickly than lighter ones. He imagined sawing a cannonball in half and dropping a half of it. According to Aristotle, the half cannonball would not fall as fast as the whole cannonball. And yet, if the two cannonball halves were joined together with a thin thread and dropped, they would not fall more slowly than the whole cannonball, which contradicted Aristotle's view.

Galileo learnt of the work by the Dutch-Belgian scientist Simon Stevin, who had dropped balls from a tower to check on Aristotle's view that heavier balls fell more quickly than lighter ones. Legend has it that in 1590, Galileo repeated Stevin's experiment by dropping a musket ball and a cannonball from the Leaning Tower of Pisa and confirming that they fell together. Aristotle was wrong!

Galileo reasoned that since a body started with zero velocity when it was released, it must accelerate as it started to fall; but how could he measure the acceleration? Timing methods in those days were very crude, so timing the fall of a speeding body as it fell over fixed distances was difficult. Then Galileo learnt of another of Simon Stevin's experiments; that of rolling a ball down an inclined plane where, since the ball did not move so swiftly, its progress could be timed. In 1603, using a water clock, Galileo carried out a careful experiment, timing the distance a bronze ball moved as it rolled down a grooved inclined plane. The results showed that the ball accelerated. Moreover, balls of different weight had the same acceleration. By extension, Galileo concluded that a freely falling body must accelerate and, moreover, the value of the acceleration was independent of the body's weight. All falling bodies fell with the same acceleration, whatever their weight. The cannonball and the musket ball fell together in his Pisa Tower experiment.

Nowadays, we recognise that the Earth attracts all objects towards its centre with a gravitational attraction we label as **g**. We are all stuck on the Earth's surface, like an iron nail stuck to the surface of a magnet. If a raised object is released, it will accelerate downwards. Near to the Earth's surface, at roughly zero height, the acceleration due to the Earth's gravitational attraction is labelled as $g_0$.

Scales for weighing objects, especially valuable items such as gold, existed at least 5,000 years ago, and they appear on wall paintings in ancient Egyptian tombs. If the weight of an object is known and it is divided by $g_0$ we obtain a property of the object called its mass. When we hold an object we can feel its weight trying to force our hand downwards. So, weight is a force equal to the object's mass times the gravitational acceleration $g_0$. In the International System (SI) of units, force is measured in Newtons (N), mass is measured in kilograms (kg), and acceleration is measured in metres per second per second (m/s²). At the Earth's surface, the downward acceleration $g_0$ experienced by all free falling objects is about 9.8 m/s².

For historical reasons, we usually measure weight in terms of mass units. So, a person weighing 70 kg (11 stone) on the Earth's surface actually has a weight of 70 × 9.81 = 686 Newtons. On the International Space Station (ISS), the same person would still have a mass of 70 kg but would have zero weight.

Galileo maintained an extensive list of contacts with scientists interested in making progress in physics, both in Italy and abroad. He made use of other scientists' experiments, always checking on the veracity of the claimed results and absorbing any helpful theories. This applied to the telescope. He didn't invent it, but he quickly understood the principle of its working, improved on its design and used it to great purpose in his astronomical studies. Galileo was a good teacher and had a talented group of students to help him. Most importantly, he wrote a number of books on physics (some published abroad), in which he challenged some of the accepted ideas and discussed new ideas, showing that the way forward to a better understanding was through experimentation.

Thus, Galileo's special contribution to scientific progress is that he stressed the need for experiments to test all ideas about the workings of natural physical phenomena and the need for the results to be made available for others to scrutinise and check for themselves.

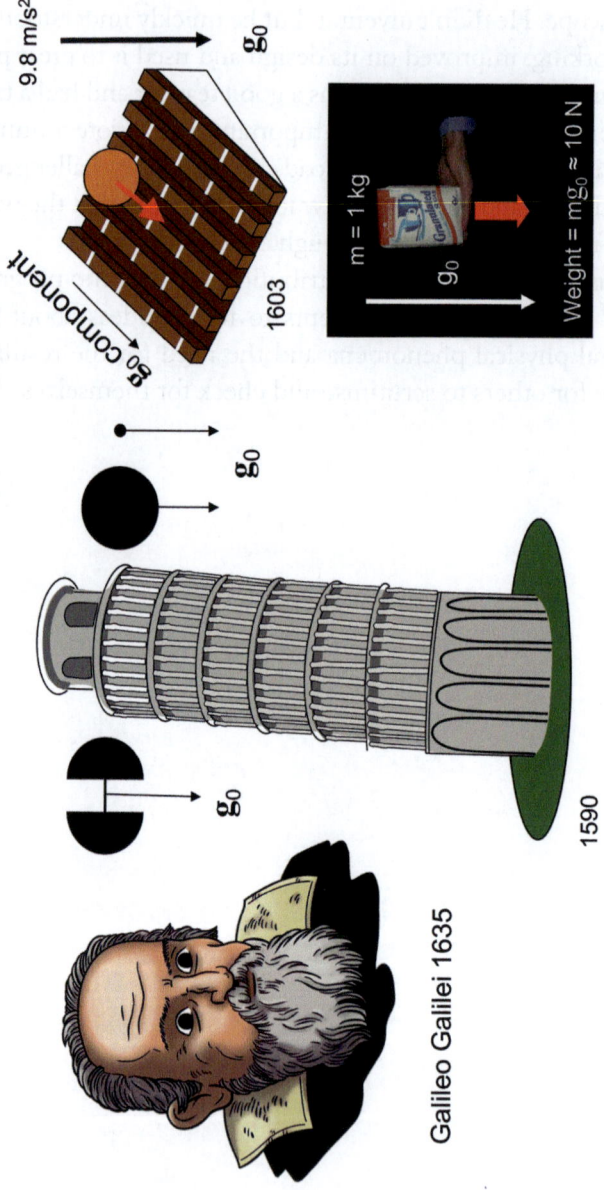

CHAPTER 4

# KEPLER'S ELLIPTICAL PLANETARY ORBITS

Galileo challenged Aristotle's view of an Earth-centred universe. In 1597, Galileo received a book, entitled *Mysterium Cosmographicum*, written in Latin by Johannes Kepler, a German mathematician, which promoted the Copernican view of a Sun-centred system with the planets moving round in circular orbits. Moreover, Kepler showed that the planetary spheres, whose circumferences contained the planetary orbits, could be nested within the five regular solids discovered by the ancient Greeks. Space, apparently, had a geometrical structure. Galileo wrote back to Kepler to say that he was also a supporter of the heliocentric Solar System. After that, there followed a fairly lengthy correspondence between the two. However, Galileo's outspoken views on the subject got him into trouble with the religious and academic authorities at home, in Italy, who adhered to the Aristotelian view.

Tycho Brahe, a Danish astronomer, spent many years collecting naked eye data on the position of the planets in the night sky. For two short years, Kepler was Brahe's mathematical assistant. After Brahe's death in 1601, Kepler carried out a long, painstaking study converting Brahe's planetary data from Earth-based measurements to Sun-based measurements, in the search for clues to explain some of the strange planetary motions.

Eventually, in 1609, Kepler discovered two empirical laws of planetary motion. The first law stated that the planetary orbits were elliptical with the Sun as one focus. If a circle is squashed to form an ellipse, the centre point splits into two focuses. Thus, Copernicus' idea of a Sun-centred Solar System was right, but his idea that the planetary orbits were circular was wrong. The second law stated that a planet's radius from the Sun swept out equal areas in equal times. It is speculated that Kepler hit on this law because of his knowledge of the ancient Greek form of infinitesimal calculus. Archimedes derived the area of a circle of radius r by filling the circle with many triangles with equal areas, all with their vertices at the centre of the circle and each base on the circumference. The area of a triangle is half base length times the height. The measured length of the circumference is $2\pi r$ and this must be equal to the sum of the base lengths of the triangles. When very many triangles are used, the height of each triangle is equal to the radius. So, summing the area of all the triangles gives the area of a circle as $\pi r^2$.

Finally, in 1619, Kepler discovered a third law connecting the square of the time (T) taken for a planet to complete its orbit with the cube of the half-length (a) of the major axis of its elliptical orbit. However, Kepler only had a relative distance (a/AU) of each planet from the Sun, based on a poor estimate of Earth's distance AU (one Astronomical Unit) from the Sun.

When we whirl a mass around on the end of a string, we are all familiar with that outward radial tug. (Certainly, boys with conkers are!) This is called the centrifugal force. It has been known about for thousands of years. As a weapon, the whirling mass is the crux of the slingshot. David killed the giant Goliath with a stone from his slingshot. When the stone is released, it shoots off in a straight line tangentially, not outwards in a radial direction. What is the invisible sling that holds the orbiting planets to the Sun? This invisible force must exactly balance the planets' centrifugal force.

Kepler speculated that the planets might be attracted to the Sun by a magnetic-like force. From his work on optics, he knew that the intensity of light at a point some distance from a light source depended on the inverse square of that distance from the source. The light spreads out from the source forming a light sphere, so that at a distance r from the source, the surface area of the sphere is $4\pi r^2$. Thus, at any point on this sphere, the intensity is the source power divided by $4\pi r^2$. Based on that knowledge, he suggested that if the Sun did attract the planets across space then the attractive force might depend on an inverse square law, but he did not pursue his speculations. Kepler also developed a mathematical model showing that the Moon was responsible for the two Earth tides per day.

Galileo didn't believe in elliptical planetary orbits and thought that Kepler was mistaken. He remained a supporter of the Copernican idea that the planets moved in everlasting circular orbits around the Sun. Also, Galileo dismissed Kepler's speculation that the Sun might possess an attractive force which pulled on the planets across space as mystical nonsense. Given that Galileo knew that masses accelerated towards the Earth, this was a rather strange attitude to take. Clearly, he couldn't see the parallel. Furthermore, Galileo rejected Kepler's mathematical model showing that the Moon caused the Earth tides. In all cases, Galileo was wrong and Kepler was right.

During the decade of the 1630s, we know that a young Cambridge student, Jeremiah Horrocks, interested himself in Kepler's work and that he was very proficient in using Kepler's tables to predict the positions of the planets in the sky. He determined the date of the next transit of Venus across the face of the Sun and was the first to observe it in 1639. He also showed that Kepler's laws applied to the Moon's orbit around the Earth and pointed out that the motion of a disturbed pendulum bob performed an elliptical orbit analogous to those performed by the planets. Horrocks died in 1641, but some of his papers were posthumously published by the Royal Society in 1672.

Galileo was the first scientist to explore the motion of pendulums and to determine their periods for small oscillations. In 1673, the Dutch mathematical physicist Christiaan Huygens described his mathematical model for a conical pendulum, where the bob rotated in a circle. His mathematical expression for the centrifugal force experienced by the bob was $mr\Omega^2$, where m was the mass of the bob, r the radius of the circle and $\Omega$ the angular velocity, or rate of change of angle during the circular motion. He sent a copy of his work to Newton. Like others, Huygens also saw the conical pendulum analogy with the rotating planets and could see that the component of the tension in the string was analogous to the Sun's attractive centripetal force, but he could not imagine what form such a force could take.

With hindsight, we know that the secret of Kepler's second law is that the angular momentum of each planet remains constant. The angular momentum H of a body is the product of its mass m times the radial distance r of the mass from the centre of rotation times the velocity v of the mass perpendicular to the radius. Mathematically, this is written as $H = m(r \times v)$.

If you sit in a nearly frictionless swivel chair, holding a 1kg bag of sugar, and get someone to rotate you and then leave you alone, you will have nearly constant angular momentum H. As you spin, you can feel the centrifugal force on the bag of sugar pulling it radially away from you. If you think of the bag of sugar as a planet and yourself as the Sun, then the force provided by your arms holding on to the bag of sugar is akin to the mysterious force of gravity holding on to the planet. When you hold the bag of sugar way out in front of you, you and the bag of sugar slow down. When you clutch the bag of sugar tightly to your chest, you and the bag of sugar speed up. The planets do the same thing. As they move further from the Sun, they slow down; as they move nearer, they speed up.

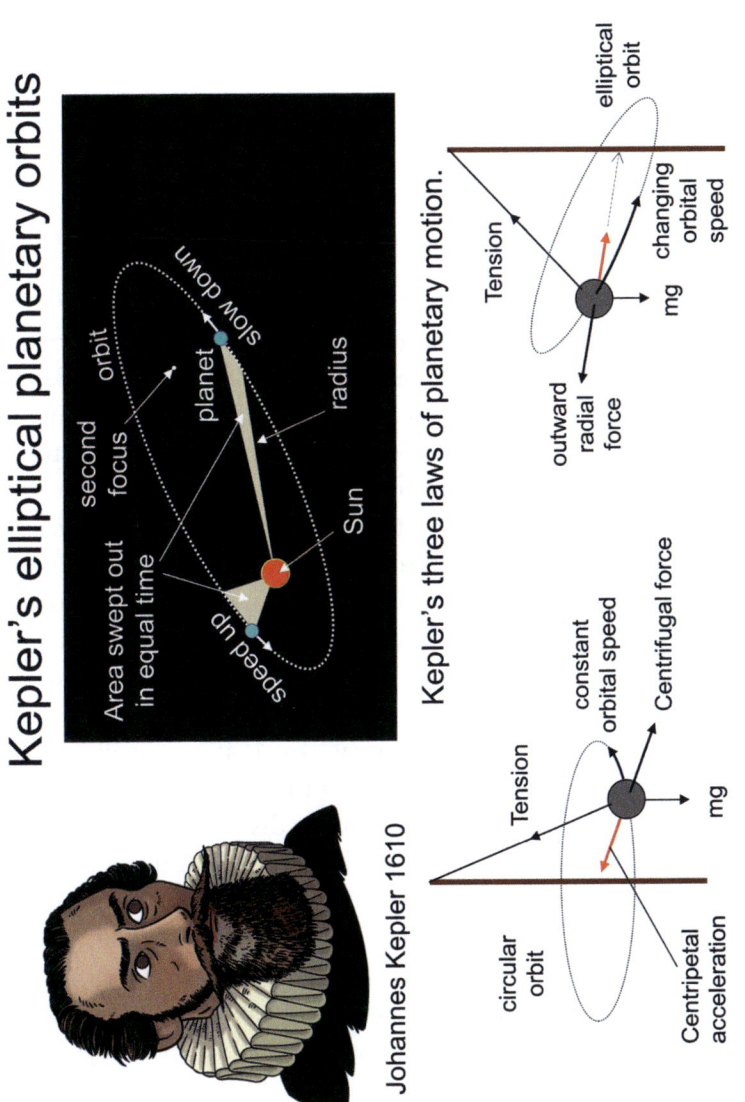

CHAPTER 5

# NEWTON'S MASTERPIECE

At the beginning of the 17th century, Sir Francis Bacon, a Cambridge graduate of Trinity College, held very high public office in England. As a part-time activity, Bacon interested himself in the up-and-coming field of science. In 1620, he wrote a book on science in Latin with the title *Novum Organum Scientiarum*. In the book, he described how scientific studies should be conducted, with great emphasis on the need for experimentation. The book appealed to many well-off gentlemen of learning at that time, who were conducting scientific experiments at home as an enjoyable pastime. Following a disastrous scandal in 1621, Bacon was forced to leave the field of politics, and he devoted the rest of his life to writing about science, dying in 1626.

Bacon's book was very influential. In 1645, a small group of scientifically inclined gentlemen used to meet together in London, at what they called their Invisible College, to discuss matters of science. These were dangerous times, as the Civil War was at its height. These low-key science meetings continued during Cromwell's time, until the demise of the Commonwealth period in 1659. Then, with the restoration of the monarchy in 1660, a gentlemen's club or, rather, a Society, was opened in London, devoted to scientific exploration. The new king, Charles II, had his own laboratory and supported the scientific endeavours of the Society, granting it a Royal Charter in 1662. The purpose of the Royal Society was to promote mathematical physics through experimental learning. It took as its Latin motto *Nullius in verba*, meaning don't take anyone's word for it; do the experiment yourself and be convinced. In 1664, the Royal Society began issuing its Transactions, publicising the results of experiments. Clearly, Galileo's and Bacon's views had been taken to heart.

Isaac Newton was born in 1642. His youth corresponded with a turbulent period of British history, with the Civil War beginning in 1641, the execution of Charles I in 1649, followed by the establishment of the Commonwealth under Oliver Cromwell.

Newton entered Trinity College at Cambridge University in 1661, just after the restoration of the monarchy. Nominally, Newton undertook a Law degree but, in today's terms, it was more a General Studies degree. Apart from Law, he studied ancient Greek and Latin, Theology, Astronomy, Geometry and, probably, the work of Aristotle. Although the Royal Society was embracing renaissance physics, Cambridge University courses remained rather backward, so it is unlikely that Newton's degree contained anything on the latest advances made in physics. But, the truth is, little is known about Newton's studies during this period. In 1663, Newton's favourite teacher, Isaac Barrow, arrived to be the first Lucasian Professor of Mathematics. Henry Lucas was the Member of Parliament for Cambridge University before the Civil War and in his will he left funds to establish the Chair. In 1665, Newton gained an undistinguished degree.

In view of Newton's later rise to fame, there are some intriguing facts worth noting about the people who may have influenced him while he was at university. John Wallis was a Cambridge University mathematician who published books on Algebra, Arithmetic, Mechanics and translations of ancient Greek works. Moreover, Wallis was a founder member of the Royal Society. While at Cambridge, Wallis worked with Jeremiah Horrocks, so they presumably discussed the Renaissance works of Copernicus, Kepler and Galileo. Although Wallis had moved to Oxford University by the time Newton arrived at Cambridge, we do know that Newton interested himself in Wallis' studies, including his algebraic study of the binomial expansion of $(1 + x)^n$, where n is a positive integer, called the exponent. The exponent determines the number of times that the expression $(1 + x)$ is multiplied by itself. The history of the binomial expansion goes back to ancient Indian and Chinese mathematicians.

For $n = 0$, the expansion of $(1 + x)^0 = 1$, so there is just one coefficient, namely 1. For $n = 1$; $(1 + x)^1 = 1 + 1x$, so there are two coefficients. For $n = 2$; $(1 + x)^2 = 1 + 2x + 1x^2$, so there are three coefficients. For $n = 3$; $(1 + x)^3 = 1 + 3x + 3x^2 + x^3$. For $n = 4$; $(1 + x)^4 = 1 + 4x + 6x^2 +$

$4x^3 + 1x^4$ and so on. The coefficients are symmetrical, the first and last always being 1.

On stacking the coefficients in rows, a triangular pattern is revealed:

|  |  |  |  |  |  |  |  |  |
|---|---|---|---|---|---|---|---|---|
| n = 0 |  |  |  | 1 |  |  |  |  |
| n = 1 |  |  | 1 |  | 1 |  |  |  |
| n = 2 |  |  | 1 | 2 | 1 |  |  |  |
| n = 3 |  | 1 |  | 3 |  | 3 |  | 1 |
| n = 4 | 1 |  | 4 |  | 6 |  | 4 |  | 1 |

This triangular pattern was written about, in 1100 AD, by the Persian mathematician and poet Omar Khayyam. The pattern existed in printed form in China by 1300 AD. The inner coefficients for the exponent n + 1 are obtained by adding the two coefficients for the exponent n immediately above it. By the Middle Ages, the pattern was also well known in Europe. Later, it was called Pascal's triangle, being named after the French mathematician Blaise Pascal, who made much use of the binomial expansion in his development of probability theory in the early 17th century. Pascal's triangle is used by the insurance industry to set their rates for life insurance. It is used in the study of heredity and even for predicting election results from a sample of voters. Pattern spotting in mathematics is very useful.

Wallis investigated infinite series, too. It was Wallis who introduced the symbol $\infty$ for infinity. In developing Wallis' ideas, Newton eventually produced his general binomial expansion for any exponent n (sometimes n is called the index), where n can be positive, negative and even a fraction. For negative and fractional exponents, the binomial expansion results in an infinite series. Wallis was also interested in the early form of infinitesimal calculus developed by Archimedes and other ancient Greek philosophers. Isaac Barrow also had an interest in infinitesimal calculus. Both Wallis and Barrow would have read the book on the subject written by Kepler. Kepler also wrote his book on the cosmos and a book on optics, too, as well as publishing his tables and laws of planetary motion.

Presumably, Newton had access to Wallis', Galileo's and Kepler's books, via the library. We know that Newton read books by the famous French philosopher René Descartes, where mention was made of Bacon's principle that scientific advances must be based on experiment. Descartes wrote about the Copernican system in terms of the ether and planets being carried around by vortices. In his work on geometry,

Descartes introduced the use of algebra. Newton also read Descartes' book on optics, which described ideas about light and colour. Newton bought a prism and tried out some of Descartes' ideas experimentally, to see whether he believed them. So Newton spent much time while at university learning about the new ideas in physics. However, these were extra-curricular studies, not part of his degree. So just gaining a pass in his degree is not surprising.

The rapid spread of the Coronavirus throughout the world in early 2020 (the Covid-19 pandemic) was aided by the very many forms of transport which now exist. In many countries, schools and universities were closed, large gatherings were prohibited and borders were closed, with complete lockdown of movement in some countries. Suddenly we all became acutely aware of the scare of a modern-day plague. With the arrival of the plague in Britain in 1665, London suffered very badly. To be on the safe side, Cambridge University was closed. In the 17$^{th}$ century, the means of travel was fairly limited to most of the population in England, which helped to slow the spread of the disease. Newton returned home to the relative safety of the Lincolnshire countryside. At the time, there was nothing to suggest that Newton might have the makings of a brilliant scholar. According to Newton himself, the 2 years that he spent at home to avoid the plague were the most intellectually fertile of his life. He carried out more optical experiments, formulated the general form of the binomial theorem for infinite series which he used in his development of fluxions, his form of infinitesimal calculus, and he began his investigation of gravity.

When Newton returned to Cambridge University in 1667, he was a different man. Isaac Barrow recognised that something had happened to Newton in the 2 years that he had been away. He was now an accomplished experimenter and mathematician. In 1669, Barrow resigned his Lucasian professorship and Newton succeeded him. In 1671, Newton built a very compact reflecting telescope and lent it to the Royal Society. They were fascinated by it and published details of it, making Newton's name known throughout the scientific circles in Europe. In 1672, Newton became a fellow of the Royal Society and began to correspond with some of the members, notably with Robert Hooke, the curator of experiments for the Royal Society. A visit to the Royal Society at that time must have come as a shock to Newton because much of the inner City had been destroyed during the four days of the Great Fire in 1666, so London would have been a vast building site.

By the early 1670s, a number of British scientists accepted that the Sun's attraction was due to the gravitational influence of its mass which spread out in all directions, in the same way that Earth's mass had a gravitational influence, attracting masses to its surface. In 1674, Hooke published a paper in which he theorised about the balance between a planet's gravitational attraction by the Sun and its centrifugal force. By 1679, Hooke and others had shown that Kepler's third law as applied to circular orbits could be rewritten to show that the square of a body's radius from the Sun times the radial acceleration (given by Huygens) was equal to a constant. Due to the balance between the centrifugal force and the centripetal force of the Sun's attraction, this meant that the Sun's attraction was equal to a constant divided by the radius squared; the inverse square law no less. In 1679, Hooke wrote to Newton and explained his approach but said that he couldn't show that the inverse square law applied to elliptical orbits.

Hooke's letter renewed Newton's interest in gravity. Although Newton later claimed that he had mostly completed his gravity studies during his 2 years at home from university, between 1665 and 1666, there is no evidence to show that he had actually made much progress. One important aspect related to gravity that Newton proved was that the mass of a body could be represented by a point mass at its centre of gravity, an idea introduced by Archimedes. This gave added support to the idea that the intensity of the Sun's attraction at some point depended on the inverse square of the distance of that point from the Sun's centre. Legend has it that when Newton saw an apple fall from a tree in his orchard, he wondered whether the Moon was falling towards the Earth, too, but was prevented from doing so by its orbital motion.

The Moon follows a nearly circular orbit around the Earth, making one circuit every 27.3 days. So, the rotational speed (called the angular velocity) of the Moon, labelled $\Omega_m$, is known. The fact that the Moon doesn't actually fall is due to the balance between the Earth's attractive gravitational force on the Moon and the centrifugal force ($M_m D_{Em} \Omega_m^2$) experienced by the Moon in orbiting around Earth. Using this fact and the latest French measurement of the distance $D_{Em}$ between the centres of the Earth and the Moon, Newton could calculate the acceleration due to Earth's gravity at the Moon's centre without knowing the mass of the Earth $M_E$ or the mass of the Moon $M_m$. Then, assuming an inverse square law and using the latest French measurement of the Earth's radius $R_E$, he could calculate the acceleration due to Earth's gravity at the surface of the

Earth first measured by Galileo. His estimate of $g_0$ was about 9.0 m/s², close to the measured value of 9.8 m/s², which he concluded was pretty good experimental evidence to confirm the truth of the inverse square law. Hooke had not done this.

Newton then set to work to show what the other scientists couldn't show, namely that the inverse square law applied to elliptical orbits, too, and he used a geometrical method to do it. These days, it is usual to use an algebraic method, called Central Orbits, to model the motion of the planets.

In 1687, Newton, with support from his friend Edmund Halley, published his masterpiece, *Philosophiae Naturalis Principia Mathematica*, generally known as the *Principia*. The book, written in Latin and separated into three sub-books, took Newton 3 years to write.

It is interesting to see the historical backdrop against which Newton's book was written and published. Charles II had died in 1685 and his brother had become King James II of England, Ireland and Wales and King James VII of Scotland. It was an unsettling time as James II tried to impose his authority. He interfered in the religious aspects of running Cambridge University and Newton was one of the prominent academics who opposed the interference. The year following Newton's publication of the *Principia*, 1688, was that of the so-called Glorious Revolution, when James II was deposed by his daughter Mary and her husband, the Dutch Prince William of Orange. A Dutch naval invasion led by Prince William, far larger than that threatened by the Spanish Armada of 1588, landed an army at Brixham in Devon and marched towards London. James II fled to France and Parliament proclaimed William and Mary to be the joint rulers of Great Britain and Ireland; the reverberations of which persist even today. There is a later parallel with Albert Einstein's work on his theory of general relativity, updating Newton's ideas on gravity, written and published in Germany during the stressful period of the First World War. Perhaps for Newton, publishing *Principia,* and for Einstein, publishing his paper on general relativity, it was a form of escapism from the tumultuous goings-on during those times. Certainly, writing this book gave me something else to think about while the daily news was all about the political shenanigans involving Brexit, about which I can do nothing and, then, Coronavirus, which was scary.

The first book in Newton's *Principia* described the three laws of motion which were partly based on Galileo's earlier observations and took into account the French mathematician René Descartes' view of inertia. The first law states that a mass moving with uniform speed in a straight line would continue to do so, unless acted upon by a force. This is sometimes called the law of inertia. Inertia is the resisting force that we experience when we try to start a body moving, or try to stop a moving body. Inertia arises during periods of acceleration or deceleration. The second law defines force. In its simplest form, we write force F equals mass m times acceleration a.

$$F = ma \quad \text{where } a = \frac{d^2x}{dt^2} \text{ and x is the distance moved in time t}$$

This law probably stemmed from the observation that the force that we call weight equals mass times the acceleration due to gravity. Now in the weight case, the mass is stationary and is called a gravitational mass. In the case of an accelerating mass, the mass is called an inertial mass. In his first book, Newton claimed that gravitational mass and inertial mass were identical, or equivalent. In his second book, Newton described an experiment to demonstrate this equivalence. He used a simple pendulum with a hollow bob which could be filled with different substances, each having the same weight and, therefore, the same gravitational mass. He then raised the bob through a small angle and released it, so that the pendulum performed small oscillations. The swinging bob, accelerating in the Earth's gravity field, had inertial mass, but in all cases the period of swing was the same. Therefore, Newton concluded that gravitational mass and inertial mass were equivalent. This is called the principle of equivalence. In 1889, the Hungarian scientist Roland von Eötvos, using a torsion balance, confirmed the equivalence result with much greater accuracy. In my controversial view, I suspect that inertia gives rise to an induced gravity field to which the accelerating mass responds. So, whether the mass is stationary in a gravitational field or is accelerating, in either case, the mass responds to gravity. Consequently, for this situation, we would expect the gravitational mass and the inertial mass of a body to be the same. The third law states that for every action, there is an equal and opposite reaction. That is, for every force, there is an equal and opposite force.

The second book deals mostly with fluid mechanics, particularly bodies moving through a resisting medium, including air. Newton continued to make use of a pendulum in his experiments. For example, he clapped his hands and used a pendulum to time how long it took for an echo to return from the end wall of a long corridor. Knowing the length of the corridor, he then worked out the speed of sound.

The ancient Greek philosophers had introduced the idea of the invisible ether (sometimes spelt aether) as the fifth element that filled all of space. In 1644, René Descartes suggested that the ether of space might be an extremely dense, fluid-type substance. A local disturbance of the ether would form an ether vortex. Assuming that mass was less dense than ether, then mass particles would congregate at the vortex centre. The laboratory centrifuge works in this way. Ether vortexes would lead to the formation of astronomical bodies of various sizes. Descartes proposed that each star was at the centre of a gigantic ether vortex. In the case of the Sun, the bodies at the centres of smaller ether vortexes were swept around the main vortex to form the planetary Solar System. The planets, being less dense than ether, would also be attracted towards the Sun at the centre of rotation.

In his search to detect the resistive presence of the ether, Newton carried out pendulum tests in a vacuum, but he was unable to reach any conclusion from his experiments. Newton argued that since there was no sign of the planets slowing down, this ruled out the existence of a dense resisting medium. Consequently, he dismissed Descartes' idea of a dense fluid-type ether, concluding that if the ether existed it must be extremely fine. However, in 1692, Newton commented that the idea that the ether was a vacuum containing nothing was absurd, since it was the ether that supported the gravitational influence of a mass throughout space. Nevertheless, he confessed that he did not know what the ether was made of. Newton's view of the ether as an extremely fine resistant-less medium through which gravitational influence permeated, prevailed for several centuries.

With hindsight, perhaps it's worth mentioning the phenomenon of superconductivity, where for some conducting materials near to absolute zero temperature ($0 \text{ K} \approx -273°C$), electric currents flow without resistance. So, the gravitational analogue might be that a dense ether does exist which, near to absolute zero, allows mass currents to flow through space without resistance.

Gravitation is like magnetism in that it wields its influence across space, without any apparent direct contact between bodies. It is worth pointing out that Newton thought of gravity in terms of a region of influence around a mass. The idea of a field was introduced, nearly two centuries later, by Michael Faraday when he studied the regions of influence around magnets, initially using iron filings to make the magnetic influence visible and then using a small compass to plot out the directions of lines of magnetic force. The observation that there was a direction associated with magnetic influence led to the field concept. Any entity with a magnitude (of influence in the case of magnetism) and a direction is called a vector. A single bold letter in the text signifies a vector, for example, **g**. Every point of a field is represented by a vector. Faraday extended the idea to the region of influence surrounding an electric charge. The similarity between mass as a source of gravity and charge as a source of electricity quickly led to the assumption of the existence of a gravity field around a mass. The field vectors for any point source are radial.

Scientists later realised that the energy density of a gravitational field is negative, meaning that the region of space around a mass must contain negative energy. But this didn't fit in with the idea that energy was a positive quantity. Where did all the positive energy reside to keep the overall energy of any volume of space positive? This was pre-Einstein ($E = mc^2$), when the idea that inert mass might contain enormous amounts of positive energy did not occur to scientists. They pondered on the idea that space might contain a hidden source of positive energy. In more recent times, scientists have introduced a new form of ether, called the quantum vacuum, full of virtual particles with short-lived positive energy. But, post-Einstein, a hidden source of positive energy in space implies the existence of unseen mass (since $m = E/c^2$), raising the possibility that the ether is dense but does not resist the movement of mass. The possible existence of an invisible ether with quantum properties remains a modern-day mystery.

The third book, subtitled *The System of the World*, describes Newton's theory of gravity. When Hooke claimed that he deserved some credit for developing the theory, Newton's response was to remove any mention of Hooke's name from *Principia*. The law of gravity is that the force **F** of attraction between two masses M and m is equal to the product of the two masses divided by the square of the distance r between their centres multiplied by a constant known as Big G.

Newton never gave an equation for the inverse square law for gravity in his book, which may come as a surprise to you, as it did to me. Suppose we surround the mass M with a sphere of radius r. The surface area of the sphere is $4\pi r^2$ and at any point on the sphere the gravitational influence **g** of the mass M will be reduced to $M/4\pi r^2$. So, the force **F** of attraction experienced by a mass m located at a point on the sphere surrounding M will be:

$$\mathbf{F} = m\mathbf{g} = \frac{mM}{\gamma 4\pi \mathbf{r}^2}(-\hat{\mathbf{r}}) \text{ Newtons}$$

$\hat{\mathbf{r}}$ is the outward unit radial vector

The factor $\gamma$ is due to the gravitational permittivity of the ether of space. Newton's universal gravitational constant $G = 1/4\pi\gamma$.

Newton wasn't able to say what the value of G was. Newton explained how his theory of gravity fitted in with Kepler's laws. The motion of the Moon around the Earth obeys Kepler's laws. The book also contained Newton's theory of the oceanic tides, which depended on the gravitational fields of the Earth, the Moon and the Sun.

Newton made use of a quantity (mv) associated with a mass m moving with velocity v. He called the quantity momentum. During Newton's time, billiards became a very popular indoor game for the gentry. Perhaps the junior common room at Trinity had a billiard table and Newton experimented with colliding billiard balls. By whatever means, he discovered that linear momentum was conserved. That is, the total momentum of the balls before a collision was equal to the total momentum of the balls after the collision.

In his 2nd law of motion, Newton defined the force **F** experienced by a mass m moving with velocity **v** in terms of the rate of change of its momentum.

$$F = \frac{d(m\mathbf{v})}{dt}$$

When the mass m is constant, the two forms of Newton's 2nd law agree, since

$$\frac{d\mathbf{v}}{dt} = \frac{d}{dt}\left(\frac{d\mathbf{x}}{dt}\right) = \frac{d^2\mathbf{x}}{dt^2}$$

The German mathematician Gottfried Leibnitz (Newton's rival in the derivation of calculus) thought that a moving billiard ball contained a living force, which he termed in Latin, *vis viva*. In fact, Leibnitz's *vis viva* was equal to $mv^2$, related to kinetic energy, but this concept was unknown to Newton.

In modern terms, we say that a body of mass m has energy E if it can do work W. Although engineers throughout the ages were familiar with the idea of work, it had not been defined mathematically. Nowadays, work W done by a body of mass m is defined as the product of the applied force F times the distance x moved by the body. Mathematically, using increments (tiny amounts) of work dW and of movement dx, we write

$$dW = F.dx$$

We now state that the increment of work done dW is equal to the incremental change of energy – dE, so that

$$dE = -F.dx$$

This leads us to a new definition of force that was unknown to Newton, namely that force **F** arises when there are gradients in energy E.

$$F = -\frac{dE}{dx} = -\nabla E \quad \text{where} \quad \nabla = \frac{d}{dx} \text{ is called the gradient.}$$

It turns out that energy is also conserved, but this wasn't realised until the mid-18$^{th}$ century, with the work of Joule, Mayer, Von Helmholtz and Thomson (see Chapter 29).

We can now explore the concept of energy in relation to gravity. Suppose we have a fixed mass M which creates a gravitational field **g**. The external energy of a test mass m can arise in two ways. Firstly, consider the test mass placed in the gravity field **g**. It has potential energy to do work. Secondly, if the test mass moves with velocity v relative to the mass M, it has kinetic energy.

With our new definition of force, we can examine the ideas of potential and kinetic energy.

In the static case, for a test mass m in a gravity field **g**, we have

$$F = -mg = -\nabla E = -\frac{dE}{dr}\hat{r}$$

Integrating (using Leibnitz's form of calculus), the potential energy is

$$E = mgr \quad \text{where r is the radial distance of the test mass from M.}$$

For r fixed, the potential energy of the mass m is fixed. There is a sphere, of radius r, surrounding the mass M on which the potential energy is constant. The surface of this sphere is called an equi-potential surface. We write the potential energy as

$$E = m\phi \quad \text{The function } \phi \text{ is called the gravitational potential.}$$

Since $\phi = gr$, we see from the inverse square law that

$$\phi = -\frac{M}{\gamma 4\pi r} \quad m^2/s^2$$

Taking the gradient of the potential, we find that

$$\mathbf{g} = -\nabla\phi$$

For the dynamic case, using Newton's 2$^{nd}$ law, if the test mass is set free it will accelerate towards the source M.

$$F = m\frac{d^2r}{dt^2} = -\nabla E$$

Now we can write $\dfrac{d^2r}{dt^2} = \dfrac{d}{dt}\left(\dfrac{dr}{dt}\right) = \dfrac{dr}{dt}\dfrac{d}{dr}\left(\dfrac{dr}{dt}\right) = v\dfrac{dv}{dr}$

Thus we have

$$mv\frac{dv}{dr} = -\frac{dE}{dr}$$

Integrating gives

$$\frac{1}{2}mv^2 = -E$$

This energy of movement is defined as kinetic energy.
 We now know that energy is conserved so that, ignoring friction,
 Potential energy + kinetic energy = a constant.
So, as a mass falls in a gravity field, it exchanges potential energy for kinetic energy and vice versa. This result is the basis of classical mechanics.
 In using the kinetic energy approach, we have avoided the need to

consider the effect of a moving gravity field associated with the moving mass.

One aspect of Newton's gravity theory that seems strange to us today is that he accepted that when a mass moved its gravitational pattern of influence (its gravity field) moved with it, re-establishing itself instantaneously throughout the Universe. This idea even extended to events, so that lightning striking Trinity College Cambridge would be known throughout the Universe as it happened. Scientists refer to this as Newton's absolute time. In the case of electromagnetism, James Maxwell, a Trinity College graduate like Newton, showed (nearly 200 years later) that disturbances in the electromagnetic field move about at the speed of light c. Nowadays, it is generally assumed that gravitational changes are also broadcast at the speed of light.

When we examine starlight, we are looking millions of years back in time. In fact, the stars won't be where we see them to be. Similarly, aliens able to view the Earth from another galaxy might conclude that life forms have not started on our planet, yet.

# Newton's masterpiece

Big G

Inverse Square Law

potential surface
gravity field

angular velocity = $2\pi/27.3$ days

Centrifugal force

$M_M$

Gravity force

$D_M$

$R_E$

$M_E$

Inverse square law

$m_i = m_g$

Equivalence

Isaac Newton 1687

PRINCIPIA
Book 1
Book 2
Book 3

## CHAPTER 6

# MEASURING VAST DISTANCES AND THE SPEED OF LIGHT

At some time during the 2nd century BC, a curious observation was reported to Eratosthenes, the Chief Librarian at the university in Alexandria. It concerned a deep well in the important town of Syene on the River Nile. The town is now called Aswan and is better known for the dam there. It was reported that at noon in mid-summer, the Sun's rays shone straight down the well and lit up the water at the bottom. This observation showed that at noon in Syene, which lies on the latitude of the Tropic of Cancer, the Sun's rays arrived exactly perpendicular to the Earth's surface. Now, the ancient Greek philosophers assumed that the Earth was a giant sphere. Moreover, Alexandria and Syene lay on roughly the same circle of longitude. The distance between Syene and Alexandria, which was to the north, was known to be about 800 km (in modern units). This measurement was quite a feat of surveying for those days. Eratosthenes erected a vertical pole in the university grounds and at noon he observed the small shadow cast on the ground. Knowing the height of the pole and measuring the length of the shadow, he determined that the Sun's rays were 7.2° (seven degrees and 12 minutes) away from the perpendicular at Alexandria. Thus, the distance of 800 km between Alexandria and Syene formed the arc length of a sector of the circle of longitude, of angle 7.2°, implying an Earth radius $R_E$ of about 6,300 km.

The ancient Greeks also interested themselves in astronomical measurements. In the 3rd century BC, during a lunar eclipse, Aristarchus used a method of similar triangles to estimate the Moon's diameter as a fraction of the Earth's diameter and then to estimate the Moon's distance

$D_{Em}$ from the Earth. He got a value for $D_{Em}$ of about ten times the Earth's diameter, or twenty times the Earth's radii. It was a good first try to measure the distance $D_{Em}$ but was not a good estimate.

Aristarchus also attempted to measure the Earth's distance from the Sun, designated as AU, for one Astronomical Unit. A time was chosen when the Moon was in its first or third quarter, as seen from the Earth. For this condition, it was assumed that the angle between Sun-Moon sightline and the Moon-Earth sightline was a right angle. Next, the angle between the Sun-Earth sightline and the Earth-Moon sightline was measured. It was measured as 87°. Unfortunately, this angle was quite wrong. The light from the Sun bathing both the Earth and the Moon is virtually parallel, but not quite. A modern value for the angle is 89.85°. The wrong angular measurement and the poor estimate of the Moon's distance $D_{Em}$ meant that the calculated Earth-Sun distance AU was wildly inaccurate (about 7 million kilometres) and wasn't corrected for many centuries.

In the 2$^{nd}$ century BC, Hipparchus, the founder of trigonometry, used his expertise to devise a method called Parallax to re-estimate the Moon's distance $D_{Em}$ from the Earth. During a lunar eclipse in 190 BC, the angle of elevation of the Moon's centre was measured at Alexandria and at the Hellespont (on the Dardanelles, now in Turkey). The difference in latitude $\lambda$ between the Hellespont and Alexandria (virtually due south) was known to be about 10°, so the baseline distance (the chord of a circle) of a gigantic triangle with an apex at the Moon could be calculated. From the two angles of inclination, Hipparchus was able to determine the two base angles. Knowing that the sum of the angles in a triangle is 180° he then calculated that the apex angle, or angle of parallax, was about 0.2°, an amazing feat given the simple apparatus used. He then calculated that the Moon's distance $D_{Em}$ from the Earth was about fifty-nine Earth radii, or 370 thousand kilometres. It was a remarkably accurate value. Ptolemy of Alexandria repeated the measurement in the 2$^{nd}$ century AD and confirmed the value, as did French scientists in the 17$^{th}$ century.

Better measurements of astronomical distances had to wait until the 17$^{th}$ century, when French scientists, with improved methods of measuring tiny angles, tackled the job. The Earth's northern angles of latitude can be accurately determined by measuring the elevation of the Pole star above the horizon. The elevation is equal to the latitude. Then, using the surveyor's method of triangulation, the French scientists accurately measured the surface arc length of 1° of the Earth's latitude, roughly along the circle of longitude from Paris to Amiens. From this, they obtained an accurate value of the Earth's radius $R_E$, being 6,370 km, which turned out to be very near to the value measured by the ancient Greeks. Using the method of Parallax, they repeated Hipparchus' measurement of the distance $D_{Em}$ of the Moon from the Earth and found it to be about 370 thousand kilometres, in agreement with the ancient Greek measurement made nearly 2,000 years earlier.

The next step was the big one; to measure the distance $D_{EM}$ from Earth to Mars at its nearest approach (every 15 to 17 years), again using the method of Parallax. It meant measuring the angles of a giant triangle with a known base length on Earth and with Mars at the vertex. To do this, Mars had to be visible at the same time to two, very widely spaced apart, observers, who had to make angular elevation measurements of Mars at their location at the same time. This involved knowing the exact position on the Earth's surface of the two observation points and clocks that could be synchronised. In 1671, an expedition sailed to Cayenne, an Atlantic coastal town in the French colony of Guiana, in South America. On the day of nearest approach in 1672 and at a specified time, the angles of elevation of the sightline Paris to Mars and the sightline Cayenne to Mars were measured. When the data was collected together in Paris, use of spherical trigonometry showed that the vertex angle at Mars was 23 seconds (about 0.0064 of a degree). The baseline was the distance through the Earth between Paris and Cayenne in French Guiana. With these values, the distance $D_{EM}$ from Earth to Mars was estimated to be about 55 million kilometres.

A way now became possible to estimate the distance AU of the Earth from the Sun. Input from Kepler's third law gave the ratio of the Earth-Mars radii in terms of their periodic times. The Earth's orbital period is 365 days and that for Mars is 687 days. Assuming circular orbits, a calculation showed that the Earth was about 105 million kilometres from the Sun. This was a better estimate than the value calculated by the ancient Greeks, but still quite wrong. Today's accurate figure for AU is 149 million kilometres (93 million miles). The distances of the other planets from the Sun have since been determined, and the immense size of the Solar System is now clear.

Christiaan Huygens, the Dutch mathematical physicist, attempted to determine the distance from the Earth to Sirius, the brightest star in the heaven of the Northern Hemisphere. Huygens assumed that Sirius was another sun like our own with the same brightness. He knew that the intensity of a point light source followed an inverse square law of

# MEASURING VAST DISTANCES AND THE SPEED OF LIGHT

distance. So, the faintness of the star compared with the Sun was due to its increased distance away. Huygens directed the light from the Sun through a tube onto a screen with a small hole in line with a glass bead and made refinements until he got an image the same (from memory) as that of Sirius. Being an expert at optics, he was able to calculate that he had stopped down the Sun's brightness by an amount 27,664 times. So, he reasoned, Sirius must be 27,664 times further away from the Earth than the Sun was. It was a poor first estimate, but it showed a rough way to measure the distance to Sirius. Several scientists of that time repeated the experiment. Newton estimated the distance to Sirius to be nearer to 1 million AU. The modern value of the distance of Sirius from Earth is 549,000 AU, or $0.821 \times 10^{17}$ m. The size of the Solar System in comparison with the vast distance to Sirius makes one realise how small a part of space the Solar System occupies.

Measurements of the distances of the planets from the Sun were made using the geometry of the ancient Greeks coupled with better angle-measuring apparatus. But it assumed that the planets were where they were observed to be by telescope. In other words, it was assumed that the sunlight reflected by the planets reached the Earth instantaneously. If light took time to travel across space then the modelling was not quite right and small errors would creep in.

Galileo had tried to determine the speed of light with an Earth-based experiment exposing lanterns on hilltops several miles apart, but the measurement technique was far too crude and he was not successful. In 1610, Galileo had observed the large moons of Jupiter through his telescope as pinpricks of light. The moons passed into Jupiter's Sun-shadow and were eclipsed with clockwork regularity, so much so that tables of the moons' eclipses were made. In 1676, the young Danish astronomer Ole Römer was employed at the Paris Observatory to improve on the eclipse tables. For Io, the innermost moon of Jupiter, Römer noticed that the interval between successive eclipses increased when the Earth was moving away from Jupiter and shortened when the Earth was moving towards Jupiter. Since the orbits of the planets lie in the same plane and are roughly circular, Römer realised that the reason for the differences in successive eclipse times was due to the change in the distance between the Earth and Jupiter. Using the latest French estimates of the distance of the Earth and Jupiter from the Sun, Römer calculated the speed of light to be $c = 2.4 \times 10^8$ m/s. Given all the estimates involved, Römer's value is remarkably near to today's known value of $c = 3 \times 10^8$ m/s.

Knowing AU, we can calculate that the light from the Sun takes about 9 minutes to reach Earth. The time for the Sun's reflected light from Jupiter reaching the Earth depends on whether the Earth and Jupiter are on the same side of the Sun in their orbits (conjuction) or on opposite sides (opposition). When the two planets are on opposite sides there is the time taken for light to cross the diameter (2 AU) of Earth's orbit to be taken into account. This is about an extra 16 minutes.

The first terrestrial measurement of the speed of light was made by Armand Hippolyte Fizeau, a French experimental physicist, in 1849. Fizeau was one of the last gentlemen scientists. Having been left a fortune by his father, he was able to work on his research interests and carry out experiments without having to worry about getting funding for them. The crux of Fizeau's apparatus was a cogwheel rotating about a horizontal axle. The wheel had n equally spaced cogs. A focussed beam of light was shone perpendicularly onto the periphery P of the wheel so that as the gaps and cogs passed by, a series of light pulses was transmitted to a mirror M a distance $D_{PM}$ away and the light pulses were reflected back to the wheel periphery. Fizeau increased the angular velocity ω (radians/s) of the wheel until a cog first totally obscured the reflected pulse of light.

Suppose the cogwheel angular velocity is $\omega_1$ for the 1$^{st}$ light obscuration. For this rotational speed, a gap lets the light signal through and a cog blocks, or obscures, the reflected light signal. From this, Fizeau calculated that the speed of light c was

$$c = 2 \times D_{PM} \times 2n \times \frac{\omega_1}{2\pi} \text{ m/s}$$

Fizeau carried out his experiment in what is now a western suburb of Paris. He set up his rotating cogwheel in a house, and the pulses of light were directed towards a mirror set up far away on the side of a hill called Montmartre. The success of Fizeau's experiment was due to his expertise in optics and his ability to collimate the pulses of light. For Fizeau's experiment, $D_{PM} = 8.633 \times 10^3$ m and n = 720. The observed light pulses were seen as a star. The first obscured light pulses (blackness) occurred when $\omega_1/2\pi$ = 12.6 revs/s. This gave a slightly high value for the speed of light of

$$c = 3.1 \times 10^8 \text{ m/s.}$$

# MEASURING VAST DISTANCES AND THE SPEED OF LIGHT

One of the difficulties with Fizeau's method was determining the angular velocity $\omega_1$ for exact obscuration of the light pulse. Another was that the experiment extended over a long distance which had to be measured accurately. For many years, Fizeau worked with Léon Foucault, but after 1847, the two scientists went their separate ways. Over a number of years, from 1849 to 1862, Foucault designed and refined a compact experiment to measure the speed of light. His apparatus, involving a rotating mirror, was confined to about 20 m. Foucault obtained a value of the speed of light $c = 2.98 \times 10^8$ m/s.

The US scientist Albert Michelson spent many years on experiments to measure the speed of light. His apparatus was similar to that of Foucault's, but he replaced the rotating mirror with a rotating octagonal block with each face containing a plane mirror. The principle behind the experiment was the same as that of Fizeau's and Foucault's in that a collimated sequence of light pulses was transmitted to a mirror a long distance away and reflected back and the time taken measured. This simple description does not mention the intricacies needed to make the experiment, carried out during the steam age, such an outstanding success. The octagonal block was rotated using an air compressor powered by a steam engine, and the frequency of rotation of the octagonal block was measured by shining the light pulses onto a tiny mirror at the end of a tuning fork vibrating at a known frequency. Modern electronic methods have greatly simplified the measurement of the speed of light, which is generally taken to be $c = 3 \times 10^8$ m/s.

We live in the Solar System and our Sun is in an outer spiral arm of the Milky Way Galaxy. The Milky Way Galaxy is about 53 light years across and contains billions of stars. And the Universe contains billions of galaxies each containing billions of stars. The latest estimate of the distance from the Earth to the edge of the observable Universe is 46.5 billion light years. This translates to a distance of $0.45 \times 10^{27}$ m. The vast numbers of stars in the Universe probably means that the Earth is not the only planet where life exists. However, the vast interstellar distances involved would seem to rule out the possibility of any direct contact with extra-terrestrials. But, of course, that assumption is based on our current understanding of physics and its possible applications. There might be a way we don't know about, yet.

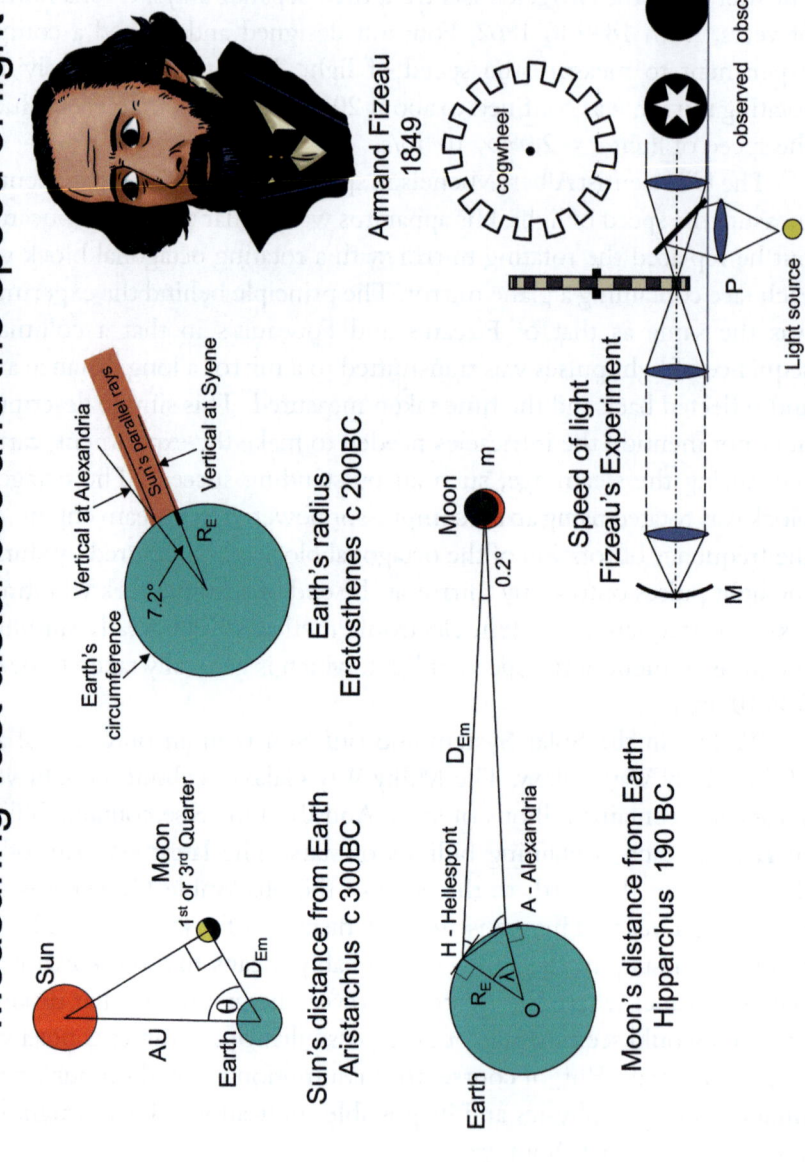

## CHAPTER 7

# GRAVITY IS A UNIVERSAL FORCE

From the inverse square law, the acceleration due to gravity at the Earth's surface is given by

$$g_0 = G \frac{M_E}{R_E^2} \quad m/s^2$$

Newton knew the value of $g_0$ and he had a good estimate of the Earth's radius $R_E$, but he didn't know the Earth's mass $M_E$, so he was unable to calculate the value of Big G, the universal gravity constant. Clearly, he was unable to think of an experiment to determine Big G. He probably felt that such an experiment would have to involve an astronomical-sized body which had a noticeable gravitational influence.

On the surface of the Earth, the weight of a body equals its mass m multiplied by $g_0$. So, by weighing a body and knowing $g_0$ we can determine its mass m. The idea that the gravitational force between two small spherical bodies with known masses, say m and M, placed in close proximity might be used in an experiment to determine the value of Big G seems ludicrous at first sight. Surely, their gravitational force of attraction is far too small to cause an interaction. Strangely, this is not true. Although the attractive force F between ordinary-sized bodies (including humans) is exceedingly weak, it is enough to cause an effect that can be detected.

The first apparatus designed to measure Big G was built in 1793 by John Michell, a professor of Geology at Cambridge University. It made use of the interacting gravity fields between two small masses, but the ingenuity of the apparatus was in its use of a torsion balance. Sadly,

Michell died before completing his experiment. Henry Cavendish, a rich aristocratic amateur scientist, had been a close confidant of Michell. Fortunately for science, Cavendish acquired and refurbished Michell's apparatus. In 1798, he successfully carried out the experiment and was the first to measure Big G.

To understand how the torsion balance works, we first consider the see-saw. Suppose we have a see-saw, of length $2\ell$, with a child of the same mass m sat at each end. Then the see-saw will balance about its centre, or fulcrum. If one of the children pushes down on the ground with a force F, then an equal and opposite vertical force F will act (Newton's $3^{rd}$ law) on the child, causing it to rise and the see-saw to start to rotate about a horizontal axis. In technical terms, the length $\ell$ from the fulcrum, or centre of rotation, to a child is called the moment arm of the see-saw. When a force is applied to the end of a moment arm causing it to rotate, the effect is called a couple, often labelled as C. In mathematical terms $C = \ell F$.

The torsion balance is a see-saw which rotates about a vertical axis. A dumbbell, consisting of a lightweight rod of length $2\ell$ with a mass m at each end, is hung from a vertical wire attached to the rod's centre. The wire-suspended horizontal dumbbell forms the torsion balance. A horizontal force F applied to one mass will cause the dumbbell to rotate about a vertical axis and we have the effect of a couple $C = \ell F$. As the wire twists, it reacts with a resisting couple $C = k\theta$, where $\theta$ is the angle through which the dumbbell rotates and k is called the torsional stiffness of the wire. Thus $\ell F = k\theta$. The natural period of oscillation of the torsion balance is used to determine the torsional stiffness k. So, knowing $\ell$ and k and measuring $\theta$, we can determine F.

In Cavendish's torsion balance, the masses m at the end of the dumbbell are made of iron and are about the size of tennis balls. The horizontal dumbbell is allowed to settle down to the wire's untwisted condition, where movement of the dumbbell ceases. At this stage, two larger iron balls of equal mass M, about the size of cannonballs, are introduced on opposite sides of the small masses also at a radial distance $\ell$ from the balance centre, so that the distances d apart of their mass centres are the same at either end. Thus, a horizontal X-pattern is formed, with the torsion balance rod forming, say, the forward stroke / and a fixed large ball sited at each end of the backward stroke \. Each large fixed ball gravitationally attracts its neighbouring small ball with a force F given by Newton's inverse square law. The gravitational attraction of the small balls causes the dumbbell rod to twist fractionally about its centre. A mirror attached to the wire and a light beam reflected back to a scale marked out horizontally, is used to measure the angle θ that the wire is twisted. Thus, we can determine the force F experienced by each small mass. Apart from Big G, the values of all the parameters in Newton's inverse square law are known, so Big G can be determined. Big G equals 0.00000000006673 m³/kg.s², so it is very small. To make the value of Big G more manageable, scientists write it as $G = 6.673 \times 10^{-11}$ m³/kg.s². Thus, the gravitational permeability $\gamma = 1.193 \times 10^9$ kg.s²/m³.

Knowing the values of γ, or Big G, the acceleration due to gravity at Earth's surface $g_0$, and the radius $R_E$ of the Earth allows the mass of the Earth $M_E$ to be determined. A calculation shows that $M_E$ = 5,980,000,000,000,000,000,000,000 kg. Again, to make the number more manageable, scientists write $M_E = 5.98 \times 10^{24}$ kg.

Assuming a circular orbit, then Earth's orbital velocity $v = 3 \times 10^4$ m/s. Since the Sun's gravitational attraction just balances the Earth's centrifugal force, and knowing the distance AU of the Earth from the Sun, we can determine an approximate value for the Sun's mass as $M_S = 1.989 \times 10^{30}$ kg. This ignores any effect that the other planets might have on the calculation. So the mass of the Sun is around 300,000 times bigger than the mass of the Earth.

It is not easy to determine the mass of the Moon $M_m$, because the acceleration due to gravity at the Moon's surface and the Moon's Radius $R_m$ are both unknown. To estimate the Moon's radius $R_m$, its diameter ($2R_m$) is compared with the known diameter ($2R_E$) of the Earth. The ancient Greeks did this but their estimate was poor. A modern estimate is that the radius of the Moon $R_m = 0.3\ R_E$. On that basis, if we assume that the density of the Moon is the same as the averaged density of the Earth, the mass of the Moon $M_m = 16.124 \times 10^{22}$ kg. Having now been to the Moon and carried out gravitational experiments there, we know this to be an overestimate. The suggested explanation is that the planets have metallic cores, whereas the Moon has a rocky core, with a density about half that of the Earth's.

Jupiter's periodic time is 11.86 years. Assuming a circular orbit, then from Kepler's third law, the distance $D_{SJ}$ of Jupiter from the Sun is $7.75 \times 10^8$ km. The mass of Jupiter is $1.898 \times 10^{27}$ kg.

There has been some debate as to what part the gravity fields of Jupiter and Venus play, when they align on either side of the Earth. Could they act as a tipping point in the movements of Earth's tectonic plates?

The increase in stretching and compressing of the Earth's mass by these external gravity fields gives rise to temperature changes in the Earth's crust which may be responsible for an increase in volcanic activity. This highlights the link that exists between frictional heat and gravity. Also, planetary alignment may affect the Sun's atmosphere, with the rebound effect on the Earth's atmosphere.

We see that the gravitational fields of the planets and the stars permeate every corner of the Solar System and of the Universe. A typical human of mass 70 kg experiences a force (weight) of about 700 N pulling them down to the surface of the Earth. We know that, as well as the Sun, the Moon's gravity causes ocean tides on Earth, but it also has a minor effect on us, too, of about 2 mN. Whether this affects our bodies in any way is not clear. Even our nearest star, Alpha-Centauri, 4.36 light years away, has a pull on us (of about $10^{-11}$N), but it is far too small to have any effect on us, as far as we know. But astrologers assume otherwise.

# Gravity is a universal force

| Astronomical Body | Mass of Body kg | Nearest distance to Earth m | Attractive force between Earth and body N |
|---|---|---|---|
| Earth | $5.972 \times 10^{24}$ | | |
| Sun | $1.989 \times 10^{30}$ | $1.495 \times 10^{11}$ | $3.541 \times 10^{22}$ |
| Moon | $7.348 \times 10^{22}$ | $3.636 \times 10^{8}$ | $2.215 \times 10^{20}$ |
| Jupiter | $1.898 \times 10^{27}$ | $6.640 \times 10^{11}$ | $1.716 \times 10^{18}$ |
| Venus | $4.867 \times 10^{24}$ | $0.380 \times 10^{11}$ | $1.343 \times 10^{18}$ |
| Mars | $0.639 \times 10^{24}$ | $0.546 \times 10^{11}$ | $8.541 \times 10^{16}$ |
| α-Centauri | $2.572 \times 10^{30}$ | $0.412 \times 10^{17}$ | $6.342 \times 10^{11}$ |
| Sirius | $4.018 \times 10^{30}$ | $0.821 \times 10^{17}$ | $2.375 \times 10^{11}$ |

Henry Cavendish 1798

## CHAPTER 8

# RELATIVITY – AN IMPORTANT DIVERSION

Galileo asked the question, "If a ship is sailing along at constant speed through a calm sea and a stone is dropped from the masthead, where does the stone land?" The answer is at the foot of the mast. Ignoring any wind, the stone is not left behind as it falls. What Galileo was implying is that the laws of physics are the same in any system moving with constant velocity.

Galileo also drew attention to the relative motion of bodies moving with uniform speeds. He asked the question. "If two ships pass each other on the high sea, can a sailor on board, with no reference to land, say which one is moving?" The answer is no. Either ship could be taken as stationary, with the other moving at the relative velocity between them. Time t would be the same on both ships. Most of us have had that curious experience at a railway station while waiting in our carriage for the train to start. Suddenly there's movement, but is it our carriage or the carriage of the train alongside that's moving? It's only when we look at the station building that we can tell which is moving.

A long train moves with speed v along a straight track in the x-direction. It passes through a station and the time is set as t = 0 as the last carriage leaves the platform. We can choose two reference systems. The stationary (0) reference system has its origin at the end of the platform. The coordinates $x_0$ measure the distance along the track from the end of the platform to some point in the train. The moving (1) reference system has its origin at the end of the last carriage. The coordinates $x_1$ measure the distance to some point in the train from the end of the last carriage. At time t = 0, the two origins coincide.

At time t, a ticket inspector on the train is at a distance $x_1(t)$ from the end of the last carriage. At the same time, he will be at a distance $x_0(t)$ from the end of the platform. Thus

$$x_0(t) = x_1(t) + vt \text{ and } x_1(t) = x_0(t) - vt.$$

These two transforms of viewpoint of the position of the ticket inspector are called Galilean transforms.

From our perspective, the world has three dimensions: up, down; left, right; front, back. We choose a set of 3 perpendicular axes, labelled x, y and z to represent 3-D. This is a Cartesian frame of reference. A frame of reference moving with uniform speed is called an inertial frame. It does not involve acceleration. Galileo's ideas apply to inertial frames. The train moving along the track in the x-direction is in 1-D.

The ancient Greek philosophers thought that the Universe, or free space, was filled with a heavenly substance they called the ether. It was assumed to have fluid-type properties, so was able to swirl about in space. For several thousand years, it has remained a mysterious substance. In Newton's view, ether existed to support the force of gravitational influence between masses in the Universe, since force needed a medium in which to convey its effect. So, in modern terms of gravitational fields, ether flow was the flow of gravitational fields between masses in space. But the flow had to allow resistance-free movement of such masses in space. The flow of gravitational influence is analogous to fluid flow, but it is not a fluid. In the absence of sources of gravitation, the ether is stationary.

During the Victorian era, when electromagnetism was developed, it was assumed that the ether supported electric fields, too. So the ether had two properties. In the absence of electrical sources, the ether was assumed to be stationary. Maxwell was interested in the possibility of electromagnetic waves moving through the ether, like water waves, created by disturbing the surface of a pond by throwing in a stone. The wave disturbance is at right angles to the direction that the wave moves in. Maxwell described a mechanical model of an electromagnetic wave moving through the ether. His illustration contained an array of rollers with their axles perpendicular to the page. The direction of the angular velocity of each roller represented a line of magnetic force. Threaded vertically between and in contact with the rollers was a strip to represent the electric force field. Tugging downward on a strip gave the direction of the electric field and caused the magnetic rollers to rotate in opposite directions either side of the strip. The adjacent strip on the other side of the rollers moved upwards, and so on, alternating along the array. In this way, an instantaneous electromagnetic wave was generated across the array. It wasn't quite right; the model needed a bit of slack, as Maxwell was convinced that electromagnetic waves moved through the ether at the speed of light c.

Maxwell thought that light itself was an electromagnetic ether wave and suggested that the term luminiferous ether be used when dealing with such waves (to distinguish it from gravitational ether effects). With the discovery of electromagnetic waves by Hertz, in 1888 (see Chapter 26), most scientists accepted Maxwell's view that such waves were a disturbance in the ether.

Newton had dismissed Descartes' idea of a fluid ether swirling around the Sun. So, scientists assumed that the ether was stationary with respect to the Sun and that the planets moved through it. Many scientists, including Maxwell, then asked, "Is it possible to detect the Earth's averaged orbital speed, $v = 3 \times 10^4$ m/s, moving through the stationary ether?" We must remember that at the time the question was asked, the Solar System and the Universe were one and the same. The realisation that the Solar System was a tiny part of a galaxy and that the Universe was filled with innumerable galaxies all moving relative to one another, thus making it impossible to define a reference point to which a stationary ether could be tied to, was yet to come.

At this time, it was thought that the speed of light c was just like the speed of sound $c_s$, only their speeds were vastly different. If the speed of sound was measured in the direction of a wind moving with a steady speed v, it was known that the overall speed of sound was $c_s + v$. If the speed of sound was measured in the opposite direction then, from relativity, the overall speed of sound would be $c_s - v$. Applying relativity to the Earth moving with orbital speed v, the speed of light c moving through the stationary ether when measured on the Earth in the direction of motion was expected to be $c - v$ and in the opposite direction to be $c + v$.

Maxwell's prediction, in 1864, of the existence of electromagnetic waves travelling through the stationary luminiferous ether at the speed of light c, had aroused a great deal of interest in the scientific community. Hermann von Helmholtz, in Berlin, was particularly interested in the prediction and discussed the idea with his research student Heinrich Hertz. However, Hertz felt that the topic was too difficult to take on as a PhD study.

During the years 1880 to 1882, the US physicist Albert Michelson studied at several universities in Germany. Michelson had been born in Germany but he had emigrated with his family, aged just 2 years old, to the US in 1854. No doubt he had learnt German from his parents. Michelson arrived in Berlin and worked as an assistant in von Helmholtz's laboratory.

Assuming a stationary luminiferous ether, Maxwell had raised the question about using the speed of light c + v in the direction of Earth's orbital motion and comparing it with the speed of light c − v measured in the opposite direction to determine the Earth's orbital velocity v. While in Berlin, Michelson pondered on Maxwell's idea. Michelson was an expert in measuring the speed of light but he realised that since the value for c lay between $2.9 \times 10^8$ m/s and $3.1 \times 10^8$ m/s, the error band for measuring c was far greater than the Earth's orbital speed $v = 3 \times 10^4$ m/s. So, a straightforward way of using light to detect the difference between c + v and c − v was not possible. He needed to think of another way to detect the passage of light through the ether and hit on the ingenious idea of using light interference. Based on the wave theory of light, two superimposed, slightly out of phase, light waves produce interference fringes, enabling distance measurements to be made of the order of the wavelength of light, or $10^{-6}$ m.

The US scientist Alexander Graham Bell, born in Scotland, who invented the telephone, provided Michelson with the necessary funding to enable him to build his interferometer in Berlin and carry out the experiment there, which he did during 1881–1882.

The interferometer is formed by two interchangeable arms of length D at right angles. A single frequency light source is used to send a beam of light into the two L-shaped arms at the point O. This is done by using a 45° partially silvered mirror at O. Half the light beam passed straight through the 45° mirror along the arm $OM_1$ towards the mirror $M_1$. The other half of the light beam was reflected from the face of the 45° mirror and travelled along the perpendicular arm $OM_2$ towards the mirror $M_2$. The light remains in-phase after splitting at the 45° mirror.

The two light signals reflected back from the mirrors $M_1$ and $M_2$ were combined as they passed back through the 45° mirror and were then shone onto a viewing screen. For perfect rays of light, the two signals remain in-phase, exactly overlapping, so the intensity is a maximum and a bright spot of light would show on the screen. For any other condition, where the two signals were out-of-phase, the intensity would be reduced and the spot of light would be dimmed. For slightly imperfect beams, the spot of light becomes a series of concentric circular fringes.

The basis of the experiment was that the ether was stationary and that the Earth moved through it. Light was assumed to be a disturbance of the stationary ether and moved through it with a speed c. At the start of the experiment, the arm $OM_1$ of the interferometer was arranged pointing eastwards, and the arm $OM_2$ pointed northwards. The two arms were of equal length D but, to distinguish between the two arms, we let length $OM_1 = D_1$ and length $OM_2 = D_2$. At midnight, it was assumed that the eastward arm $D_1$ and the northward arm $D_2$ both moved eastwards with a speed equal to the Earth's orbital speed v.

Time $t_1$ is the time it takes for a light wave to travel from O to $M_1$ and back to O. Since light was assumed to be a disturbance of the stationary luminiferous ether, the speed of light moving along the arm from O to $M_1$ was expected to be $c - v$, while the speed of light back along the arm from $M_1$ to O was expected to be $c + v$. The time taken is given by the distance travelled divided by the speed.

$$t_1 = \frac{D_1}{(c-v)} + \frac{D_1}{(c+v)}$$

Rearranging, gives

$$t_1 = \frac{2D_1}{c} \frac{1}{\left(1 - \frac{v^2}{c^2}\right)}$$

Time $t_2$ is the time it takes for a light wave to travel from O to $M_2$ and back to O. In time $\tfrac{1}{2}t_2$ the mirror at $M_2$ has moved eastward a distance $\tfrac{1}{2}vt_2$. In the same time, the light wave travels a distance $\tfrac{1}{2}ct_2$ from O to $M_2$. Using Pythagoras' theorem,

$$\left(\tfrac{1}{2}ct_2\right)^2 = D_2^2 + \left(\tfrac{1}{2}vt_2\right)^2$$

Rearranging, gives

$$t_2 = \frac{2D_2}{c} \frac{1}{\sqrt{1 - \frac{v^2}{c^2}}}$$

# RELATIVITY — AN IMPORTANT DIVERSION

The difference in time $t_1 - t_2$ taken by the two light waves (the same wave before being split) to strike the viewing screen resulted in an interference pattern of circular fringes, which was used as the reference condition. Turning the interferometer by 90° about a vertical axis through O, so that the former eastern arm became a northern arm, was expected to cause a double fringe shift due to Earth's movement through the stationary ether.

However, within experimental error, no fringe shift was observed. The initial fringe pattern remained unchanged. The level of accuracy provided by Michelson's apparatus was more than sufficient to be able to detect any effect on the speed of light as it passed through the luminiferous ether, in any direction. But he failed to detect any effect. He published his negative results and scientists were puzzled.

In 1882, Michelson returned to the US where he had the good fortune to meet the US chemist Edward Morley. They discussed the Michelson interferometer and together they built an improved version. In 1887, they carried out a new experiment. But, again, no fringe shift was observed. The experiment was repeated at noon. Nothing! The experiment was repeated at different times of the day and year and with the arms in different directions. Nothing! The experiment has been repeated with better measuring techniques, but still nothing. It could be that the luminiferous ether moved with the Earth, but this possibility made no sense. Why should the Earth be special?

A few scientists examined the theory for the interferometer, searching for the cause of the null experimental result. For there to be no change in fringe pattern when the interferometer was rotated by 90°, suggested that light took the same amount of time to traverse each arm of the interferometer (arriving at the viewing screen in phase), regardless of the velocity v of the Earth. That is, $t_1 = t_2$. In other words, the speed of light c remains unchanged, regardless of the motion of the light source.

But if $t_1 = t_2$, this means that

$$D_1 = D_2 \sqrt{1 - \frac{v^2}{c^2}}$$

So, although the arms were measured to be the same length when laid side by side, it seemed that the length $D_1$ parallel to the direction of motion contracted, while the length $D_2$ perpendicular to the motion remained unaffected. This result was derived theoretically by the Irish physicist George FitzGerald in 1894. The length contraction is now known as the FitzGerald contraction. Generally, the contraction will not be noticeable unless the velocity v approaches light speed c. What was the cause of the apparent contraction? Was there something about light that made its use unsuitable in an experiment aiming to detect the ether?

Although the Michelson–Morley interferometer experiment of 1887 failed to detect the existence of the luminiferous ether, the implication of the failure was so important that Michelson (the inventor of the interferometer) won the Nobel Prize in Physics in 1907 for his work.

# Relativity – An important diversion

1632 Galileo Passing ships

1887 Michelson — Morley experiment fails to detect the ether

Albert Michelson 1887

1892-3 FitzGerald's length contraction

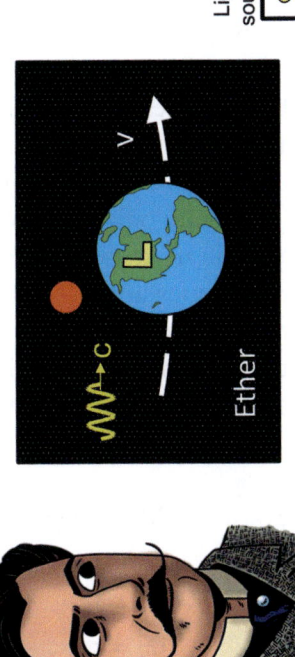

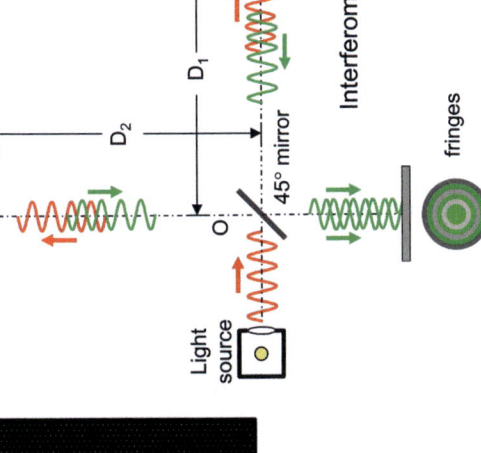

Michelson – Morley experiment

## CHAPTER 9

# THE LORENTZ TRANSFORM AND SPECIAL RELATIVITY

Special relativity is the study of events as they appear to observers moving with uniform speeds relative to one another. It formalises the question asked by Galileo about which ship is moving when two ships pass at sea. A sailor on board one ship observes that the other ship has energy due to its motion. That is, the motion of its mass. And the same view is taken by a sailor on the other ship. Where does this (kinetic) energy reside? Our study of gravity, so far, has really only dealt with static mass. When we look at mass dynamics, we will need to consider the energy question.

Let us update our 1-Dimensional Galilean transformation to a 3-Dimensional transformation using x, y and z perpendicular coordinate axes. The 3-D Galilean transforms of the coordinates used to change from the stationary (0)-frame to the (1)-frame moving with velocity v are given by

| Stationary (0)-frame | Moving (1)-frame |
|---|---|
| $x_0 = x_1 + vt_1$ | $x_1 = x_0 - vt_0$ |
| $y_0 = y_1$ | $y_1 = y_0$ |
| $z_0 = z_1$ | $z_1 = z_0$ |
| $t_0 = t_1$ | $t_1 = t_0$ |

Note that the time t in both frames is the same. This is Newton's Absolute time.

In 1904, the Dutch physicist Hendrik Lorentz proposed a theory to

# THE LORENTZ TRANSFORM AND SPECIAL RELATIVITY

account for the null result of the Michelson–Morley experiment, which incorporated the FitzGerald contraction and introduced the idea of local time for a frame of reference.

To modify the Galilean transform, Lorentz knew that he needed to incorporate the length contraction in the x-direction; the direction of relative velocity v. To make the mathematics less laborious, it is usual to introduce the factor γ (not to be confused with γ used for gravitational permeability) as

$$\gamma = \frac{1}{\sqrt{1-\frac{v^2}{c^2}}}$$

Also, because the speed of light c remained the same in both inertial frames, the time t had to change, too, between the stationary (0)-frame and the moving (1)-frame. Lorentz's new coordinate transforms were

Stationary (0)-frame        Moving (1)-frame

$x_0 = \gamma(x_1 + vt_1)$         $x_1 = \gamma(x_0 - vt_0)$

$y_0 = y_1$                $y_1 = y_0$

$z_0 = z_1$                $z_1 = z_0$

$t_0 = \gamma\left(t_1 + \frac{x_1 v}{c^2}\right)$       $t_1 = \gamma\left(t_0 - \frac{x_0 v}{c^2}\right)$

The above set of four coordinate equations is known as the Lorentz transformation. Note that the lengths measured perpendicular to the direction of motion in the two reference frames remain unaffected, but the time measured in the two reference frames is now different.

The stationary (0)-frame has perpendicular coordinate axes $x_0$, $y_0$, $z_0$. Suppose that there is a fir tree at the point P with coordinates $x_0(1)$, $y_0(1)$, $z_0(1)$ and an oak tree at the point Q with coordinates $x_0(2)$, $y_0(2)$, $z_0(2)$. We can calculate the distance between P and Q from Pythagoras' theorem. If we want to determine the distance PQ in terms of the moving (1)-frame coordinates then we must transform the $x_0$ coordinates into $x_1$ coordinates. We don't have to bother with transforming the $y_0$ and $z_0$ coordinates as they remain unchanged. Then we can use Pythagoras' theorem again to measure the length PQ in the moving (1)-frame.

In the stationary (0)-frame, the length $D_0$ measured along the $x_0$ axis is given by $D_0 = x_0(2) - x_0(1)$. When measured at time $t_1(1)$ in the moving

(1)-frame, the length along the $x_1$ axis is $D_1 = x_1(2) - x_1(1)$. We can determine the length $D_1$ in terms of the $x_0$ coordinates by transforming the $x_1$ coordinates into $x_0$ coordinates using the Lorentz transform. We find that

$$D_1 = D_0\sqrt{1 - \frac{v^2}{c^2}} = \frac{D_0}{\gamma}$$

confirming that $D_1$ is subject to the FitzGerald contraction.

Let us now introduce the idea of an event E; something that happens at a particular place in a coordinate reference frame at a particular time in that frame. For example, suppose in the (0)-frame the fir tree is struck by lightning at time $t_0(1)$. Suppose that the oak tree is struck simultaneously, at the same time $t_0(1)$. Due to the differences in the measurement of time between the (0)-frame and the (1)-frame, a simultaneous event in one frame involving two separate places will not be a simultaneous event in the other.

However, a simultaneous event involving one place in the (0)-frame (for example, a walker arriving at the fir tree just as the lightning struck it) will be a simultaneous event in the (1)-frame.

The Lorentz transform of time leads to the strange effect of time dilation. Suppose a time interval $T_1 = t_1(2) - t_1(1)$ is measured at a fixed point $x_1(1)$ in the moving (1)-frame and that the time interval observed in the stationary frame is $T_0 = t_0(2) - t_0(1)$. Applying the Lorentz transform to the time intervals, we find that

$$T_1 = T_0\sqrt{1 - \frac{v^2}{c^2}} = \frac{T_0}{\gamma}$$

This means that $T_1$ is less than $T_0$, written mathematically as $T_1 < T_0$.

Now, the time interval $T_1$ could be the periodic tick-tock of a clock at rest at the point $x_1(1)$ in the (1)-frame. Seen from the (0)-frame, the clock is moving with speed v and has a periodic tick-tock $T_1$. But $T_1 < T_0$, so an observer in the (0)-frame concludes that time passes more slowly when measured by a clock in the (1)-frame. This is encapsulated in the statement that to a stationary observer, 'Moving clocks run slow'. The increase in $T_0$ relative to $T_1$ is the effect called time dilation.

Time dilation leads to a twin paradox. Suppose that a space-travelling twin, wearing a wristwatch, leaves the Earth, while his twin, also wearing a wristwatch, remains behind. When the spaceship returns, the non-spacefaring twin checks the astronaut's watch and his own watch expecting the time difference to show him that by staying on Earth he had aged more. Similarly, the astronaut checks his brother's watch and his own watch expecting the time difference to show him that he had aged more during his space flight. Thus, we have a paradox. Based on symmetry, one might expect that both times are the same, but using the astronaut's watch is invalid because it doesn't take into account the effect that periods of acceleration have had on it during the space flight.

The French mathematician Henri Poincaré was the first scientist to realise what the failure of the Michelson–Morley experiment to detect the ether implied. There was no unique ether tied to one particular frame of reference. There were only relative inertial frames of reference and the ether for each inertial frame was the same; hence the null reading. No special inertial frame of reference against which all other inertial frames could be referred to meant that it was impossible to detect absolute motion. Furthermore, claimed Poincaré, the laws of physics must be the same in all relative inertial frames of reference. In 1904, Poincaré called this hypothesis the 'Principle of relativity'.

It was Albert Einstein, a German physicist working in the Swiss Patent Office, who pulled all the strands together and formulated the theory of special relativity, which he published in 1905.

Einstein ignored the ether; he did not dismiss its existence. If the ether does exist, its properties must be the same in every inertial frame, so a Lorentz transform must leave it unchanged. Scientists now view the ether in terms of quantum physics.

Based on the idea of length contraction in the direction parallel to the velocity v, Lorentz suggested that a speeding object would increase its mass. The British mathematician Oliver Heaviside had already come to the same conclusion. If $m_0$ is the mass in the stationary (0)-frame then in the moving (1)-frame the mass is $m_1$, where

$$m_1 = \frac{m_0}{\sqrt{1-\frac{v^2}{c^2}}} = \gamma m_0$$

This formula was tested by the German experimental physicist Walter Kauffman in 1900. From observations of the deflections of electrons moving at various speeds through a magnetic field, he found the results were in good agreement with the change in mass with velocity formula.

In 1905, Einstein introduced the idea that a mass at rest has intrinsic energy $E_0 = m_0c^2$ and that the energy-mass relationship, $E = mc^2$, holds in the general case.

The mass-velocity relationship stems from relativity, but I feel that the formula is a misrepresentation of what happens. For moving electric charge, the assumption is made that charge remains unchanged and is conserved. But, in the moving mass case, it is assumed that the mass changes. In my view, both charge and mass are conserved. In the moving charge case, the increase in energy is stored in the magnetic field. I believe that in the moving mass case, the increase in energy is stored in the gravitomagnetic field. This predicted but little-known field was first detected by US scientists in 2011 (Chapter 35). The relativistic mass-velocity formula denies the existence of the gravitomagnetic field. For moving conserved mass m, the classical kinetic energy of $½mv^2$ is the non-relativistic part of the gravitomagnetic energy.

As the speed v of a mass approaches that of light c, the mass-velocity formula predicts that the mass m becomes infinite. The conclusion is, then, that nothing can go faster than the speed of light. But remember that c is the terminal speed of an electromagnetic wave in the ether. The terminal speed of an acoustic wave in air is Mach 1, the speed of sound. But we are all familiar with supersonic aircraft which travel faster than the speed of sound. I suspect that the same applies for light speed. In my view, as a mass' speed approaches light velocity c its mass doesn't change, but the gravitomagnetic field surrounding the mass builds up like the shock waves in the acoustic case. So, the mass-velocity formula must be treated with caution. Although we have absolutely no idea how to propel a spacecraft at superluminal speed today, in the distant future, we may discover a way to do it.

In 1900, the German physicist Max Planck introduced the idea that the energy contained in electromagnetic wave radiation of frequency f is composed of discrete lumps of energy E, or quanta, given by $E = hf$. The SI unit of energy, or work, is the Joule (J). The tiny constant $h = 6.626 \times 10^{-34}$ J.s, called Planck's constant, is the smallest amount of angular momentum that can exist. These days, electromagnetic quanta are called photons.

Combining Einstein's $E = mc^2$ with Planck's $E = hf$, we see that photon energy $E = hf = mc^2$, indicating that a photon has effective mass given by $m = hf/c^2$. The higher the frequency f, the greater the energy E of a photon and the greater its effective mass, giving it more inertia and hitting power. In his miraculous year of 1905, Einstein used Planck's concept of radiation quanta to explain the photoelectric effect; why some radiation of frequency f, however intense, could not dislodge surface electrons, while less intense but higher frequency radiation could. It was all down to the individual hitting power of a photon.

With hindsight, we can use the ideas of Planck and Einstein to look at the Michelson–Morley experiment from a quantum viewpoint. This allows us to see why the interferometer of Michelson and Morley gave its strange result. Although light speed c remains constant, the wavelength of light does not; it changes with velocity. So, we cannot use the wavelength of light as a universal measure. The FitzGerald contraction is a correction factor needed to account for the change in wavelength. Contrary to what many scientists would have you believe, in my opinion, the FitzGerald contraction is an illusion.

If you are happy to forego my mathematical explanation of why I think that the Lorentz contraction is an illusion, then move on two paragraphs.

The wave formula, linking wave speed c with frequency f and wavelength $\lambda$, is $c = f\lambda$. Since a photon of frequency f has an effective mass $m = hf/c^2$, we see, from the wave formula, that it also has an effective mass $m = h/(\lambda c)$. Using the above mass–velocity formula (in place of gravitomagnetism) and replacing the mass m with the effective mass $m = h/(\lambda c)$ of a photon, we get $\lambda_1 = \lambda_0/\gamma = \lambda_0 \sqrt{(1 - v^2/c^2)}$. That is, the wavelength $\lambda_1$ in the moving (1)-frame appears to be shorter than the wavelength $\lambda_0$ in the stationary (0)-frame. In the stationary (0)-frame, the number of wavelengths for the arm $D_0$ is given by $n = D_0/\lambda_0$, where n may be a fraction. Thus, the length of the moving arm $D_1$ is $D_1 = n\lambda_1 = (D_0/\lambda_0)\lambda_1 = D_0\sqrt{(1 - v^2/c^2)}$. That is, the arm appears to suffer a FitzGerald contraction. The contraction term $\sqrt{(1 - v^2/c^2)}$ should be treated as a correction factor, needed because a light source in the stationary frame is used to make measurements in the moving frame.

Since $c = f\lambda$, we also have $f_1 = \gamma f_0 = f_0/\sqrt{(1 - v^2/c^2)}$. That is, in the moving (1)-frame, the frequency appears to increase (perhaps beyond light frequency). Since $E = hf$, it means that the photons in the (1)-frame have greater energy. Finally, $f = 1/\tau$, where the interval of time $\tau$ is the wave period. Consequently, we have the result $\tau_1 = \tau_0/\gamma = \tau_0\sqrt{(1 - v^2/c^2)}$, which means that $\tau_1 < \tau_0$. So, using light to measure time gives rise to time dilation. When v tends to c then $\tau_1$ tends to zero, so that time stands still for a photon. A stellar photon as it travels across the vast distances of space does not age. When a photon of starlight strikes our retina, we are looking back in time by an infinite amount, to when the photon left the star and began its vast journey across space. Note that $c = \lambda_0/\tau_0 = \lambda_1/\tau_1$, showing that the speed of light remains the same in any frame moving with uniform velocity.

The null result of the Michelson–Morley experiment did not prove the non-existence of the ether! But it did indicate that the speed of light remains constant when changing from one frame of reference moving with uniform speed to another moving with a different uniform speed. But, the frequency and wavelength both change! This is seldom mentioned.

We now consider the acoustic analogue of the Michelson–Morley experiment. For light waves moving through the ether, there is no unique reference frame. All motion is relative. For acoustic waves, stationary air provides us with a unique medium and a reference frame. There are two cases to consider:

1. A moving source of sound and a stationary observer. (To the observer, the acoustic waves are squashed or stretched.)
2. A stationary source of sound and a moving observer. (To the observer, the acoustic waves are speeded up or slowed down.)

For case 1, consider an observer on the platform waiting for a train. The observer hears the warning whistle of an approaching non-stop express train rise in pitch, or frequency. The sound waves emitted by the whistle have frequency $f_E$ and are unaltered in the (1)-frame of the moving engine. But to a stationary observer (receiver) on the platform, the approaching sound waves bunch up in the (0)-frame, squashing the wavelengths. According to the wave formula $c_s = f\lambda$, where $c_s$ is the speed of sound, a decrease in wavelength $\lambda$ means an increase in frequency $f$. So, in the (0)-frame, there is a rise in frequency to $f_R$.

In 1842, very early on in the history of railways, the Austrian scientist Christian Doppler carried out experiments using a trumpet (rather than a whistle), emitting a single note of constant frequency $f_E$ on a train moving with speed v to determine the change of frequency $f_R$ received by a stationary observer as the sounding trumpet approached. According to Doppler, the frequency increase of the approaching trumpet note is given by

$$f_R = \left(\frac{c_s}{c_s - v}\right) f_E$$

As the express train leaves the platform behind, the wavelengths of the trumpet note get stretched, so the frequency drops. The frequency decrease of the departing trumpet note is given by

$$f_R = \left(\frac{c_s}{c_s + v}\right) f_E$$

To consider case 2, we assume that the trumpeter is on the platform, the (0)-frame, and the observer is a passenger on the express train, the (1)-frame.

As the train approaches the platform, the speed of sound in the (1)-frame has increased to $c_s + v$, although the wavelengths haven't changed. The trumpet note received by the passenger leaning out of the carriage window has an increasing frequency $f_R$ given by

$$f_R = \left(\frac{c_s + v}{c_s}\right) f_E$$

As the train leaves the platform behind, the trumpet note received by the passenger has a decreasing frequency $f_R$ given by

$$f_R = \left(\frac{c_s - v}{c_s}\right) f_E$$

Note that case 1 and case 2 give different results because we have a unique medium against which to analyse results. In case 1, the air is squashed or stretched by the acoustic signal, while in case 2, we have a reference frame tied to stationary air, against which we can measure the velocity of the observer.

If the wind blows in the direction of motion with speed w, we must replace $c_s$ with $c_s + w$. If the wind blows in the opposite direction, we must replace $c_s$ with $c_s - w$.

We now consider the Doppler Effect for light. There is no unique ether and, therefore, no unique reference frame for the ether. Suppose we choose our light source, of frequency $f_E$, to be at the origin in the (0)-frame and that the observer, at the origin in the (1)-frame, moves away from the source with speed v. The change in frequency $f_R$ of the light source seen by the observer follows the same pattern as that for Case 2 in the acoustic case. In the Doppler formula, the c now refers to the speed of light and from special relativity we must account for time dilation by introducing the factor γ. Thus, the Doppler Effect seen by an observer ((1)-frame) moving away with relative speed v of a light source with frequency $f_E$ (in the (0)-frame) is given by

$$f_R = \left(\frac{c-v}{c}\right)\gamma f_E = \sqrt{\frac{c-v}{c+v}}\, f_E = \left(\sqrt{\frac{1-\frac{v}{c}}{1+\frac{c}{v}}}\right) f_E$$

Since velocities are relative, it applies to a fixed observer and a source moving with speed v, too. So a light source moving away has a lower frequency. In other words, the light signal is red-shifted in frequency.

It wasn't until the late 1920s that the US astronomer Edwin Hubble realised that the spiral nebulae seen in space with giant telescopes were actually galaxies, containing millions of stars, separate from the Milky Way Galaxy. Our Sun is a small star in the Milky Way Galaxy. Within a few years, spectrographic studies of many of these other galaxies showed that most had a red shift, or a reduction, in the frequency of their starlight as seen from Earth. However, the Andromeda Galaxy, the nearest galaxy to the Milky Way Galaxy, had a blue shift. Based on the Doppler formula for light, this suggested that the blue-shifted Andromeda Galaxy is on a collision course with the Milky Way Galaxy. But most of the galaxies are red-shifted, which suggested that most galaxies are moving away from us at speeds which are estimated to be a significant fraction of the speed of light c. The further away these galaxies are from us, the faster they are moving away from us. This led to the Big Bang Theory; that the birth of the Universe began with a cataclysmic explosion of a single mass of unimaginable concentration. Originally, it was thought that gravitational attraction would eventually bring the galaxies to rest, followed by a period of contraction, leading to the Big Crunch, where all the galaxies were compressed into a single point of mass. The Big Bang theory would then undergo a cyclic repetition of an expanding and a contracting Universe. But this idea has been thrown into doubt because we have recently learnt that many galaxies are not just moving away from us but are actually accelerating away from each other, too. There appears to be some form of anti-gravitational effect in space speeding up the separation of galaxies relative to one another. If we could understand and make use of this effect, it would allow us to control gravity.

# Special relativity

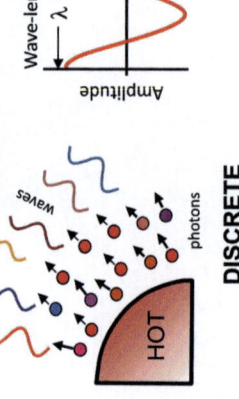

Max Planck 1900

Energy quanta

$E = hf$

Planck's constant h
Frequency f

| 1900 | Lorentz and Heaviside | Mass increases with velocity |
| 1904 | Lorentz | Transforms |
| 1904 | Poincare | Principle of Relativity |
| 1905 | Einstein | Photons and $E = mc^2$ |
| 1905 | Einstein | Special relativity |

CHAPTER 10

# GRAVITY AND ACCELERATION ARE EQUIVALENT

We have examined frames of reference moving with uniform velocities relative to one another and concluded that any frame can be treated as a stationary frame. Such frames of reference are called inertial frames. We now look at the case where frames of reference accelerate. This includes linear acceleration and radial acceleration experienced during rotation.

Most of us have had the experience of standing in a lift and feeling heavier as it starts to smoothly accelerate upwards. Accelerating upwards is equivalent to an increase in the force of gravity acting downwards and we experience a change in weight. Once a uniform speed is reached, we are travelling in an inertial reference frame, so we feel no change of weight. As we smoothly decelerate, to stop at an upper floor, momentarily, we feel lighter. Deceleration is equivalent to a force of gravity acting in the same direction. Similarly, on going down in the lift during the initial downward acceleration, we feel lighter, as though the force of gravity has been reduced. Then, momentarily, we feel heavier as we decelerate to stop at a lower floor.

Acceleration is equivalent to gravity operating in the opposite direction. Deceleration is equivalent to gravity operating in the same direction.

Although acceleration and gravity are equivalent, they are different phenomena. Acceleration has to be applied directly to a mass to force it to move. Gravity, on the other hand, exerts its influence on mass across space, forcing it to move without any apparent contact. Gravity has field properties but acceleration does not.

When a car accelerates forward, we are pressed back into our seat.

When the car's brakes are applied, we are thrown forward during the deceleration. Everyone knows this, but why does it happen? "Oh, that's just inertia," you might say. Yes, but what is inertia? "Oh, inertia is the resisting force experienced when a mass accelerates, or decelerates," you might reply. Yes, but what causes inertia? "Mmm."

Let's start by saying that mass responds to gravity. Then we might surmise that the cause of inertia is that when a mass accelerates in one direction, an equal and opposite gravity field is induced, or generated, in the mass in the opposite direction. The mass experiences this induced gravity field as a resistance to its acceleration.

When a mass is set free in a gravity field it responds by accelerating in the direction of the field. From our supposition, in accelerating, an equal and opposite gravity field is induced in the mass. The result is the two gravity fields cancel. The mass is in free fall. From Galileo's experiments, we know that if a number of objects are dropped together, whatever their mass, they will all fall together. They are all in free fall and weigh nothing.

Most countries with space programmes have drop shafts to carry out microgravity experiments on the surface of the Earth. In a drop shaft, experiments are carried out under free-fall conditions equivalent to near zero-g, referred to as microgravity. Probably the best known drop shaft in the USA is the one at the NASA Glenn Research Center at Cleveland, Ohio, which provides about 5 seconds of microgravity. The Japanese drop shaft is situated at the IHI facility, on Hokkaido Island. It provides a period of 10 seconds of microgravity. The European drop shaft, or Fallturm, is based at the ZARM facility in Bremen, in Germany, where 5 seconds of microgravity is available. The Indian drop tower is in Madras and provides 2.5 seconds of microgravity. Less well known are the huge drop tower at Chelyabinsk, in the foothills of the Urals in Russia, near the border with Kazakhstan, and the Chinese drop tower at the National Microgravity Laboratory in Beijing.

Even a photon responds to gravity. According to German physicist Max Planck, a photon has energy $E = hf$, where f is its frequency of vibration and h is a constant, called Planck's constant ($h = 6.626 \times 10^{-34}$ J.s). According to Einstein, the relationship between mass m and energy is $E = mc^2$, where c is the speed of light. Combining the two equations, we see that a photon has effective mass $m = hf/c^2$. So, a photon will be affected by gravity. It was Albert Einstein who pointed out that stellar light rays should bend if they passed close to the Sun. This prediction was confirmed experimentally by the English astronomer Arthur Eddington during a solar eclipse in 1919. The result created a great amount of publicity and made Einstein famous.

But what happens if you try to accelerate a photon? A photon cannot go faster than the speed of light. Trying to accelerate a photon (or its source) results in the photon changing its frequency.

During curved motion, a mass experiences radial acceleration. Again, from equivalence, this shows up as a gravity effect. When we corner to the left in a car, the radial acceleration is directed inwards to the left and we are forced in the opposite direction, outwards to the right.

When Galileo used his inclined plane to examine balls falling in Earth's gravity field, as well as letting balls just roll down, he started some rolling sideways across the plane and he noticed that the curve they made was a parabola.

Newton went a step further and, in Book 3 of his *Principia*, he imagined firing a cannonball horizontally from a cannon gun mounted on a hilltop. He assumed that after an initial period of acceleration, the ball would have a uniform horizontal speed (ignoring wind resistance) but would continue to accelerate vertically downwards, in free fall, due to Earth's gravity $g_0$. During this time, the ball would follow a parabolic curve, before hitting the Earth.

What Galileo and Newton both assumed was that near to the Earth's surface, the acceleration due to gravity $g_0$ was roughly constant. But Newton predicted that the Earth's gravity field followed an inverse square law of distance r from the Earth's centre. So, as the distance from the Earth increases, there is a rate of change of gravity. From equivalence, this means as a test mass increases its distance from Earth it will experience a steady rate of change of acceleration. This is called jerk. There is also a rate of change of jerk called jounce, and so on. For the Earth's gravitational field alone, the effect due to gravity is

$$\text{Acceleration} = \frac{GM_E}{r^2} \text{ m/s}^2 \quad \text{Jerk} = \frac{(2GM_E)^{\frac{3}{2}}}{r^{\frac{7}{2}}} \text{ m/s}^3 \quad \text{Jounce} = \frac{7}{2}\frac{(2GM_E)^2}{r^5} \text{ m/s}^4$$

These are smooth and continuous values. Astronauts experience increasing acceleration and jerk during the launch phase of their rocket. Similarly, an oscillating mass on a spring experiences acceleration, jerk and jounce but, being smooth, jerk and jounce are not obvious effects.

A more common example of jerk is experienced by a driver who pulls away very quickly in their car when the traffic lights turn green. While the acceleration is increasing, jerk is present but, again, it is not that obvious. A better example, where jerk is abrupt and very noticeable, is with a whirling mass at the end of a string, which is suddenly let go. The radial acceleration experienced by the whirling mass almost instantaneously vanishes and during this very brief instant of time the mass experiences jerk.

The old-fashioned roller coaster is a good example of the conservation of energy, providing we ignore friction. An unpowered upright car with passengers is raised to a high point and then let go down an undulating track with dips and humps. As it drops down a dip, it exchanges gravitational potential energy for kinetic energy. As it climbs up a hump, it exchanges kinetic energy for potential energy. Passing over the top of a hump with any speed leads to one feeling lighter due to radial acceleration acting in the same direction as gravity. If the downward track is steep, then any acceleration by the descending car in the same direction as gravity leads to one feeling lighter. Passing round a dip with speed leads to radial acceleration in the opposite direction to gravity and the feeling of being squashed in one's seat. Coming out of a curved section of a roller coaster to a straight, or level, section, there is a change from radial acceleration to zero acceleration. This sudden change in acceleration results in an abrupt, discontinuous spike-like jerk which can cause passenger discomfort. Modern designs of roller coaster tracks include complete loops where the car travels on the inside of the loop. Rather than circular loops, teardrop-shaped loops (called clothoids) are used, so that the transition to a level section is smoothed and abrupt jerk avoided.

To give would-be astronauts some experience of the weightlessness they would encounter in space, an extreme form of roller-coaster ride was conceived. Russia and the USA have led the way with designated aircraft used to perform a series of hump-backed flight paths, providing periods of weightlessness for the occupants of the planes. The plane's horizontal speed is maintained but on the upward parabolic flight path the plane decelerates, until at the top of a hump it reaches zero upward velocity. This period of upward deceleration is equivalent to an upward gravity force which cancels with Earth's gravity force, so the plane's occupants are weightless. After passing the top of the hump, the downward flight continues on its parabolic curve during which it is in free fall, so the plane's occupants remain weightless. The period of weightless-parabolic flight lasts for about 20 seconds. During each flight, the plane performs a number of hump-backed motions providing a series of weightless periods for the plane's occupants. Outside of the weightless zones, the plane's occupants experience an enhanced gravity field of 1.8$g_0$.

To earn their keep, these aircraft are now used in commercial ventures, offering the public the chance, for a fee, to experience weightlessness. The strange up and down sensations of the aircraft often results in nausea and sickness by the passengers. The crew of the Russian aircraft refer to their plane as the Vomit Comet. In the USA, such flights are offered by the Zero Gravity Corporation. One of the most famous participants of their flights was the late Professor Stephen Hawking.

Newton was one of the first scientists to realise that if a cannonball was given a big enough impetus, it would continue to fall around the Earth, never actually impacting with it, just like the Moon. In such a case, the curved flight path would become circular, or slightly elliptical. This is a special case of curved motion giving rise to radial acceleration. The International Space Station (ISS) is an artificial moon. For it to remain in orbit, its radial acceleration must equal that of the Earth's gravity field. The ISS orbits at an altitude of 400 km. From Newton's inverse square law we can calculate the force of gravity at this distance from the Earth's centre. The radial acceleration is given by $v^2/R$, where v is the orbital speed of the ISS and R is its distance from the Earth's centre. For the two to balance, this requires v = 7.66 km/s.

Interestingly, the mass of the ISS is not a factor for it to stay in correct orbit. So, attaching visiting spacecraft or adding crew members doesn't make any difference. This reflects the difficulty that we had in trying to determine the mass of the Moon. At first sight, this seems odd, given that the centrifugal force on the ISS increases with its mass. But the clue is that the ISS is in free fall and the acceleration is the same for all bodies, whatever their mass, as Galileo was one of the first to observe.

Astronauts in the International Space Station (ISS) usually require three, or four, days to acclimatise to weightlessness. Lack of gravity causes a major health problem for astronauts. In deep space, where gravity is much reduced, astronauts will float about. Only if their spaceship accelerates in some way, either linearly or in a curved manner, will they experience a force akin to gravity.

In the MGM film *2001: A Space Odyssey*, the wheel-shaped space station between the Earth and the Moon is shown rotating smoothly so as to provide artificial gravity (the effect of radial acceleration), allowing the inhabitants to walk around the inside rim. It's a space version of the circus wall-of-death motorbike ride. In the film, even the spaceship going to Jupiter has a rotating section to provide artificial gravity for the crew.

In view of Einstein's subsequent use of the equivalence between gravity and acceleration in his model of gravity, it is worth noting that the linear acceleration of mass is directly tied to gravity, whereas the radial acceleration of mass is linked, via path curvature of the mass, with mass velocity and a new force field called gravitomagnetism. But details of this must wait until later.

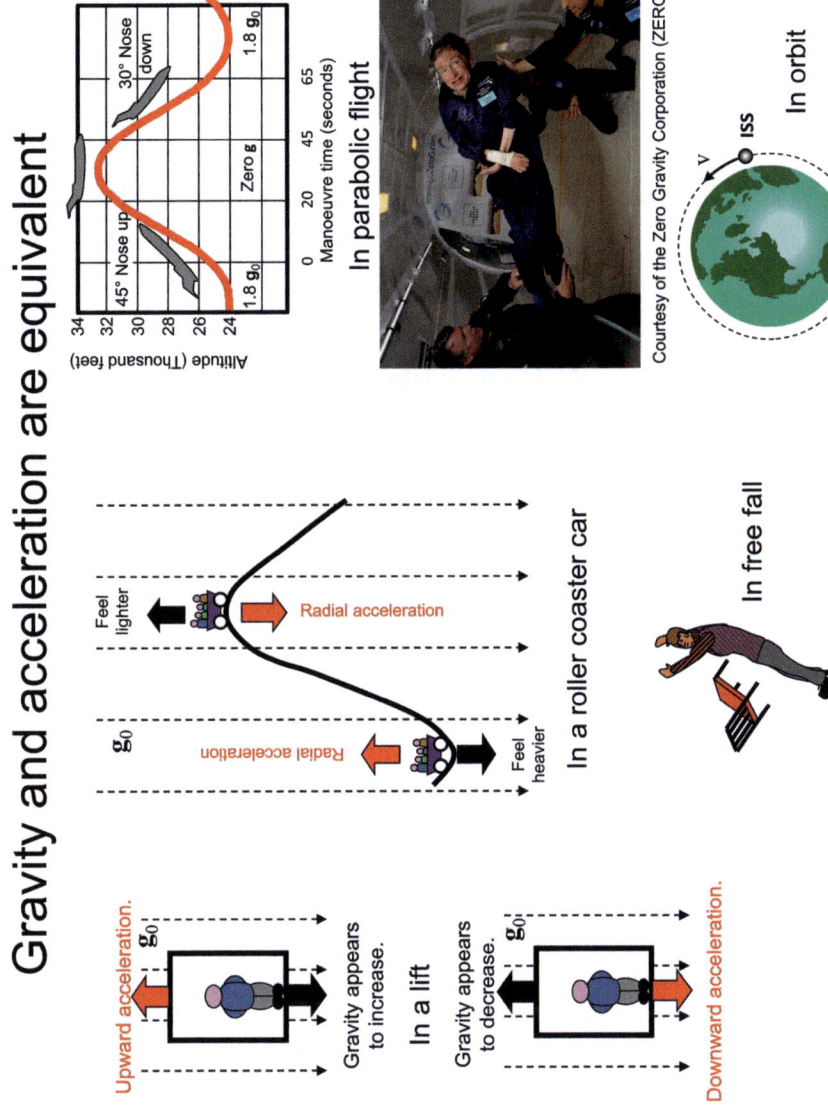

CHAPTER 11

# SPACE AND MINKOWSKI'S SPACE-TIME

Euclid taught mathematics at the University of Alexandria some time during the 3rd century BC. His most famous book, in thirteen parts, is *The Elements,* in which he collected together all the known work on geometry and arithmetic. *The Elements* is the oldest scientific textbook in history still in use today. The first part starts by defining things that geometry deals with, such as points, straight lines, plane (flat 2-D) surfaces, angles, circles, rectilinear figures, and so on. Then comes a number of postulates based on the definitions. Euclid deals with many plane geometrical shapes, such as triangles, rectangles, parallelograms, circles, and so on, as well as solid 3-D shapes, such as oblongs, cubes, pyramids, spheres, and so on. He states geometric propositions, or theorems, and gives their proof. The best known is Pythagoras' theorem for a right-angled triangle in 2-D, where he proves that the square on the hypotenuse is equal to the sum of the squares on the other two sides. Pythagoras' theorem can be extended to 3-D, where the square of the hypotenuse is equal to the square of three sides of a right-angled pyramid. Some readers will remember, as schoolchildren, working through one of Euclid's geometric proofs and writing QED at the end of it. QED stands for the Latin phrase *Quad erat Demonstrandum,* meaning *That which was to be proved has been proved*.

If we accept that a circle has 360° (based on the ancient Babylonian idea, from around 2,400 BC, that a complete circuit of a year took 360 days), then crossing perpendicular lines make four right-angles of 90° and a straight line is 180°. On this basis, the interior angles of a triangle add up to 180°. In Euclid's plane geometry, the shortest distance between two points is a straight line, and parallel straight lines never cross.

A scalar is just a number; so, examples are mass (kg), length (m), speed (m/s), temperature (°C), and so on. A vector is a scalar with an associated direction, and the symbol representing it is often printed in bold type. Some examples are distance **d** (1m, eastward), velocity **v** (50 km/hr, north-westwards) and the acceleration due to gravity at Earth's surface $\mathbf{g}_0$ (9.8 m/s², vertically downwards).

The French philosopher René Descartes initiated the use of algebra in conjunction with Greek geometry. Assuming that space is a 3-D form of Euclidean geometry, Descartes introduced the idea that a 3-D framework, made of 3 perpendicular axes originating from a point in space called the origin, could be used to determine all points in space relative to the origin. The 3 coordinate axes of this Cartesian framework are represented by x, y and z directions, with three corresponding unit vectors **i**, **j**, **k**. Unit vectors are signposts of direction. Thus, the vector **i** points in the x direction and the length of the signpost is 1 unit, etc. We can draw the vector **d** in Euclidean space as an arrow from the origin to the point with coordinates (x, y, z) and write it symbolically as

**d** = x**i** + y**j** + z**k**

Sometimes, to make things easier, we might replace the vector components x, y, z with $x_1, x_2, x_3$ and the unit vectors **i, j, k** with $\mathbf{e}_1, \mathbf{e}_2, \mathbf{e}_3$ and introduce the Greek capital letter S, or Sigma, to sum and write

$$\mathbf{d} = \sum_{p=1}^{3} x_p \mathbf{e}_p = x_1 \mathbf{e}_1 + x_2 \mathbf{e}_2 + x_3 \mathbf{e}_3$$

Or, even simpler, we might just write

**d** = [$x_1$ $x_2$ $x_3$]

This horizontal line is called a row vector. Or, transposing, we can write the components in a vertical line to give us a column vector:

$$\mathbf{d}^T = \begin{bmatrix} x_1 \\ x_2 \\ x_3 \end{bmatrix}$$

Pythagoras' theorem in 3-D space can be written as

$$(d)^2 = \mathbf{d}.\mathbf{d} = \mathbf{d}\,\mathbf{d}^T = [x_1\ x_2\ x_3]\begin{bmatrix}x_1\\x_2\\x_3\end{bmatrix} = (x_1)^2+(x_2)^2+(x_3)^2$$

We slide the line vector down over the column vector and multiply the components.

In 1906, the French mathematician Henri Poincaré introduced the idea of an imaginary fourth dimension of length, involving time. The new dimension was at right angles to the other three dimensions in space. An imaginary length measured in the new dimension was ict, where i stands for imaginary and ct is the distance moved by a light pulse in time t. Any point in Poincaré's 4-D dimensions of space-time has coordinates (x, y, z, ict). For ct = 1, we get an imaginary length, or number, i, in the new dimension. The magical property of the imaginary number i is that i × i = $i^2$ = -1.

A rectangular array of numbers is called a matrix. In modern mathematical form, the array is enclosed in brackets [ ]. The matrix was first developed to solve a set of linear equations and has a long history stretching back to ancient Babylon and ancient China, where 2,000-year-old examples have been found of their use. Following the vector terminology, the matrix is made up of rows of numbers or columns of numbers. Matrix mathematics has its own rules of operation. The advent of the electronic computer is ideally suited for dealing with vast arrays of numbers, so the use of matrices (the plural of matrix) has burgeoned.

As an example, think of the matrix as an operation, or set of instructions, which take something from a starting place and time to a finishing place and time. That something might be a passenger on a train. The rail network has a set of instructions to get the train from the start of its journey to its final destination. The origin of the (0)-frame is the station platform of departure. The origin of the moving (1)-frame is the guard's van, or caboose, at the end of the train. A passenger can move about the train. The distance-time coordinates of a passenger from the start of the journey are $(x_0, y_0, z_0, t_0)$. The distance-time coordinates of the movement of the passenger on the train are $(x_1, y_1, z_1, t_1)$.

Mathematicians love to establish patterns. The equations for the Lorentz transformation in 4-D given in Chapter 9 can be written in matrix form.

$$\begin{bmatrix} x_0 \\ y_0 \\ z_0 \\ ict_0 \end{bmatrix} = \begin{bmatrix} \gamma & 0 & 0 & -i\beta\gamma \\ 0 & 1 & 0 & 0 \\ 0 & 0 & 1 & 0 \\ i\beta\gamma & 0 & 0 & \gamma \end{bmatrix} \begin{bmatrix} x_1 \\ y_1 \\ z_1 \\ ict_1 \end{bmatrix} \quad \text{where } \beta = \frac{v}{c} \text{ and } \gamma = \frac{1}{\sqrt{1-\beta^2}}.$$

The 4 × 4 array of numbers on the right-hand side is the Lorentz transform matrix **L**. At zero time in both frameworks, the two origins and the coordinate axes of both frameworks coincide.

On the left-hand side of the equals sign, the column vector represents the distance $s_0$ moved in time $t_0$ in the 4-D space-time (0)-framework. The column vector $s_1$ on the right-hand side represents the distance moved in time $t_1$ in the 4-D space-time (1)-framework, moving with speed v relative to the (0)-framework. To regain the Lorentz transform equations, we multiply the Lorentz matrix **L** by the 4-D space-time (1)-coordinates.

For example, to obtain the $x_0$ coordinate in terms of the (1)-coordinates, we slide the first row of the matrix over the end and down the (1)-coordinate column and sum the multiplied terms; $\gamma$ times $x_1$ plus 0 times $y_1$ plus 0 times $z_1$ plus $-i\beta\gamma$ times $ict_1$. You can check (also see Chapter 9) that this gives

$$x_0 = \gamma x_1 + \gamma v t_1 = \gamma(x_1 + vt_1)$$

The other (0)-coordinate values are obtained using the same method of sliding the opposite row of the matrix over the end and down the (1)-coordinate column, multiplying the terms and summing the result.

In terms of line vectors and the Lorentz matrix, we can write

$$\mathbf{s}_0 = \mathbf{L}\mathbf{s}_1$$

Suppose a light bulb is switched on at the origin in the 3-D (0)-frame of Euclidean space. At any time $t_0$ later the radius of the light sphere centred on the origin is $ct_0$. If $(x_0, y_0, z_0)$ are the 3-D coordinates of a point on the light sphere then, from Pythagoras' theorem in 3-D, $(x_0)^2 + (y_0)^2 + (z_0)^2 = (ct_0)^2$.

In the 4-D (0)-frame, the distance from the origin to the point $(x_0, y_0, z_0, ict_0)$ is the vector $\mathbf{s}_0 = [x_0, y_0, z_0, ict_0]$. The square of the distance $s_0$ is given by the scalar product $\mathbf{s}_0 \cdot \mathbf{s}_0 = (s_0)^2 = (x_0)^2 + (y_0)^2 + (z_0)^2 + (ict_0)^2$, which is zero.

Poincaré showed that the square of the 4-D distance is invariant.

That is, for the (0)-frame and the (1)-frame, $(s_0)^2 = (s_1)^2$. Nought equals nought is always true.

We can see this with the Lorentz transform.

$$(s_0)^2 = s_0 s_0^T = (s_1 L)(s_1 L)^T = s_1 L L^T s_1^T = s_1 s_1^T = (s_1)^2$$

where

$$LL^T = \begin{bmatrix} \gamma & 0 & 0 & -i\beta\gamma \\ 0 & 1 & 0 & 0 \\ 0 & 0 & 1 & 0 \\ i\beta\gamma & 0 & 0 & \gamma \end{bmatrix} \begin{bmatrix} \gamma & 0 & 0 & i\beta\gamma \\ 0 & 1 & 0 & 0 \\ 0 & 0 & 1 & 0 \\ -i\beta\gamma & 0 & 0 & \gamma \end{bmatrix} = \begin{bmatrix} 1 & 0 & 0 & 0 \\ 0 & 1 & 0 & 0 \\ 0 & 0 & 1 & 0 \\ 0 & 0 & 0 & 1 \end{bmatrix}$$

Suppose in the (0)-frame of space-time that an event $E_0(1)$ has coordinates $(x_0(1), y_0(1), z_0(1), ict_0(1))$ and that an event $E_0(2)$ at a different place and different time has coordinates $(x_0(2), y_0(2), z_0(2), ict_0(2))$. If the events $E_0(1)$ and $E_0(2)$ occur simultaneously in the (0)-frame, when $t_0(1) = t_0(2)$, they will not occur simultaneously in the (1)-frame, moving with relative velocity v, unless they are located at the same place, too.

Suppose that the events $E_0(1)$ and $E_0(2)$ in the (0)-frame are not co-located, but are nearby; a short (incremental) distance apart. The vector joining the two events may be represented by a 4-D line increment $[ds_0] = [dx_0\ dy_0\ dz_0\ icdt_0]$, where $dx_0 = x_0(2) - x_0(1)$, etc.

Following the same pattern as before, the scalar product gives the square of the short distance between the events:

$$ds_0 \cdot ds_0 = (ds_0)^2 = (dx_0)^2 + (dy_0)^2 + (dz_0)^2 - (cdt_0)^2$$

The square $(ds)^2$ of the incremental length ds is called the space-time metric. In matrix form, we have

$$(ds_0)^2 = \begin{bmatrix} dx_1 & dy_1 & dz_1 & icdt_1 \end{bmatrix} \begin{bmatrix} 1 & 0 & 0 & 0 \\ 0 & 1 & 0 & 0 \\ 0 & 0 & 1 & 0 \\ 0 & 0 & 0 & 1 \end{bmatrix} \begin{bmatrix} dx_1 \\ dy_1 \\ dz_1 \\ icdt_1 \end{bmatrix} = (dx_1)^2 + (dy_1)^2 + (dz_1)^2 - (cdt_1)^2$$

In 1908, Hermann Minkowski, a German mathematician from Lithuania, at that time a province of Russia, took up and expanded Poincaré's 4-D space-time idea. However, he chose to make the 4$^{th}$ space-time coordinate real, using ct instead of ict. An event in Minkowski's space-time has coordinates (x, y, z, ct). Although Minkowski dispensed with the imaginary space-time coordinate, the space-time metric remains unchanged.

$$(ds_0)^2 = \begin{bmatrix} dx_1 & dy_1 & dz_1 & cdt_1 \end{bmatrix} \begin{bmatrix} 1 & 0 & 0 & 0 \\ 0 & 1 & 0 & 0 \\ 0 & 0 & 1 & 0 \\ 0 & 0 & 0 & -1 \end{bmatrix} \begin{bmatrix} dx_1 \\ dy_1 \\ dz_1 \\ cdt_1 \end{bmatrix} = (dx_1)^2 + (dy_1)^2 + (dz_1)^2 - (cdt_1)^2$$

If the 4-D incremental length ds is small, we can follow Poincaré's invariance idea that $(ds_0)^2 = (ds_1)^2$. The metric remains the same in whatever inertial frame it is measured.

It is difficult to imagine the four axes in 4-D space-time all perpendicular to one another. However, most of us, especially hill walkers, will have seen a contour map. This is a 2-D plane form of Earth's 3-D surface, where the height has been suppressed and replaced with contour lines. Where the contour lines are close together, the land rises steeply. Where the land is nearly flat, the contour lines are far apart. So, a 2-D map gives us a view of 3-D topography. But we are dealing with the surface of the spherical Earth, which is not a flat surface (although it may appear to be so, locally), so our 2-D map suffers from some distortion. For Euclidean space, a 2-D surface is flat. Consequently, we can suppress the height (say z-axis) in Euclidean 3-D space and introduce the 4th vertical coordinate ct in its place without introducing any distortion, so that we have a 3-D form of 4-D.

For this 3-D version of space-time, Minkowski introduced the concept of the light-cone. At any event E with coordinates (x, y, ct) in 3-D space-time, he formed a double right-circular cone, with the cone-axes parallel to ct, and with both cones sharing a common apex at the event. The apex is treated locally as an origin. Letting ct = 0, the initial event at the apex happens at the present time with coordinates (0, 0, 0). The upper inverted cone spreads into the future and the lower cone shrinks from the past. Suppose the event is switching on a light bulb at the 3-D space-time point (0, 0, 0). In 3-D x-y-z space at time t the light wave-front forms a sphere of radius ct, but in 3-D x-y-ct space-time at time t the light wave-front spreads out in the 2-D x-y plane to form a circle with a radius equal to ct. At any point (x, y) on the circle, Pythagoras' theorem gives $x^2 + y^2 = (ct)^2$. In 3-D flat space-time, as the circle moves upwards with distance ct, its radius ct increases by the same amount and it traces out a conical surface on which $x^2 + y^2 = (ct)^2$. The apex angle between the vertical ct axis and a straight line in the cone surface is 45°.

From symmetry, we can extend the cone backwards in time. Starting at some time -t in the past, as the space-time -ct reduces to zero, the light-wave front shrinks to zero and the light cone shrinks to the apex at the point where the event happens.

On the double cone surface, we see that $s^2 = x^2 + y^2 - (ct)^2 = 0$ for all ct. The cone surface is called light-like as it represents the distance reached by the wave-front of light in the x-y plane at that particular time t. On any x-y plane, an observer at a point lying outside the circle, where $s^2 > 0$, will be unaware of the approaching light signal. This region is called space-like. While for a region on the x-y plane within the cone, where $s^2 < 0$, the light signal will have already passed by and the region is called time-like. The movement of any particle, passing through the origin at local time t = 0, must always stay within the cone, because nothing can go faster than the speed of light. In 3-D space-time, the path of such a particle will follow a world-line within the double cone. Any point on the world-line will have its own light-cone.

The future cone shows the region of space-time where a later event can be influenced by the initial event E at the cone apex. The past cone shows the region of space-time where an earlier event may influence the initial event E at the cone apex.

The world-line of a stationary particle is a straight vertical line. The world-line of a particle moving with constant velocity v is a straight line with constant slope. A photon moving at the speed of light c moves along a straight line with a slope 1 at an angle of 45°. The world-line of an accelerating particle is curved.

For two nearby events occurring at the same time (simultaneously) in 3-D space-time, $(dt)^2 = 0$, so $(ds)^2 = (dx)^2 + (dy)^2$ is positive. For two events which occur at the same place in 3-D space-time but at slightly different times, $(dx)^2 = (dy)^2 = (dz)^2 = 0$, so $(ds)^2 = -(cdt)^2$ is negative.

Suppose the (1)-frame moves with uniform speed v in the x-direction as measured in the stationary (0)-frame. Suppose two events occur at the same place in the (1)-frame. If $dt_1 = \tau_1$ is the time interval between the two events measured in the (1)-frame and $dt_0 = \tau_0$ is the time interval between the events measured in the (0)-frame, we have already seen that $\tau_1 = \tau_0/\gamma$. The time measured by the clock at rest at the place where the events occur is called the proper time. Often, the proper time $\tau_1$ is just printed as $\tau$. We can use the invariance of the space-time metric to get the same result. Travelling on the train, your wristwatch measures your proper time, not the time measured by a passing clock tower.

Having established how the coordinates of position and time change between inertial frames, the next step was to find out how velocities and accelerations transformed. Then the transform of momentum of a particle of mass between inertial frames could be determined. Then, using Newton's 2nd law of motion, where force is equal to the rate of change of momentum, the way force transforms between inertial frames could be resolved. For force acting in the x-direction on a particle of mass, it is found that the force is the same for all inertial frames.

Minkowski carried out a very important transformation, showing that Maxwell's equations describing the phenomenon of electromagnetism (see Chapter 26) are invariant. Maxwell had assumed that electric and magnetic fields resided in the ether at rest and that disturbances were broadcast as electromagnetic waves through the stationary ether. Minkowski showed that Maxwell's equations were the same for all inertial frames; there was no absolutely stationary ether which could be used as a fundamental frame of reference. Everything was relative!

Special relativity can be applied directly to electrodynamics. A stationary line-charge lying along the x-axis in the (1)-frame has a radial electric field. If the line-charge moves with speed v in the x-direction relative to the (0)-frame then, viewed in the (0)-frame, the line-charge develops a circular magnetic field about it. Although charge is unchanged, due to the Lorentz contraction, the density of the line-charge is changed.

In 1902, Minkowski moved to Göttingen University, as a professor of Mathematics. Prior to this move, Minkowski had been the professor of Mathematics at the Zurich Polytechnic, a Federal Swiss Technical Institute located within the University of Zurich. While there, Minkowski had taught the young Albert Einstein, who attended mathematics classes there at the turn of the 19th into the 20th century. No doubt, as his former student, Minkowski would have been interested in Einstein's 1905 paper on special relativity. Einstein's paper was backed up by thought experiments (see Chapter 12). Minkowski's 1908 paper on special relativity introduced space-time and gave Einstein's thought experiments a geometrical basis. Unfortunately, Minkowski did not have the opportunity to develop his ideas about space-time any further as he died a year later, in 1909. However, the seeds of space-time had been well and truly sown.

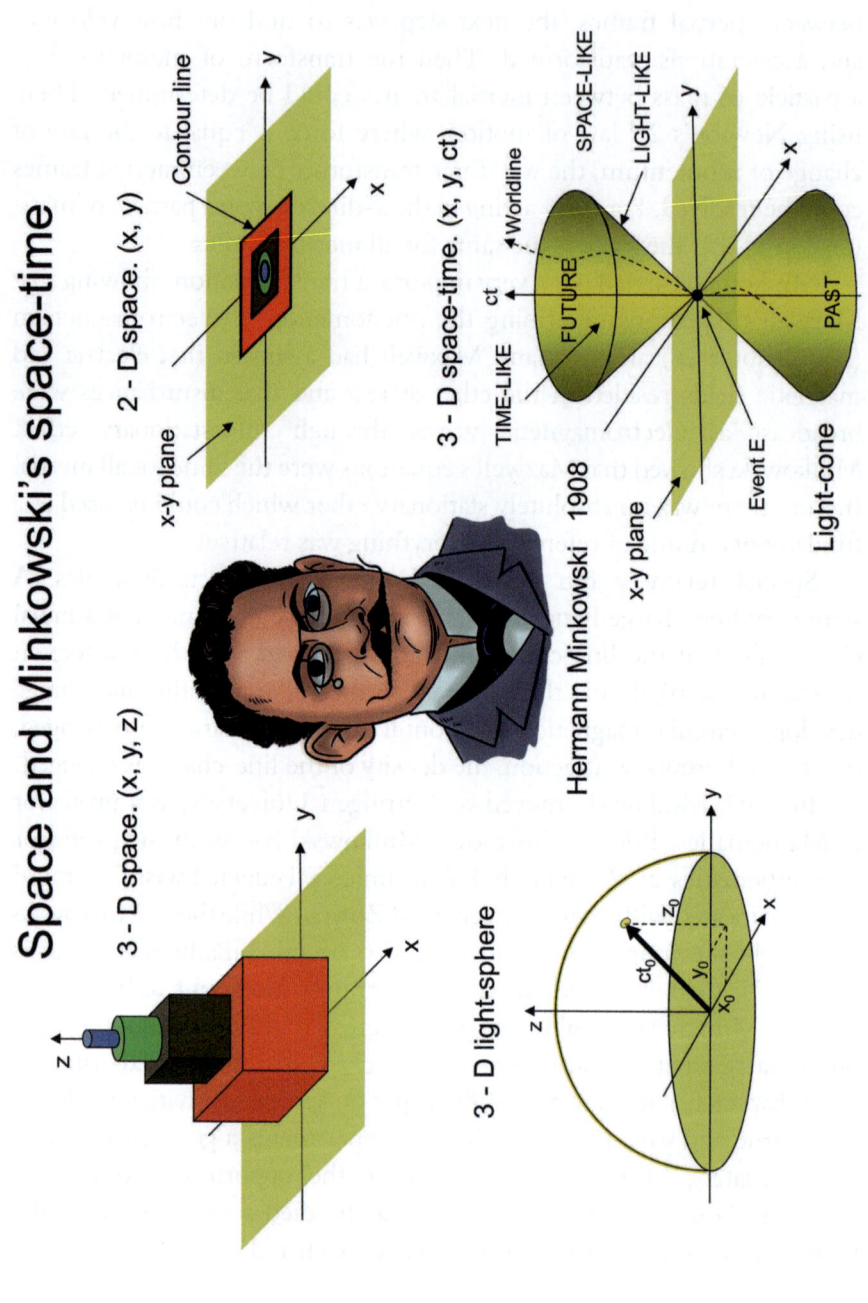

## CHAPTER 12

# EINSTEIN'S EARLY TRIUMPHS

Albert Einstein's mind was working in scientific overdrive during his miraculous year of 1905, leading to the publication of his papers on

1. The explanation of the photo-electric effect (June).
2. The measurement of the size of molecules (July).
3. Special relativity (September).
4. $E = mc^2$ (November).

The photo-electric story started with Heinrich Hertz's search for electromagnetic waves (Chapter 26) at Kiel University. He discovered that bathing the electrodes of a spark gap in ultraviolet light reduced the voltage at which a spark jumped across the gap. Hertz's laboratory assistant was Wilhelm Hallwachs, who showed, in 1888, that it was the illumination of the cathode (negative electrode) that was the important factor. The effect became known as the photo-electric effect. After the death of Hertz in 1894, Philipp Lenard, an Austro-Hungarian scientist, became the Professor of Experimental Physics at Kiel. Lenard studied cathode rays, for which he won the Nobel Prize in Physics in 1905. He also continued the work on the photo-electric effect and proved that a cathode, such as zinc, emitted electrons when illuminated in light.

In 1902, Lenard showed that increasing the intensity of incident radiation, although resulting in an increase in the number of emitted electrons, made no difference to the maximum velocity that any of the emitted electrons could attain. However, keeping the intensity fixed but increasing the frequency of the incident radiation did cause an increase in the maximum velocity of some of the emitted electrons. This did not seem to make sense. Surely, increasing the intensity of radiation, so that more electromagnetic wave energy hit the metal surface, should increase the overall kinetic energy of the freed electrons and, therefore, the maximum velocity of some of them. It was like shining a brilliant (high-intensity) red light laser onto a metal surface but getting no release of electrons. And then shining faint (low-intensity) torchlight with a green filter onto the same metal surface and getting electron emission. What had changing from lower wave frequency red to higher wave frequency green got to do with it?

Einstein pictured the situation in a new way. He made use of Planck's concept of a quantum particle of radiation, with energy $E = hf$. The quantum was later named the photon by the US scientist Arthur Crompton. Rather than the uniform spread of energy in the beam of incident radiation, Einstein saw it as a stream of electromagnetic particles, or photons, each containing a quantum of energy. It was rather like a stream of grapeshot fired from a cannon gun. When a photon hit an electron in the metal surface, if it had enough energy, it could free it, or knock it out of the surface. The energy of the photon became the kinetic energy of the emitted electron. (It's not quite as simple as that, as some incident energy is retained by the metal, the amount depending on the work function.) Increasing the intensity of the radiation, but keeping the frequency the same, meant more photons striking the metal surface, but each one still had the same amount of incident energy. More photons meant more emitted electrons, but no change in the distribution of their kinetic energies. However, increasing the radiation frequency meant that the energy of each incident photon was increased. Now when the photons struck the surface electrons, they stung more and, if freeing them, gave them a greater kinetic energy, increasing the maximum velocity of some of them.

Einstein got the Nobel Prize in Physics in 1921, for his explanation of the photo-electric effect. His explanation also gave a huge boost to Planck's idea of the quantum of radiation, and the two scientists were at the forefront of the development of Quantum Mechanics.

In 1827, the Scottish botanist Robert Brown had sprinkled pollen grains onto the surface of a beaker of water and noticed that the grains jiggled about. This jiggling became known as Brownian motion. Earlier, in the 17th century, the amateur Dutch naturalist Antonie van Leeuwenhoek had built several of the world's first microscopes and had, among various discoveries, seen bacteria in water. It was something quite unknown; 10,000 bacteria would occupy a grain of sand. He sent drawings of his results to the Royal Society and he was elected a fellow in 1680. Initially, Brownian motion was thought to be due to unseen creatures in the grain pollens.

Eventually, it was realised that the pollen grains were not inhabited by tiny creatures. It was guessed that, instead, the pollen grains were jiggled by water molecules in a manner analogous to Maxwell's kinetic theory of gases, where gas pressure is due to the interaction between gas molecules. Count Amedeo Avogadro was the Professor of Physics at the University of Turin at the beginning of the 19$^{th}$ century. In 1811, he hypothesised that 'Equal volumes of gases (at the same temperature and pressure) contain the same number of molecules'. At the time, few scientists paid any attention to Avogadro's idea. Fifty years later, sadly, after his death, further research confirmed the brilliance of Avogadro's hypothesis. His original idea for gases was extended to all substances and led to the introduction of Avogadro's number $N_A$, which defines the number of molecules in a mole of substance. The mole is the SI unit for an amount of matter. If $\rho$ is the density of a substance of known atomic weight then the number N of molecules per unit volume is given by

$$N = \frac{\rho N_A}{\text{Atomic Weight}}$$

Einstein developed a mathematical model for Brownian motion which showed how to determine the size of a water molecule, something that cannot be seen. The width of a water molecule is about $3 \times 10^{-10}$ m. At the time, the idea of atoms and molecules was still in its infancy. From his theoretical model, Einstein was one of the first scientists to be able to calculate Avogadro's number $N_A$.

$$N_A = 6.02 \times 10^{23} \text{ molecules/mole}$$

The number of water molecules in a litre of water, which occupies a volume of $1 \times 10^{-3}$ m$^3$, is about $3.346 \times 10^{25}$.

Einstein's published paper formed the basis of his 21-page PhD thesis submitted to the Zurich Polytechnic in 1905 which, after minor changes, was accepted. In 1908, the French scientist Jean Perrin used Einstein's work to model the settlement of particles in water falling under gravity. This formed part of his research work and led, later, to his being awarded the Nobel Prize in Physics in 1926.

In his 1905 paper entitled, *On the Electrodynamics of Moving Bodies*, Einstein stated his idea of special relativity.

1. The laws of physics are the same in all inertial frames. There is no preferred inertial frame.
2. The speed of light is the same in all inertial frames and is independent of the motion of the source.

In his paper, Einstein described a thought experiment involving lightning striking the front (A) and back (B) of a fast-moving train, moving with linear velocity v. This is slightly different to the Michelson-Morley experiment in that it involves two light signals travelling in opposite directions. But, the Lorentz transforms apply to both signals.

An observer at the mid-point (M) of the station platform ((0)-frame) saw the two lightning strikes at the front and back of the train happen simultaneously at points A and B on the railway track. At the instant the lightning strikes happened in the (0)-frame, the midpoint of platform M was coincident with the mid-point M' of the train where an observer was seated. Although the train may appear to undergo a FitzGerald distortion, a conductor on the train with a tape measure would declare that in the (1)-frame AM' = BM', no matter in which direction he made the measurement.

Given that the speed of light is finite, common sense tells us that the seated passenger will see the lightning strike hit the front of the train (A) before the back of the train (B), by a nanosecond, or so. So, the seated passenger said that the lightning strikes were not simultaneous. For both observers to be right, Einstein argued that the times measured by the two observers moving relative to one another must be different.

In the (0)-frame, the time for the light signal from the lightning strike at the front of the train A to reach the middle of platform M is $t_0(MA) = AM/c$. In terms of the Lorentz transform for time

$$t_0(M_A) = \gamma\left[t_1(M'_A) + \frac{AM'v}{c^2}\right]$$

Similarly, the time for the light signal from the lightning strike at the end of the train B to reach M is $t_0(MB) = BM/c$. Again, in terms of the Lorentz transform for time

$$t_0(M_B) = \gamma\left[t_1(M'_B) + \frac{BM'(-v)}{c^2}\right]$$

Comparing the total times measured for the two light pulses in the (0) and (1)-frames, we find

$$t_0(MA) + t_0(MB) = \gamma[t_1(M'A) + t_1(M'B)] \text{ since } AM' = BM'$$

Thus, time in the stationary (0)-frame is equal to $\gamma$ times the time in the moving (1)-frame. Since $\gamma$ is greater than 1 ($\gamma > 1$) then, to the stationary observer on the platform, clocks and watches on the train appear to run slow.

But, as Einstein stated, there is no preferred inertial frame; everything is relative. The passengers on the train could equally argue that from their point of view they were stationary and that the platform was moving with speed $-v$. Moreover, their version of the Lorentz transform for time would confirm that the station clock on the platform was running slow.

In his 1905 paper entitled, *Does the Inertia of a Body Depend on its Energy-Content?*, Einstein derived the world's most famous equation $E = mc^2$ linking mass m with energy E. Since publication, experts have pointed out that Einstein's derivation of the equation is subject to the restrictive conditions of a point particle of mass m moving at low, non-relativistic speed.

As derived by theory and confirmed by experiment

$$m = \frac{m_0}{\sqrt{1 - \frac{v^2}{c^2}}}$$ where $m_0$ is the mass of the body at rest.

The term in the square root may be expressed (using Newton's binomial expansion) as an infinite series. If terms higher than $v^2/c^2$ are discarded then, on multiplying both sides of the subsequent equation by $c^2$, we have

$$mc^2 = m_0c^2 + \frac{1}{2}m_0v^2$$

The kinetic energy is $\frac{1}{2}m_0v^2$, so $m_0c^2$ is taken to be the rest energy of the mass $m_0$. However, the assumption that mass increases with speed is tantamount to assuming that $E = mc^2$, so the derivation is not a proof.

Since the publication of Einstein's paper, experts, using more complex mathematics, have shown that the equation can be extended to structured bodies and, therefore, applies under all conditions.

Newton's 1$^{st}$ law of motion defines a force F as that which causes a mass $m_0$ to veer off course. From Newton's 3$^{rd}$ law, we see that inertia is the reaction to forcing a body to veer off course. From Newton's 2$^{nd}$ law, we see that inertia occurs when a body accelerates and that the inertial back reaction is equal to mass $m_0$ times an acceleration a.

We can write the back reaction force of inertia as F, where

$$F = m_0 a = m_0 \frac{dv}{dt} = m_0 \frac{dx}{dt}\frac{dv}{dx} = m_0 v \frac{dv}{dx} = m_0 \frac{v\,dv}{dx} = \frac{\frac{1}{2} m_0 v^2}{x} = \nabla E$$

From the conservation of energy, we can replace the kinetic energy with potential energy. The term $\nabla E$ is mathematical shorthand for the gradient of energy E. It doesn't apply to rest energy.

A body subjected to an energy gradient, if free to move, will accelerate and experience inertia. So, to answer Einstein's query, since inertia is equal to the negative gradient of energy, then inertia is dependent on the spread of a body's energy content and inversely dependent on the length of spread.

In 1907, Einstein published details of a thought experiment in which he predicted that an electromagnetic wave shone vertically upwards through a planet's gravitational field would suffer a decrease in its frequency. The frequency f is the number of wave oscillations per second. For a light wave, the decrease in frequency would mean a frequency shift towards the red end of the spectrum.

The length of one complete wave is the wavelength $\lambda$. The time taken for one complete wave to occur or to pass by is called the wave period $\tau$. The period $\tau$ may be likened to the tick-tock of a clock. The speed of an electromagnetic wave is c. Since speed is equal to distance divided by time, we have $c = \lambda/\tau$. Now, the inverse of the period $\tau$ is equal to the frequency f, that is $f = 1/\tau$. This leads us to the wave formula $c = f\tau$. If the frequency f decreases then the period $\tau$ must increase. So, Einstein's gravitational red shift means that as a light wave travels upwards through a planet's gravitational field, where gravity decreases, its period $\tau$ increases. In other words, time slows down (dilates, or takes longer to pass) with altitude. Astronauts on the International Space Station age slightly less quickly than people on Earth do. When gravity increases the period $\tau$ decreases, so the time speeds up (passes more quickly). When Einstein published his idea of a gravitational red shift of light, in 1907, there was no way of experimentally testing his hypothesis for any electromagnetic wave.

One way to look at the gravitational red shift is to think about it in terms of a photon of light. Setting Planck's photon energy $E = hf$ equal to Einstein's energy of mass $E = mc^2$ we see that a photon has effective, or photonic, mass $m = hf/c^2$. As the photon moves upwards in the gravity field g, it gains potential energy. For ordinary mass moving in a gravity field, energy is conserved, so the increase in potential energy of the mass is equal to the decrease in its kinetic energy (reduced speed). But a photon always travels at the speed of light c and cannot slow down. The change in the photon's kinetic energy is achieved by reducing its frequency, leading to a red shift in the frequency of light. Thus, the change in frequency $\Delta f$ as the photon moves upwards a distance $\Delta s$ in the gravity field g is

$$\Delta f = -\frac{g}{c^2} \Delta s f$$

Consider a photon leaving the surface of a star. At infinite distance from the star's gravitational mass, a stellar photon's frequency f is zero. Since $\tau = 1/f$, this means that time is infinite for a stellar photon as it travels across the Universe. If something takes an infinite time to happen, it means that time stands still. As a photon of starlight descends through the Earth's gravity field, the reverse of the red shift occurs and the photon begets a frequency.

Although not electromagnetic, the effect of gravity on time is theoretically predicted for the simple pendulum. In the red shift, gravity acted on the effective mass of a photon, while with the pendulum, gravity acts on the mass of the bob. The period $\tau$ of the simple pendulum is the familiar tick-tock of a clock. The period of each complete swing is $\tau = 2\pi\sqrt{(\ell/g)}$, where $\ell$ is the pendulum length and g is the strength of the gravity field. Note that the bob mass m does not appear in the formula because gravity acts on all masses in the same way. That is, they all experience the same acceleration, g. With increase in height, g decreases and $\tau$ increases. Each tick-tock should take longer as the clock is raised higher in the gravity field. In his book *Novum Organum Scientiarum*, written in 1620, Bacon suggested taking a pendulum to the top of a mountain to see whether the period $\tau$ increased. But, at the time, the only comparison would have been with an hourglass, where the particles of sand are acted upon by gravity, so it's not clear whether the difference in time could have been detected. Perhaps a sundial might have been used? Nowadays, we can use spring-driven clocks and watches, but are these also affected by gravity? As the spring unwinds, energy is lost, so effective mass is lost, which is acted upon by gravity. At a great distance from the planet, g tends to zero and the period of oscillation τ tends to infinity. Thus, the frequency of oscillation of a simple pendulum tends to zero and the oscillation is frozen in time.

The general assumption is that gravity affects time, however it is measured.

In 1960, 5 years after the death of Einstein, the first successful experiment to measure the Earth's gravitational red shift was made. The experiment, by the US scientists Robert Pound and Glen Rebka, was carried out in the 22.56m high lift shaft of the Jefferson Tower in the Physics building of Harvard University.

A gamma ray transmitter (frequency $f_T = 10^{22}$ Hz) was used to beam a signal up the lift shaft towards a narrow band gamma ray receiver.

The gravitational red shift near the top of the lift shaft was of order $10^7$ Hz, putting the signal outside the bandwidth of the gamma ray receiver. To overcome this, the receiver was mounted on a mobile platform that moved downwards, with speed v, towards the stationary transmitter. The Doppler blue-shifted frequency $f_R$ of the transmitted signal at the receiver is

$$f_R = \left(\sqrt{\frac{1+\frac{v}{c}}{1-\frac{v}{c}}}\right) f_T \approx \left(1+\frac{v}{c}\right) f_T \text{ for low velocity v.}$$

The Doppler blue shift in frequency $\Delta f_B = f_R - f_T = (v/c) f_T$ was arranged to cancel the gravitational red shift in frequency $\Delta f_R = -(g/c^2) \Delta s f_T$, so that the gamma ray receiver should detect the transmitted gamma wave signal. To cancel the frequency shifts

$$v = \frac{g}{c} \Delta s \text{ m/s where } \Delta s = 22.56 \text{m}.$$

The required velocity was just over 2 mm/hour. With incredible precision, Pound and Rebka were able to adjust the velocity v so that the gamma ray absorber received the transmitted signal, thus confirming Einstein's prediction of a gravitational red shift to within 10%.

In the NASA Gravity Probe-A experiment, carried out in 1976, a rocket carrying an atomic (microwave) clock was launched into an elliptical trajectory, reaching an altitude of nearly 10,000 km. During the flight, the slight increase in clock period τ with altitude (therefore, change in g) was measured.

There is still more to be squeezed out of Einstein's red-shift. The photons (light, radar, gamma waves) in the vertical beam, by changing frequency, generate a gravity field which cancels with the Earth's gravity field. Thus, the photons are force-free and move with constant speed c.

Now a photon's frequency is linked to its temperature. So, accompanying the frequency shift is a cold-shift in temperature, creating a thermal gradient, which is not cancelled. The analogue of a photon beam in a thermal gradient suggests that a gravity field is created!

The year 1905 was, indeed, a miraculous year for Einstein and for science, too.

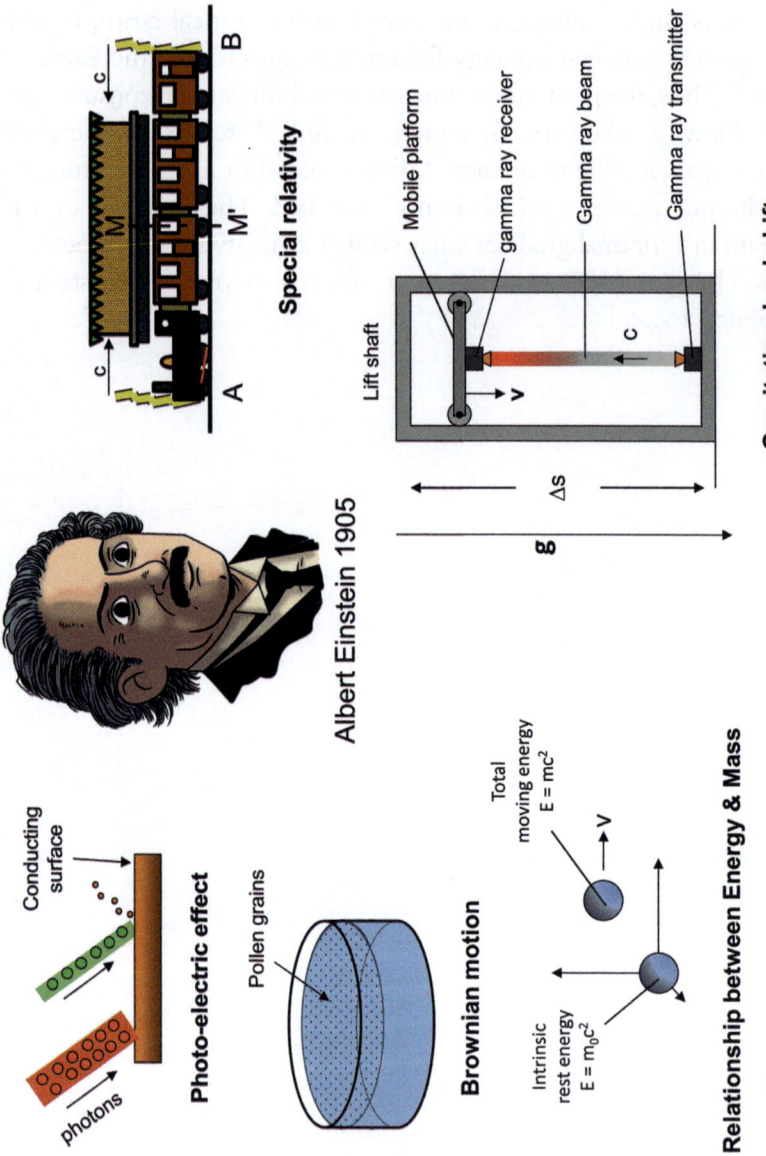

CHAPTER 13

# GRAVITY, SPACE AND SPACE-TIME CURVATURE

In the year 1907, Einstein published a paper surveying relativity and speculating on the possibility of extending it to the case of uniformly accelerating frames of reference moving relative to one another. This was the beginning of his thoughts about a general theory of relativity.

At this time, Einstein introduced his principle of equivalence which stated that at any point in space it is not possible (using a mass as a detector) to tell the difference between acceleration in one direction and a gravity field in the opposite direction. Einstein's notion is sometimes called the weak principle of equivalence and is linked with Newton's observation of the equivalence between inertial mass and gravitational mass. Light from an accelerating source also undergoes a frequency shift, as though in a gravity field. Einstein's reasoning behind his principle of equivalence was his thought that a man (mass) who had fallen off a ladder was weightless during his fall. Acceleration in one direction was equal and opposite to that of a gravity field in the opposite direction, resulting in gravity cancellation in the man's accelerating framework. On hitting the Earth, the change of the man's frame of reference back to that of the Earth's was a very painful reminder of the effect of inertia.

By virtue of the weak principle of equivalence, a theory for transforming between accelerating 3-D frames of reference moving relatively to one another would also be a theory of gravity. One way of forming a model of a uniformly accelerating 3-D framework might be to replace it with a uniform gravitational field. Then, changing gravity fields would be equivalent to changing between relatively accelerating frames of reference. But this requires the presence of mass to create the gravity field, and with it, inertia.

Einstein extended his principle of equivalence to state that in all freely falling reference systems, the laws governing physical processes must be the same, or equivalent. This is called the strong principle of equivalence to differentiate it from the weak version.

In 1908, Einstein read the paper by Hermann Minkowski which introduced the concept of 4-D space-time. Minkowski used vectors and matrices to give special relativity a geometrical footing, whereas Einstein's approach had been purely algebraic. Einstein knew Minkowski personally, having been taught by him during his student days at the Zurich Polytechnic, and he was initially concerned that a mathematician was moving into his area of scientific interest. However, Einstein was always quick on the uptake and ready to extend other scientists' discoveries and theoretical ideas. It dawned on him that Minkowski's 4-D space-time might be the approach he needed to extend special relativity to the next step; that of moving between uniformly accelerating relative frames of reference. With that thought, Einstein became very interested in the concept of space-time. In fact, the approach was an essential step along the road which led him to general relativity.

In Minkowski's 4-D space-time, when a body accelerates, it travels along a curved world-line and the body experiences an apparent force of gravity. Suppose an astronaut in free space has a rocket motor backpack. When he switches his rocket motor on (Event $E_0$), he accelerates away in the x-direction, say, and experiences an equivalent force of gravity in the minus x-direction. His world-line, starting at the apex $E_0$ of a Minkowski light-cone, is wholly contained in the flat ct-x plane; beginning as a vertical line in the ct direction (angle 0°, so infinite slope) which gradually curves over but never reaches a slope of 1 at an angle of 45°, corresponding to light speed. To simulate the existence of gravity in 4-D space-time, a body must be made to travel along a curved world-line. But when a body moves with uniform velocity (or is stationary), its world-line is straight with constant slope and there is no apparent force of gravity. A permanent background force of gravity cannot be simulated using Minkowski's 4-D space-time.

Einstein viewed the situation in reverse. To mimic a gravity field, space-time must be curved so that a body always experienced a curved world-line even when it moved with constant speed, or remained stationary, in 3-D. For example, suppose the astronaut coasts at uniform speed apparently along a straight line in the flat surface of the ct-x plane, if the surface is actually curved over then the astronaut will travel along a curved path on the surface; his mass will be subject to radial acceleration and he will feel as though he is in a gravity field.

Without knowing anything about 4-D curved space-time but just following the mathematical pattern introduced by Minkowski, Einstein could write the metric as

$$(ds)^2 = \begin{bmatrix} dx & dy & dz & cdt \end{bmatrix} \begin{bmatrix} g_{11} & g_{12} & g_{13} & g_{14} \\ g_{21} & g_{22} & g_{23} & g_{24} \\ g_{31} & g_{32} & g_{33} & g_{34} \\ g_{41} & g_{42} & g_{43} & g_{44} \end{bmatrix} \begin{bmatrix} dx \\ dy \\ dz \\ cdt \end{bmatrix}$$

On expanding out the right-hand side of the expression for the metric $(ds)^2$ we see that it contains the addition of sixteen terms.

$$\begin{aligned}(ds)^2 =\ & g_{11}(dx)^2 + g_{12}dx\,dy + g_{13}dx\,dz + g_{14}dx\,cdt \\ & + g_{21}dy\,dx + g_{22}(dy)^2 + g_{23}dy\,dz + g_{24}dy\,cdt \\ & + g_{31}dz\,dx + g_{32}dz\,dy + g_{33}(dz)^2 + g_{34}dz\,cdt \\ & + g_{41}cdt\,dx + g_{42}cdt\,dy + g_{43}cdt\,dz + g_{44}(cdt)^2\end{aligned}$$

For Minkowski's 4-D space-time, where gravity and accelerating frames of reference are absent, $g_{11} = g_{22} = g_{33} = 1$ and $g_{44} = -1$ and the other $g_{pq}$ coefficients are zero. Surfaces in Minkowski's 4-D space-time are flat, with zero curvature, so that a hyperspace version of Pythagoras' theorem applies.

Einstein assumed that the presence of gravitation or accelerating frames of reference would affect the values of the $g_{pq}$ coefficients at each point (x, y, z, ct) of 4-D space-time. Or, to put it another way, the curvature at a point of space-time due to gravity or acceleration was what determined the values of the 16 $g_{pq}$ coefficients at that point giving the metric. Fortunately, from symmetry, $g_{pq} = g_{qp}$, so that only ten independent $g_{pq}$ coefficients needed to be found. But what was the equation that linked the surface curvature of space-time with the $g_{pq}$ coefficients at every point of space-time?

To get a feel for surface curvature, let us start by looking at surfaces in 3-D space where we can envisage them. Time t is not involved. We can think of a volume of 3-D space as an oblong loaf of bread extending in the z-direction, sliced up into x-y plane surfaces, or 2-D flat layers. This is the Euclidean geometry view of space, where a point has coordinates (x, y, z). The shortest distance between two points on a 2-D x-y plane, or flat surface, is a straight line. Incremental distances in the x-y plane are dx and dy. The metric is

$$ds^2 = dx^2 + dy^2 = \begin{bmatrix} dx & dy \end{bmatrix} \begin{bmatrix} 1 & 0 \\ 0 & 1 \end{bmatrix} \begin{bmatrix} dx \\ dy \end{bmatrix}$$

Since Pythagoras' theorem is satisfied, the surface is flat. The matrix of noughts and ones is called the identity matrix. It signifies that the surface has zero curvature.

We can also think of a volume of 3-D space as being like an onion, where any point in space has spherical polar coordinates (r, θ, φ). The onion can be peeled away in spherical layers, or surfaces, each of different radius r. On each spherical surface, the angles θ and φ vary. This is the spherical geometry view of space. It has a tentative link with Kepler's idea of the structure of space and with the ancient Greek idea that the fixed stars all resided on a celestial sphere and the planets moved on equatorial circular orbits. The shortest distance between two points on a spherical surface is a circular arc. The incremental distances on a spherical surface of fixed radius r are rdθ and rsinθdφ. In spherical geometry terms of the independent variables θ and φ the metric is

$$ds^2 = (rd\theta)^2 + (r\sin\theta d\phi)^2 = \begin{bmatrix} d\theta & d\phi \end{bmatrix} \begin{bmatrix} r^2 & 0 \\ 0 & r^2\sin^2\theta \end{bmatrix} \begin{bmatrix} d\theta \\ d\phi \end{bmatrix}$$

The equation for the metric cannot be written in terms of other independent variables to make the matrix the identity matrix. So, Pythagoras' theorem is not satisfied, and the surface must be curved.

The rules for plane geometry are different to those for spherical geometry. On a plane surface, the interior angles of a triangle add up to 180° but on a spherical surface, the interior angles of a triangle, formed by three curved lines, sum to more than 180°. For a plane surface, parallel lines never cross but on a spherical Earth lines of longitude are parallel at the equator but cross at the poles, while lines of latitude always remain parallel. Spherical Geometry is a branch of Elliptical Geometry for which all surfaces have positive curvature, although the surface curvature changes with the position on the surface. On the other hand, in Hyperbolic Geometry, the interior angles of a triangle formed on the surface of a hyperboloid always sum to less than 180°. Examples of such surfaces in 3-D space are the saddle shape, the cooling tower shape and the flared horn shape of a trumpet. All surfaces of hyperboloids have negative curvature.

# GRAVITY, SPACE AND SPACE-TIME CURVATURE

Let us now consider the surface of a cylinder of radius R. In cylindrical polar coordinates, any point on the surface has coordinates (R, Rθ, z). The metric for the surface, in terms of the independent variables θ and z, is given by

$$ds^2 = (R\,d\theta)^2 + (dz)^2 = [d\theta \ dz] \begin{bmatrix} R^2 & 0 \\ 0 & 1 \end{bmatrix} \begin{bmatrix} d\theta \\ dz \end{bmatrix}$$

It is not clear whether the surface is flat, or not. Since R is constant, the equation can be rewritten in terms of the independent variables Rθ and z, thus

$$ds^2 = (R\,d\theta)^2 + (dz)^2 = [R\,d\theta \ dz] \begin{bmatrix} 1 & 0 \\ 0 & 1 \end{bmatrix} \begin{bmatrix} R\,d\theta \\ dz \end{bmatrix}$$

In this form, the matrix is the identity matrix, showing that the surface of the cylinder has zero curvature and may be taken as flat. Pythagoras' theorem holds on the surface of the cylinder. A triangle with 180° interior angles lies flat on the cylindrical surface, and parallel lines, although curved, never cross.

In the three 2-D examples above, the metric equation is of the form

$$ds^2 = [dx_1 \ dx_2] \begin{bmatrix} g_{11} & g_{12} \\ g_{21} & g_{22} \end{bmatrix} \begin{bmatrix} dx_1 \\ dx_2 \end{bmatrix}$$

The matrix determines the curvature. Since $g_{21} = g_{12}$, then three numbers determine the curvature in 2-D.

Following the same pattern for 3-D, since $g_{21} = g_{12}$, $g_{31} = g_{13}$ and $g_{32} = g_{23}$, we see that six numbers determine the curvature in 3-D. As we have already concluded above, ten numbers determine the curvature in 4-D.

Defining the curvature of a line in 3-D space is very simple; it is just the inverse of the radius of curvature of the line at that point. A line can only have positive or zero curvature. A mass travelling with constant speed along a line with curvature experiences radial acceleration which, from equivalence, feels like being in a gravity field. We can experience the equivalent gravity effect of line curvature when we travel upright in a car on the track of a roller-coaster in Earth's gravity field. If the direction of the radius of curvature (from the centre of the circle to the track) is down, we feel heavier, while if the radius of curvature is up, we feel lighter. For a level track, where the radius of curvature is infinite, the line curvature is zero and we feel no change in weight. If the mass rotates as it moves along the line in space, it is said to undergo torsion.

During the early 19th century, the famous German mathematician Karl Gauss, at Göttingen University, investigated the curvature K of 2-D surface shapes and derived a formula for them. The mathematics is difficult. In simplified terms, the curvature K at a point on a surface is obtained from the radii of curvature of two perpendicular 1-D lines drawn in the surface through the point. These are then multiplied together and the inverse gives K. Radii direction is important; if they oppose then the multiplication is negative. For the sphere of radius r, both radii of curvature are r, giving the surface curvature of the sphere as $1/r^2$. For a cylinder of radius R, one radius of curvature is R, while the other is infinite, so the surface curvature of a cylinder is zero. For the cooling tower shape, at any point on the surface, one radius of curvature points inwards, while the other points outwards, so the curvature is negative.

The surface of a toroid (doughnut or smoke-ring shape) has all three forms of surface curvature. The outer part of the toroid surface has positive curvature, while the inner surface has negative curvature. Two circles separate the outer surface with positive curvature from the inner surface with negative curvature, and so both circles must have zero surface curvature.

In the above, where we considered surface curvature in 3-D space, only the spatial coordinates (x, y, z) played a part. In 4-D hyperspace, the space-time coordinate ct can play a part in surface curvature. We know from equivalence that space-time curvature and gravity are linked and from the gravitational red shift we know that gravity and time are linked, so we see that space-time ct is linked with space-time curvature. Extreme curvature of the space-time coordinate ct hints at the possibility of time travel.

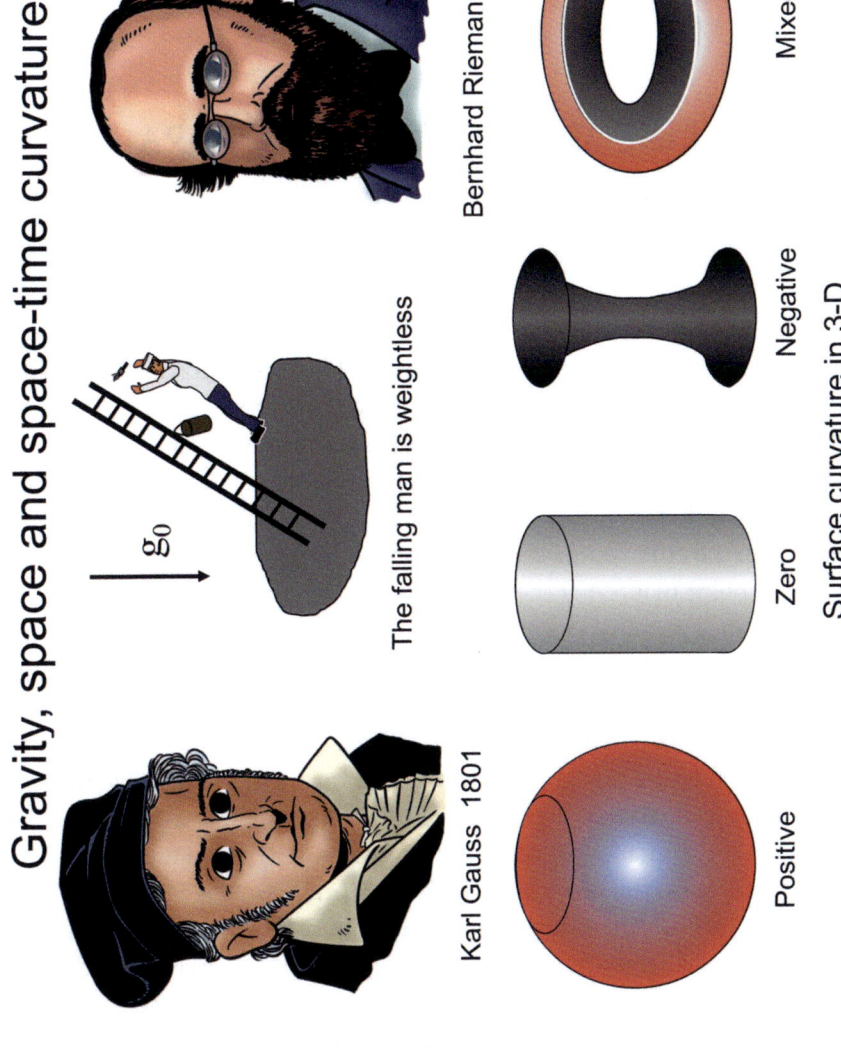

## CHAPTER 14

# AN OUTLINE OF EINSTEIN'S GENERAL RELATIVITY

In Newton's theory of gravity, the ether of space provides the medium for the transmission of gravitational influence. The gravitational permittivity γ is taken to be constant. We can compare this with fluid mechanics, where the density ρ is constant for an incompressible fluid. We assign a 3-D Euclidean nature to the ether. If we introduce a stationary point mass M into the ether, we can envisage an infinite number of plane, or flat, surfaces in the ether, containing radial gravitational **g**-field lines stemming from the point mass M at the centre (origin) of each plane.

We can also envisage an infinite number of spherical surfaces surrounding the point mass M where, for each radius r, we assign the surface with a potential φ = constant. The potential surfaces cut the flat planes orthogonally (such that **g** = – ∇φ). Let us introduce a test mass m which doesn't alter the gravity field of the mass M. If we move the test mass m over an arc on a potential surface surrounding M, there is no change in its gravitational potential energy E = mφ. Since force **F** = -∇E, then the force on the test mass m is **F** = -∇(mφ), so that **F** = m**g**.

If the path of a test mass m crosses different equipotential surfaces surrounding M, then the potential energy of the test mass will change. But, on returning to its starting position, the energy of the test mass is that with which it started out. We say that the gravitational field created by the point mass M is conservative.

Thus, the introduction of a mass M into the ether introduces a number of possible surfaces in the ether which may influence our view of the geometry of space (perhaps spherical geometry might be better than

Cartesian geometry), although space remains undistorted and essentially Euclidean in character.

As the importance of Einstein's work was gradually appreciated by scientists, so his academic career began. He left the Swiss Patent Office in 1909 to become a Physics lecturer at Berne University. However, he quickly moved on to become a professor at Prague University, then in the province of Bohemia in the Austro-Hungarian Empire. But after only a year, he returned to Switzerland to become the Professor of Theoretical Physics at the Zurich Polytechnic.

Einstein was now searching for a way to change reference frames at a point in 4-D space-time, from an accelerating reference frame (or where gravity was present), where the $g_{pq}$ coefficients of the metric matrix were unknown, to a freely falling reference frame (akin to the Minkowski condition), where the $g_{pq}$ coefficients were known. Moreover, he realised that he needed to link the change in reference frame with surface curvature in 4-D. He asked the Hungarian mathematician Marcel Grossmann for help. Grossmann had been a student with Einstein at the Zurich Polytechnic, taking the same classes, and they had remained close friends. Indeed, it was through Grossmann's father-in-law's help that Einstein had got a job at the Swiss Patent Office in Berne. Grossmann's academic career had taken off more quickly than Einstein's. In 1907, he had been appointed as the Professor of Geometry at the Zurich Polytechnic. When Einstein discussed his ideas with Grossmann, Grossmann thought of the N-Dimensional non-Euclidean geometry introduced by the German mathematician Bernhard Riemann. Grossmann was well acquainted with Riemann's work, which involved a method for transforming from one reference frame of coordinates to another. Moreover, the subject had been developed further by another German mathematician, Elwin Christoffel, the Professor of Mathematics at the Zurich Polytechnic. The extension of Riemann's work involved differential geometry and mathematical forms now known as tensors. Since Einstein knew nothing about the subject, particularly about tensors, Grossmann suggested that Einstein should start by reading the work of Riemann.

Riemann's PhD supervisor, at Göttingen University, was Gauss. After completing his doctorate in 1851, Riemann tackled a number of mathematical topics with success. But, for the story in this chapter, it is Riemann's work on extending Gauss' study of surface curvature that is of interest. Gauss had initiated the study a quarter of a century earlier but had never published any details about it.

In 1854, Riemann gave a lecture on N-Dimensional geometry with the title, 'On the hypotheses that lie at the foundation of geometry'. Riemann began with the results of Gauss' earlier study, which first mooted the idea that the Euclidean geometry developed by the ancient Greeks might not be the only form of geometry.

The presence of matter in space gives rise to curved surfaces in space; for example, equipotential surfaces. But this is not the curvature of space itself. Riemann suggested that the presence of matter might alter the property of space around matter, giving rise to space distortion. If so, gravity fields and the distortion of empty space would be linked, and the geometry of space would be non-Euclidean in character. This idea might have been why Grossmann thought of Riemann's work when Einstein came to him for help.

At the outer reaches of the Universe, an immense distance away, Riemann conjectured that the distortion, or curvature, of space there might be so intense that space coalesced to form the surface of a gigantic sphere on which the fixed stars were located. Space would be closed and it would be pointless to ask what was outside because the surface of the sphere was the outside. Riemann believed in the existence of an incompressible ether filling space. He speculated that ordinary matter was an ether sink, which drained the ether away to another dimension, outside of our three dimensions. As far as I'm aware, he gave no thought to the possibility of a compressible ether where, in a region of severe space curvature, any stretching, or compressing, of the ether would cause a regional change in the value of the gravitational permeability $\gamma$.

Riemann suggested that space containing matter, as well as being non-Euclidean, might have more than three dimensions. If space had N-Dimensions, it would require an N x N matrix to change coordinate frames of reference. Although we can easily visualise surfaces in 3-D space, trying to imagine surface shapes which exist in N-Dimensional space is impossible. Instead, we have to trust in the repeated patterns supplied by mathematics, extended from the observable known into the unobservable but predicted. Riemann's idea of the non-Euclidean geometry of N-D space was to lead to tensor calculus.

Given two points on any surface in N-Dimensional space, the shortest distance between them is called a geodesic. Following Gauss' lead, Riemann discovered that in N-D space, the surface curvature R is determined by the metric; that is, the square of an incremental (small) geodesic length. Moreover, the metric was an invariant; that is, independent of the coordinate frame of reference. The presence of matter might distort space, causing further space curvature.

In the audience of academics at Riemann's ground-breaking lecture, it was probably only Gauss who fully understood the mathematics. Riemann's lecture notes were not published until 1866. By then, Gauss was dead and so, too, was Riemann, who died in 1865 aged only thirty-nine.

By and large, the scientific community at the time viewed Riemann's non-Euclidean N-Dimensional geometry as an esoteric curiosity, with limited areas of application. However, Elwin Christoffel, appointed in 1862 as the Professor of Mathematics at the newly formed Zurich Polytechnic, had taken an interest in Riemann's work. In 1869, he published a paper explaining Riemann's geometry in more detail. More importantly, he developed a method of differentiating functions in N-Dimensional space with respect to the space coordinates. To condense the expressions, he introduced the Christoffel symbols of the first and second kind, using brackets [ ] and { }, or Γ, respectively. He also introduced a modified form for Riemann's curvature, now called the Riemann-Christoffel curvature tensor. In turn, Christoffel's work was extended by the Italian mathematician Gregorio Ricci-Curbastro and his former pupil Tullio Levi-Civita, in a paper entitled, *Absolute Differential Calculus*, published in 1901. This was the beginning of tensor calculus. Ricci formed a contraction of the Riemann-Christoffel tensor, now called the Ricci curvature tensor.

At the start of the 20$^{th}$ century, few mathematicians knew about tensor calculus, which merges algebra with geometry in N-D. A restricted form of 3-D tensors, with three perpendicular straight axes (Cartesian tensors), were used by some mathematical engineers working in stress analysis, and this is really where the word *tensor* originated. Einstein's successful use of tensor calculus to formulate a new theory of gravity certainly publicised the subject of tensors, but it remains a highly abstruse area of mathematics.

To develop the theory of general relativity, the subject of tensors cannot be avoided. Although this is not a book about tensor theory, let us remove a little bit of the mystique of what tensors are. A scalar is a tensor of rank 0. A vector is a tensor of rank 1. A matrix is a tensor of rank 2. So, we have already been dealing with tensors. Thus, given a 4-D space-time reference system, a tensor is a collection of quantities (depending on its rank) associated with a point (x, y, z, ct) whose values on changing to a different reference system are transformed in accordance with definite rules. This includes a procedure called parallel transport, where vectors can change direction as their point of origin moves along a curve. So, the main purpose of tensor analysis is to keep track of what happens when reference systems change.

Mathematicians introduce double summation to shorten the general expression for the metric $(ds)^2$ in 4-D space-time. Here, $dx = dx_1$; $dy = dx_2$; $dz = dx_3$; and $cdt = dx_4$.

$$(ds)^2 = \sum_{q=1}^{4}\sum_{p=1}^{4} g_{pq} dx_p dx_q$$

To expand this expression, we must let p vary from 1 to 4 first and then let q vary from 1 to 4. We will then get the sum for $(ds)^2$ containing sixteen terms. It is the same equation that we got on expanding the matrix form for $(ds)^2$ in Chapter 13.

In tensor calculus, the double sigma signs are dropped and superscripts are used for all coordinates $(x^p)$.

$$(ds)^2 = g_{pq} dx^p dx^q$$

Although done for brevity, dropping the sigma signs makes tensor calculus difficult to follow by the uninitiated. Tensor manipulation is achieved using indices, and there are rules which must be applied. For example, repeated indices means that the expression $a_{pq}x^p$ is the summation, $a_{pq}x^p = a_{1q}x^1 + a_{2q}x^2 + a_{3q}x^3 + a_{4q}x^4$ in 4-D space.

The common reference system for 3-D flat space is the Cartesian reference frame, where the unit vectors **i, j, k** are said to form a basis which spans space. At a point P in 3-D curved space, a reference system can be spanned by two different vector bases, given by the tangent vectors (**e**$_1$, **e**$_2$, **e**$_3$) to the curvilinear coordinate axes at P, or the gradient vectors (**E**$_1$, **E**$_2$, **E**$_3$) normal to the curved surfaces at P. They are called contravariant and covariant vector bases. Which vector base is used determines how the geometric and physical properties change when transforming to another reference system at that point. We now have to let go and extend the idea into 4-D curved space-time and rely on the mathematical pattern. Superscripts are used for contravariant vector components $A^p$, while subscripts are used for covariant vector components $A_n$. Transformed coordinates and vectors are signified with a bar ($\bar{x}^p, \bar{A}^p, \bar{A}_p$).

We have barely touched on the subject of tensor analysis. We can have covariant and contravariant tensors of rank 2 (matrices) and of higher rank. For example, the Riemann-Christoffel curvature tensor $R^i{}_{pqj}$ is of rank 4. It is a function of $g_{pq}$ (with sixteen coefficients) and the first and second derivatives of the coefficients with respect to the coordinates $x^i$ (sixteen terms), giving a total of $16 \times 16 = 256$ terms. Without the use of tensors, the mathematical model would be extremely unwieldy. We can also have mixed tensors. Raising and lowering indices links contravariant tensors with covariant tensors. However, reducing the size of mathematical expressions comes at the cost of complexity and losing sight of the physical picture.

Grossmann tutored Einstein on tensor calculus and gradually he mastered the theory. It was a mind-boggling task for Einstein, who confessed in a 1912 letter to the German physicist Arnold Sommerfeld that developing the theory of special relativity was child's play when compared with developing the theory of general relativity.

Einstein shared his thoughts with Grossmann about extending his principle of equivalence, still further, with the principle of covariance. This was the requirement that the general laws of nature must be expressed by equations which hold good for all reference systems modelled using tensors. The Lorentz transforms are a form of covariance, but they only apply to inertial systems. Einstein argued for a general form of covariance. In other words, the same laws must hold in accelerating frameworks as in inertial frameworks, and the equations expressing them must be co-variant, or transformable.

The metric $(ds)^2$ is invariant in transforming from one framework (with unbarred coordinates) to another framework (with barred coordinates). Thus, in transforming from the vector base **E**, to the vector base **Ē**, the covariant form for the metric is

$$(ds)^2 = g_{pq} dx^p dx^q = \bar{g}_{pq} d\bar{x}^p d\bar{x}^q$$

It can be shown that the tensor $\mathbf{g}_{pq}$ (that is, the metric matrix containing the coefficients $g_{pq}$) transforms as a covariant tensor. In other words, it retains the same form under a transformation. This is just what is needed for the physical laws to remain the same under a transformation and fit in with Einstein's principle of covariance. The reciprocal of $\mathbf{g}_{pq}$ is the contravariant tensor $\mathbf{g}^{pq}$, where the multiplication of matrix coefficients $g_{pq} g^{rq} = \delta p^r$, where $\delta$ is called the kronecker delta and is 1 if $p = r$, but is zero otherwise.

Einstein and Grossmann collaborated on developing a generalised theory of gravity using tensors. In 1913, Einstein and Grossmann published their first gravity paper, entitled *Outline of a Generalised Theory of Relativity and a Theory of Gravitation*, with Einstein providing the physics insight and Grossmann providing the mathematical backing. This was followed up in 1914 with another paper: *Covariance Properties of the Field Equations for a Theory of Gravitation*. Jointly, Einstein and Grossmann came very close to providing a general theory of gravity. But their field equation did not reduce to the Newtonian solution for a weak, static gravity field.

In 1914, with Max Planck's support, Einstein was offered a post at the Humboldt University in Berlin, which he accepted. This was the end of the collaboration with Grossmann but, by then, Einstein was well schooled in tensor calculus. By and large, Einstein continued to work on his general theory of relativity on his own, although others were keeping a watch on him. During this period, Einstein still made mistakes in his tensor analysis and Levi-Civita wrote to him and corrected some of them.

The German mathematician David Hilbert, at Göttingen University, took a close interest in Einstein's theory of gravity and Einstein was invited to give a seminar at Göttingen in June 1915. Hilbert took copious notes, which flattered Einstein. A few months later, Hilbert informed Einstein that he was using a variational method to derive the field equation for gravity and thought that he would have a solution fairly quickly. Having made all the effort over nearly 9 years, Einstein was very concerned that Hilbert would reap the reward of finding the correct field equation for gravity first. The threat put Einstein under a great deal of nervous strain and made him ill. But he wasn't to be beaten and he renewed his efforts. There was an element of escapism involved, too, as by concentrating solely on his gravity study he could blank out his difficult family problems and the bleak wartime situation in Germany. Finally, in an intense period of study, which ended in November 1915, Einstein managed to piece together a field equation for gravity that did satisfy Newton's theory of gravity for weak gravity fields. Einstein and Grossmann had worked with the Riemann-Christoffel curvature tensor, but it was the Ricci curvature tensor that was needed. At almost the same time, David Hilbert derived a solution for the field equations for gravity, but his result was overshadowed by Einstein's work. Synchronicity is a common occurrence in scientific development.

Since we have skimmed over Einstein's approach to general relativity, we cannot derive his set of equations for modelling the gravitational field

## AN OUTLINE OF EINSTEIN'S GENERAL RELATIVITY

in 4-D space-time. But, let us state his field equations here so that we know what they look like. There are sixteen separate equations.

$$[dx\ dy\ dz\ cdt]\begin{bmatrix} G_{11} & G_{12} & G_{13} & G_{14} \\ G_{21} & G_{22} & G_{23} & G_{24} \\ G_{31} & G_{32} & G_{33} & G_{34} \\ G_{41} & G_{42} & G_{43} & G_{44} \end{bmatrix}\begin{bmatrix} dx \\ dy \\ dz \\ cdt \end{bmatrix} = \left(\frac{-2}{\gamma c^4}\right)[dx\ dy\ dz\ cdt]\begin{bmatrix} T_{11} & T_{12} & T_{13} & T_{14} \\ T_{21} & T_{22} & T_{23} & T_{24} \\ T_{31} & T_{32} & T_{33} & T_{34} \\ T_{41} & T_{42} & T_{43} & T_{44} \end{bmatrix}\begin{bmatrix} dx \\ dy \\ dz \\ cdt \end{bmatrix}$$

In tensor form they reduce to

$$\mathbf{G}_{pq} = -\frac{2}{\gamma c^4}\mathbf{T}_{pq}$$

Einstein's tensor $\mathbf{G}_{pq}$ is composed of two parts.

$$\mathbf{G}_{pq} = \mathbf{R}_{pq} - \frac{1}{2}R\mathbf{g}_{pq}$$

$\mathbf{R}_{pq}$ = Ricci curvature tensor (4×4 matrix. Units: $1/m^2$).
$R$ = Riemann's scalar curvature (Units: $1/m^2$).
$\mathbf{g}_{pq}$ = The 4×4 metric matrix (Units: Dimensionless).
$G$ = Newton's gravitational constant = $6.672 \times 10^{-11}$ $m^3/kg.s^2$.
$\gamma$ = Gravitational permeability = $1/4\pi G$ = $1.193 \times 10^9$ $kg.s^2/m^3$.
$c$ = Speed of light in the vacuum = $3 \times 10^8$ m/s.
$\mathbf{T}_{pq}$ = Energy-momentum, or stress, tensor (4×4 matrix. Units: $kg/m.s^2$).

$R$ is the trace of the matrix $\mathbf{R}_{pq}$. That is, $R$ is the sum of the four coefficients on the leading diagonal of $\mathbf{R}_{pq}$. This is nothing like as simple as it appears to be.

On the left-hand side of the field equation for gravity, we have the geometry of space-time curvature. Einstein's tensor $\mathbf{G}_{pq}$ is a symmetric 4×4 matrix. That is, for $p \neq q$, the coefficients $G_{pq} = G_{qp}$. The $\mathbf{G}_{pq}$ matrix is split into two parts. The first part is the matrix $\mathbf{R}_{pq}$, expressing the inherent curvature of space-time introduced by matter (of any form). The second part is linked with the curvature of space-time due to the coordinate system used. Due to symmetry, only a maximum of ten of the $G_{pq}$ coefficients are different. Einstein proposed that the coefficients $G_{pq}$ should be viewed as a distribution of sixteen potentials for space-time.

We now consider the right-hand side of the equation. Einstein guessed that space-time curvature was due to the background presence of mass and energy which stressed space-time. He referred to this idea as Mach's principle. In 1863, the Austrian scientist Ernst Mach hypothesised that a mass would be inertia-less and space would be stress-free if there was no background gravity. So Einstein introduced the space-time stress tensor $\mathbf{T}_{pq}$. This is a 4×4 symmetric matrix with sixteen coefficients, where each coefficient of the matrix $\mathbf{T}_{pq}$ has units of energy density. The coefficients $T_{pq}$, for $p \neq q$, are shear stresses in space-time. The first three coefficients along the leading diagonal are pressures. The fourth coefficient $T_{pq} = \rho c^2$ is the rest energy density.

In order to balance the field equations dimensionally, each coefficient $T_{pq}$ is multiplied by a constant $2/\gamma c^4$. Note that energy has effective mass, since $m = E/c^2$, so regions of energy, for example, an electromagnetic field, create gravitational influence. This also includes gravitational energy, so mass provides two sources of gravitational influence; one from the mass itself and two from the surrounding negative energy field that it creates. This makes the model of general relativity extremely complicated and non-linear.

There is some controversy over whether Einstein's equation satisfies the conservation of energy and conservation of angular momentum. The Russian physicist Professor Anatoly Logunov has constructed an alternative theory of gravity. It runs along similar lines to Einstein's approach but retains its links with special relativity. More importantly, in Logunov's theory, both the conservation laws of energy and angular momentum are obeyed. Another difference is that while in Einstein's theory gravity appears as the result of space-time curvature, in Logunov's theory gravity retains its Faraday–Maxwell field nature with which we are familiar.

Einstein now had to determine the path of a free point mass in 4-D space-time as it moved under the influence of a gravity field. In 3-D space-time, the free-falling point mass accelerates along a world-line inside the light-cone. From the principle of covariance, the point mass is free of gravity and suffers no resistance as it travels along the world-line, so the world-line forms a geodesic. The $g_{pq}$ coefficients must be determined for a rest frame in space-time where the interval ds is a minimum distance. The method for seeking maximums and minimums, where curves reach a turning point, is called the calculus of variations. This is used in tensor analysis form and gives rise to an expression involving Christoffel symbols linked with tensor differentiation and the terms $g_{pq}$. A point mass moving in this frame would take the maximum (proper) time to do so, since all moving clocks run more slowly. In theory, the curvature of the space-time metric results in mass acceleration without resistance, implying free fall in a gravity field in 3-D. So the geometry of space-time curvature is tied to gravity. Thus the path of a free mass accelerating in a gravitational force field in 3-D space is viewed as the resistance-free passage of the same mass along a curved surface in 4-D space-time.

Turning (or stationary) points are associated with rates of change which involve differentiation. But the differential of a tensor is not a tensor, and ways have been made to add extensions to get covariant differentials of covariant and contravariant tensors. Work in this area was carried out by Christoffel during the 1860s, while at the Zurich Polytechnic. There is some debate as to whether the covariant derivative should include an extra term to model torsion, but this was ignored during the initial development of general relativity and only considered later, in the 1920s, by the French mathematician Elie Cartan.

The set of equations that describe how masses move in curved space-time is given by the Geodesic equation, where $\tau$ is the proper time. The result has links with Newton's second law.

$$\frac{d^2x^p}{d\tau^2} + \Gamma^p_{rs}\frac{dx^r}{d\tau}\frac{dx^s}{d\tau} = 0$$

$\Gamma$ is the Christoffel symbol of the 2$^{nd}$ kind, sometimes called a connection coefficient. It contains differentials of the coefficients $g_{pq}$ with respect to the coordinates $x^i$.

Einstein's field equation and the Geodesic equation, both in tensor form, are the two fundamental equations for general relativity. The results are bewildering to non-tensor experts and leave most of us in shock and awe.

Probing Einstein's theory of general relativity may provide an intellectual stimulus for the tensor experts. But, apart from a few cases for limiting conditions, solving the non-linear equations for general relativity simultaneously is virtually impossible. Nevertheless, using computers, numerical experiments have been carried out to model things involving very large gravitational fields, such as the collision between two black holes.

We need to check that Einstein's field equations reduce to Newton's form for a static gravity field. This means assuming that at any point in space-time the changes in the gravity field are very small and may be ignored and that the mass velocities there are so small that they may be taken to be zero.

The addition of the matrix $(\mathbf{R}_{pq})/R$ with dimensionless coefficients only slightly perturbs the matrix $\mathbf{g}_{pq}$. The leading coefficients of the modified matrix $\mathbf{g}_{pq}$ may be written as

$$g_{11} = 1 - \frac{\phi_{11}}{c^2} \quad g_{22} = 1 - \frac{\phi_{22}}{c^2} \quad g_{33} = 1 - \frac{\phi_{33}}{c^2} \quad g_{44} = -1 - \frac{\phi_{44}}{c^2}$$

The off-diagonal coefficients, where $p \neq q$ and symmetry applies, may be written as

$$g_{pq} = -\frac{\phi_{pq}}{c^2}$$

The modified $g_{pq}$ coefficients act like individual dimensionless gravitational potentials, so that the modified $\mathbf{g}_{pq}$ matrix forms a gravitational potential-like distribution.

If no matter is present, then $\mathbf{R}_{pq} = 0$ and $\phi_{pq} = 0$ for all p and q, so that the modified matrix $\mathbf{g}_{pq}$ becomes the Minkowski matrix $\mathbf{g}_{pq}$ for flat space-time.

The reduced set of equation in matrix form is

$$[dx\ dy\ dz\ cdt]\left(\frac{-1}{2}\right)R[\mathbf{g}_{pq}]\begin{bmatrix} dx \\ dy \\ dz \\ cdt \end{bmatrix} = \frac{2}{\gamma c^4}[dx\ dy\ dz\ cdt][\mathbf{T}_{pq}]\begin{bmatrix} dx \\ dy \\ dz \\ cdt \end{bmatrix}$$

Here, $\mathbf{g}_{pq}$ is the modified metric matrix. Expanding the expression will give us sixteen equations. But, if we divide both sides of all the equations by $(dt)^2$ and if we ignore velocity-squared terms ($v^2$) and those divided by $c^2$, we are left with the following single equation:

$$\left(\frac{-1}{2}\right)R\left(-1 - \frac{\phi_{44}}{c^2}\right)c^2 = \frac{2}{\gamma c^4}\rho c^2 c^2$$

We assume that the scalar distortion factor $R = \partial^2/\partial s^2$ along the geodesic curve. Thus, the equation becomes

$$\frac{\partial^2 \phi}{\partial s^2} = \frac{\rho}{\gamma} \text{ where Newton's gravitational potential } \phi = \frac{1}{4}\phi_{44}.$$

This is called Poisson's equation and is the result for a static Newtonian gravitational field. Poisson's equation may be written as $\nabla^2 \phi = \rho/\gamma$. We can also write the equation as $\nabla \bullet \mathbf{g} = -\rho/\gamma$, since $\mathbf{g} = -\nabla\phi$. This confirms that Einstein's field equations reduce to the Newtonian form for static gravity.

The above approach is called the linearisation of Einstein's field equations. The method can be extended to include the effect of a mass m moving with a low velocity **v** under the influence of a gravitational field **g**. Here we must defer to the tensor experts. In 3-D space, they show that Einstein's field equations become

$$\nabla \bullet \mathbf{g} = -\frac{\rho}{\gamma}$$

$$\nabla \bullet \mathbf{h} = 0$$

$$\nabla \times \mathbf{g} = \eta \frac{\partial \mathbf{h}}{\partial t}$$

$$\nabla \times \mathbf{h} = -\rho \mathbf{v} + \gamma \frac{\partial \mathbf{g}}{\partial t}$$

These equations are sometimes called the GEM (gravitoelectromagnetic) equations, as they are the gravitational analogue of Maxwell's equations for electromagnetism.

A point mass m has a radial gravity field **g**. If the mass moves with speed **v**, there is a dislocation in the radial gravity field lines, because information about the position of the mass travels with speed c radially outward along the gravity field lines. In nature, the dislocation is smoothly affected by a circular gravitomagnetic field **h** surrounding the moving mass. The permeability of the field **h** in 3-D space is labelled $\eta$. In Chapter 33, we will investigate the linearised dynamic gravity equations in more detail. The meaning of the various mathematical symbols (dot • and cross ×) will be explained in the following chapters.

Linearising the Geodesic equation and multiplying the result by mass m leads to a result of the form

$$m\frac{d\mathbf{v}}{dt} = m\mathbf{g} + m(\mathbf{v} \times \eta \mathbf{h})$$

The equation is the gravitational analogue of Lorentz's equation in electromagnetic theory. The equation can also be obtained from Einstein's special relativity, using the Lorentz transforms. For **v** = 0, it reduces to Newton's 2$^{nd}$ law of motion.

In his concluding paper on general relativity, published in November 1915, Einstein gave an approximate solution for the static case of a spherical non-rotating mass in the vacuum of space-time.

The following year, Karl Schwarzschild published an exact solution to Einstein's simple spherical mass problem. Between 1901 and 1909,

Schwarzschild had been the professor in charge of the Göttingen University Observatory. His university colleagues included Hermann Minkowski and David Hilbert. Even before then, he had queried whether space was truly Euclidean in form. From 1909 until 1914, he was the Director of the Astrophysical Observatory in Potsdam, Berlin. But, with the onset of the 1st World War, Schwarzschild had joined the German Army. By late 1915, he was an artillery officer on the Eastern front fighting the Russians. During this time, he read and understood Einstein's theory of gravity (and probably Hilbert's version, too), which was an incredible feat. Not only that, Schwarzschild's model matched the conditions for those around the Sun, providing a metric for the Solar System. His solution showed that space-time near to the Sun was distorted. But as the distance from the Sun increased, the distortion quickly disappeared, leaving space-time flat.

Schwarzschild's model is usually visualised using the analogy of a flat rubber sheet to represent a layer of space-time. A heavy ball bearing is used to represent the Sun. When the heavy ball bearing is placed at the centre of the rubber sheet, it lies in a deep dimple, representing the distortion of space, but not time, by the mass of the Sun. A light ball bearing, representing a planet, is cast onto the sheet. In circling around the dimple, it represents the motion of a planet around the Sun. Far from the heavy ball bearing the sheet is undistorted, or flat.

All the planets follow elliptical orbits around the Sun as one focus. The major axes of the planets' elliptical orbits also rotate very slowly. That is, a planet's perihelion (point of nearest approach to the Sun) rotates slightly. This slight rotation is called precession. These are incredibly difficult measurements for astronomers to make. At the beginning of the 19th century, the value for the precession of Mercury's orbit was observed to be 575 arcseconds per century. In 1859, the French astronomer Urbain Le Verrier used Newton's gravitational theory (which only applies to two masses) plus perturbations due to the other planets to calculate the precession of Mercury's perihelion. The validity of this approach remains questionable. Nevertheless, following a later correction, the Newtonian-based model predicted the precession of Mercury to be 532 arcseconds per century. Thus, there was a difference of 43 arcseconds per century not accounted for.

From the solar space-time distortion predicted by Schwarzschild, Einstein was able to calculate its effect on the orbit of Mercury. He calculated that the distortion would increase the flat space-time precession of the perihelion of Mercury predicted by Le Verrier's method by a further tiny amount of 43 arcseconds per century. This was in perfect agreement with the observation made by astronomers and gave support to his theory of general relativity.

In 1917, Einstein tried to find another solution to his field equation. He assumed, like the ancient Greeks, that the Universe was contained within a sphere. For no mass present in the sphere (including effective mass from energy fields) $\mathbf{T}_{pq} = 0$, so that $\mathbf{R}_{pq} = 0$ and space-time is flat. For the case where mass was present, he assumed that it was evenly distributed throughout the Universe. He also assumed that the Universe was static so that the radius of the Universe (however distant) did not change with time. But then he realised that he had a problem. There was a gravity imbalance across the edge of the Universe which, due to gravitational attraction, would lead to an inward collapse of the Universe with time. To keep the Universe static, Einstein introduced a cosmological constant $\Lambda$ (Lambda is the Greek capital letter L) to the left-hand side of his equation. His idea was to tweak the curvature coefficients $g_{pq}$.

$$\mathbf{R}_{pq} - \frac{1}{2} R \mathbf{g}_{pq} + \mathbf{g}_{pq} \Lambda = -\frac{2}{\gamma c^4} \mathbf{T}_{pq}$$

$\Lambda$ = Cosmological constant (units: $1/m^2$).

Einstein could have added the constant to the right-hand side of his equation to tweak the stress tensor $\mathbf{T}_{pq}$. In this case, we see that the expression $(\gamma \eta c^4/2)\mathbf{g}_{pq} \Lambda$ is a stress-like term which imposes a constant force per unit area on the surface of space-time. In other words, it acts like a pressure between the individual masses, reducing their gravitational attraction. The dimensions of the expression mean that it can also be thought of as the energy density of the vacuum of space, which modifies the energy density due to the presence of force fields. Now, gravitational energy density is negative, so introducing $\Lambda$ at every point of space means that it operates like a moderator in a nuclear reactor. If $\Lambda$ is positive, then the negative energy density of the Universe is reduced everywhere, so its presence counteracts gravitational attraction and prevents (or slows) gravitational collapse of the Universe. Too high a value for $\Lambda$ will make the Universe expand. On the other hand, if $\Lambda$ is negative, it will hasten gravitational collapse.

Astronomical observations indicate that $\Lambda$ is tiny. Averaged measurements for the Universe show that $\Lambda \approx 1.19 \times 10^{-52}$ $m^{-2}$.

No one seems to have given any thought to the idea that $\gamma$ (or G

= $1/4\pi\gamma$) might change in Einstein's equation. But this would mean introducing a parameter $\Lambda$ which was not constant, but varied, possibly with distance, time or curvature.

In case you are not happy with the idea of introducing the cosmological constant $\Lambda$ in space, think of the situation within the nucleus of an atom. Something similar must be happening there. We talk about the strong nuclear force holding the positively charged protons together. It may be that a nucleological constant creates a negative energy density within the nucleus, reducing the electrical repulsion between the protons and increasing their gravitational attraction, thereby holding the protons and neutrons together. It's the same idea but on a completely different scale.

Having introduced $\Lambda$, Einstein was dismayed to find that his modified model of the Universe was still unstable, and he suggested dropping the idea. Then, in 1929, when Edwin Hubble discovered that the visible Universe is expanding, it looked like the cosmological constant was unnecessary after all. However, the more recent discovery that this expansion is speeding up, rather than slowing down, has led to a revival of interest in $\Lambda$. This is now linked with dark energy, a speculative 'anti-gravity' phenomenon said to be responsible for boosting the expansion rate of the Universe.

Although general relativity is conceptually very different from Newton's theory, its observable consequences are almost exactly the same. The most obvious differences occur in very strong gravitational fields, which require much denser states of matter than can be found in the Solar System. This means that astronomers have to look much further afield for physical effects which might be explained by general relativity. But, as we know, this has uncovered anomalies leading to the discovery of new phenomena such as black holes, gravitational waves and dark energy.

One hundred and fifty years ago, the subject of non-Euclidean geometry was in its infancy. One hundred years ago, it developed into tensor calculus practised by about a dozen scientists. Tensor calculus involves the combining and manipulation of large arrays of data. Over the last 50 years, the development of high-speed computers, with enormous data storage capacities, has provided an ideal means of dealing with tensor operations numerically, allowing cases of strong gravity fields to be examined. Nowadays, hundreds of scientists use coded tensor calculus programs to model physical problems and perform numerical experiments. Even so, tensor calculus remains an arcane art.

A final comment on general relativity is that its derivation by

Einstein was an intellectual achievement requiring physical insight, new mathematics and enormous perseverance. Even today, only a relative few mathematical physicists fully understand the theory. Nowadays, it is accepted that general relativity is a solution for very large-scale situations, and cosmic events have provided some experimental verification of the theory. But the theory is of no practical use for experiments in a laboratory on Earth investigating the means of gravity control. Nor is the theory a complete theory, since a quantum theory of gravity is missing.

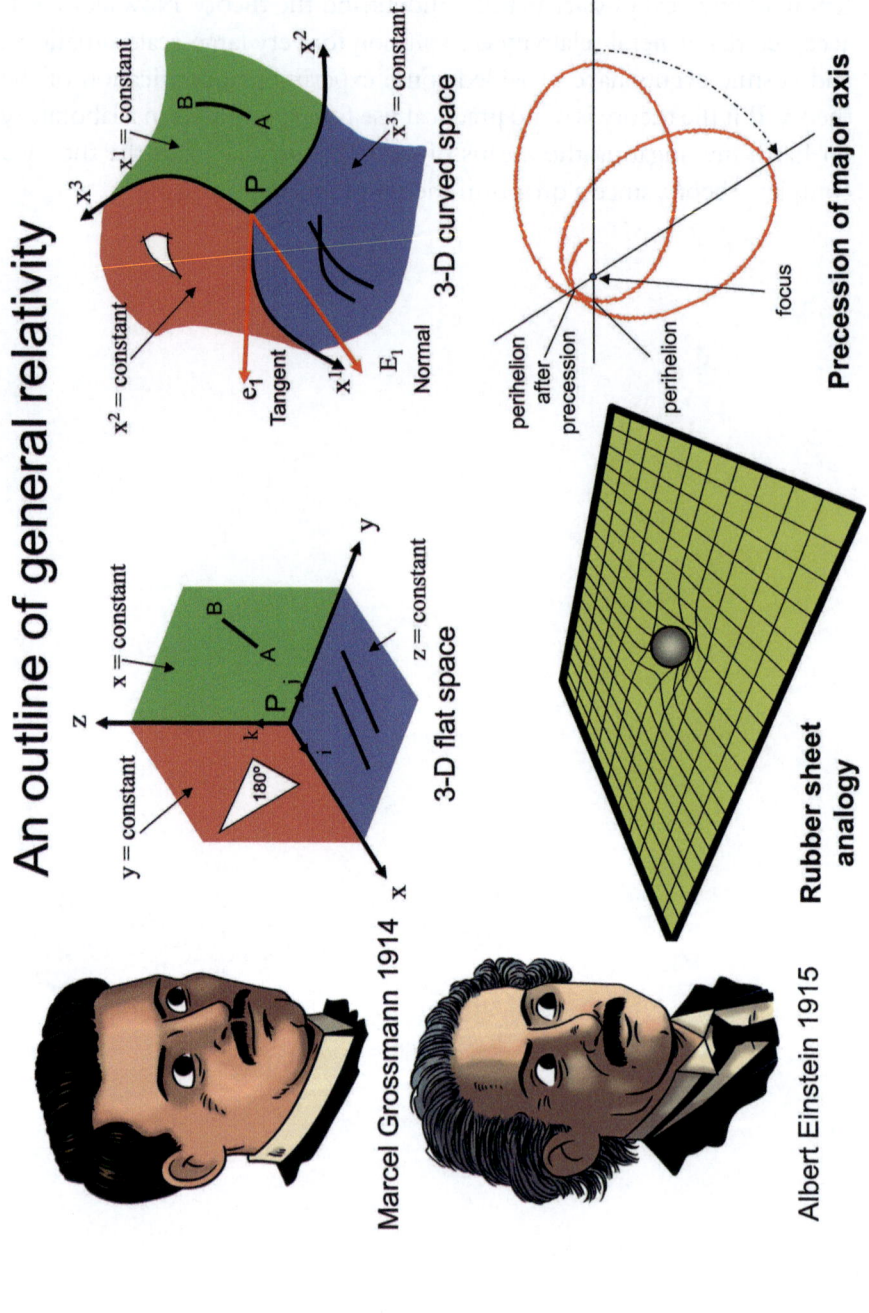

CHAPTER 15

# THE MEANS OF GRAVITY CONTROL REMAINS UNKNOWN

Following Newton's work, scientists eventually accepted that mass was a source of gravity and that a body of mass was surrounded by a region of gravitational influence; now referred to as the body's gravity field. After Einstein showed that $E = mc^2$ it was realised that an energy region of any form must possess effective mass, which would create a gravity field within and around the region. Thus, a mass creates a gravity field which contains energy which creates effective mass which creates more gravity. It's getting complicated. From Einstein's theory of general relativity we have learnt that the presence of mass in space-time distorts the surrounding 4-D geometry of space-time in such a way that surface curvature within space-time replaces the body's gravity field. Or, to put it another way, the gravitational energy associated with mass is locked up in the distortion of space-time around a mass. However, this viewpoint has not helped us to understand how to control gravity. At the moment, the way ahead in our search for gravity control seems to be stymied by Einstein's general relativity.

Knowing too much about a subject can be off-putting. In the conclusion to Ivar Giaever's Nobel Prize speech (1973 – Tunnelling in Superconductors), he said "... *The road to scientific discovery is seldom direct and does not necessarily require great expertise. In fact… often a newcomer to a field has a great advantage because he is ignorant and does not know all the complicated reasons why a particular experiment should not be attempted.*" So, perhaps, now is the time for non-experts in general relativity to propose new ideas for experiments to investigate possible ways around the current impasse.

Einstein's set of equations for gravity is non-linear, in line with many of nature's phenomena. Unravelling non-linear equations to see how the various variables interact is very difficult. Generally, to start with, mathematicians prefer to work with linearised equations where the variables are directly proportional to each other. In this way, major relationships between variables are retained but minor relationships are ignored. Therefore, as a first step in our search for gravity control, let us investigate the linearised theory of general relativity. Hopefully, this will enable us to form a simple model of the dynamical relationship between moving masses and their gravity fields. If this approach is successful, then might be the time to investigate what other secrets gravity holds with its non-linearity.

Galileo discovered that the simple pendulum could be used for time-keeping. With this device, Newton measured the speed of sound. Investigating the simple pendulum, scientists discovered the link between the period of swing of the pendulum, the length of the pendulum and the acceleration due to gravity. The simple pendulum played a major part in the growth of the clock-making industry, which had an impact in other areas, most notably in navigation. For small angles of oscillation the system is linear, but for swing angles greater than 10° the system is non-linear and the device is of limited use for studying the basic link between gravity and time.

There is an added bonus in tackling gravity force field dynamics in a more simplified way, because we will find pattern similarities with other force field dynamics which are already understood and which are based on linear theories. Mathematicians and scientists love patterns.

Let us consider how electromagnetism developed. The search started with the separate experimental studies of electricity and magnetism. Through chance, a link was discovered experimentally between current flow and magnetism. Then, again through experiment, it was discovered how a moving magnetic field gave rise to an electric field. It was only after that that a theory linking the force fields of electricity and magnetism was formulated, by James Clerk Maxwell, which described the whole phenomenon of electromagnetism. Later, the equations of Maxwell's theory were shown to satisfy the theory of special relativity. In fact, these same equations can also be expressed in tensor form, but they seldom are, as they make the subject of electromagnetism almost incomprehensible to all but a few scientists. So, even though we don't know what electricity is (perhaps it is due to some other form of space-time curvature), we do know how to control the combined phenomena of electricity and magnetism.

With the study of electromagnetism, we were fortunate that interactive effects are very large. By comparison, gravity is a very weak force. Assuming that gravity is not immutable, there are two possible reasons why we haven't been able to detect any dynamic effects. Firstly, it may be that interactions with other phenomena are very weak, so sensor detection sensitivity is a major factor. Secondly, it may be that we don't know what we are looking for (it may be a large effect, that we have not realised is linked with gravity), so we don't know what detection system to use. Because of its weakness, the study of gravity began with theoretical work and predictions checked with astronomical measurements. This was followed, later, with some laboratory experiments using sensitive detection systems. Following Einstein's work, there followed a lot of theoretical interest but little experimental progress. Today, both theory and experiment remain linked with the effects associated with astronomical bodies, again highlighting the weakness of gravity. To follow the electromagnetic route we need to devise further laboratory experiments. This means greatly improving device sensitivity (for example, using atomic gyroscopes), or searching for new, or re-interpreting misunderstood, interactions with gravity. New ideas to test are analogous to exploring new routes through the wilderness by pioneers. Having a simple guide to suggest some new ways to try would be invaluable. Such a guide does exist and has been used by scientists before. It is based on using 'read-across' between mathematical models sharing the same pattern, or form, called analogues.

In his book *Novum Organum Scientiarum*, published in 1620, Francis Bacon wrote:

> *Hence all the most Noble Discoveries have come to light, not by any gradual improvement and Extension of the arts, but merely by Chance; whilst nothing imitates or anticipates Chance but the Invention of Forms.*

By forms, Bacon meant analogues, and in another passage in his book he wrote:

> *He who knows forms grasps the unity of nature beneath the surface of bodies which are very unlike. Thus, he is able to identify things which have never been seen before and bring them about by experiment. Things which nature*

*has kept secret and humans have never dreamt of. From the discovery of forms flows true speculations and unrestricted advances.*

In an address to the Mathematical and Physical Section of the British Association given in 1870, James Clerk Maxwell said:

*The student who wishes to master any particular science must make himself familiar with the various kinds of quantities which belong to the science. When he understands all the relations between these quantities, he regards them as forming a connected system, and he classes the whole system of quantities together as belonging to that particular science. This classification is the most natural from a physical point of view, and it is generally the first in order of time.*

*But when the student has become acquainted with several different sciences, he finds that the mathematical processes and trains of reasoning in one science resemble those in another so much that his knowledge of the one science may be made a most useful help in the study of the other.*

*When he examines into the reason of this, he finds that in the two sciences he has been dealing with systems of quantities, in which the mathematical forms of the relations of the quantities are the same in both systems, though the physical nature of the quantities may be utterly different.*

*He is thus led to recognise a classification of quantities on a new principle, according to which the physical nature of the quantity is subordinated to its mathematical form. This is the point of view of the mathematician; but it stands second to the physical aspect in order of time, because the human mind, in order to conceive of different kinds of quantities, must have them presented to it by nature.*

It is noticeable that many forms, shapes or patterns in nature are repeated. It may be the patterns on animals' coats and hides and the pattern of background vegetation, or the shapes shared by various structures, or the patterns shared by the force fields in physics. It's as though nature only has a few basic designs at its disposal. Sometimes phenomena share the same pattern because there is a clear causal relationship, camouflage in the case of some animal coats, but in other cases there is no obvious connection or, perhaps, none exist. When phenomena are expressed in mathematical terms, nature's disguises are removed and the underlying form exposed. Then, any similarities shared by phenomena become very apparent. Such phenomena are called analogues.

# THE MEANS OF GRAVITY CONTROL REMAINS UNKNOWN

The ancient adage, 'As above, so below, but after another manner', which some claim stems from ancient Egyptian and ancient Greek texts, is another description of analogues linking the macroscopic shape of things in the Universe with the microscopic shape of things in the atom.

However, although mathematical models of two phenomena might be analogous, it does not necessarily mean that an underlying relationship exists between the phenomena. Furthermore, although mathematical models may appear to be very similar, they are unlikely to be exact, so interpreting what 'read-across' means needs to be treated with caution. Nevertheless, scientists have used 'read-across', looking for the analogue of effects occurring in one phenomenon which seemed to be missing from another.

We can see this book in terms of an analogy. The first part of the story, up until now, is like setting out to drive to a particular destination. We make good progress until we come to a halt at a temporary traffic light showing red, with roadworks stretching beyond into the distance. Most drivers patiently wait for the traffic light to change to amber and then green, so that they can continue their journey. The analogue for the next part of the story is that after some considerable time, when nothing has happened, a few drivers ease out of the queue and turn off down side roads just before the temporary traffic light, looking for a way around the obstruction. Having travelled this way before, making for similar destinations, these drivers have some familiarity with the geography of the area and try other roads. While they are moving they may be lucky, and a new road may lead them to their intended destination, or to some other destination just as interesting. Another aspect is synchronicity, as several drivers may arrive at the same destination at the same time. This happens with scientific discoveries, too.

We are amazingly adept at spotting patterns and noticing anomalies when the pattern is spoilt. Pattern recognition has led to the classification of science into various branches (using the analogy of the tree) such as Biology, Chemistry, Physics, etc., as we try to sort out nature's behaviour from what might appear, at first sight, to be chaos. This is essential if we are to make progress and record, for later generations, what we have discovered.

It is important that there is a common understanding of a mathematical model of a phenomenon between the theoretician and the experimenter for progress to be made. If the model is too complex, it may not help the experimenter to devise an experiment to test the theory. In 1857,

Michael Faraday, one of the world's greatest experimenters, pointed this out in a letter to James Clerk Maxwell, one of the world's greatest theoretical physicists. Faraday made the breakthrough in understanding the phenomenon of electromagnetism through experiment and found Maxwell's later theoretical model of electromagnetism difficult to comprehend. But Heinrich Hertz did understand Maxwell's theory and he used it in his experimental search for electromagnetic waves, the discovery of which has led on to the communications revolution of today.

A mathematical model starts off like a baby. To begin with, babies look alike, but as they grow they develop their own personalities and become different. The same happens with mathematical models. They are limited in what they can do to begin with and at this early stage they may share similarities with other models. But as models are developed to cover more features, they grow in complexity and, most likely, they lose their earlier similarity with other models.

So we should be aware that as mathematical models are separately developed for different phenomena, becoming more complex in the process, there is a danger that the shared patterns of simpler models become less obvious and the idea of 'read-across' is overlooked. I feel that we are in this sort of situation with gravity research today. The Einstein model replacing gravity with curved space-time is extremely complex. Nevertheless, there are a few scientists who are fully conversant with the theory as well as being experts in experimentation. But for the rest of us, perhaps now is the time to examine the analogies between simpler models to see whether 'read-across' suggests any new ideas for experiments to try in our search for the means of gravity control. In particular, let us look at potential theory and the patterns exhibited by fluid dynamics, electromagnetism and thermodynamics to see whether they offer any new insights into the phenomenon of gravity. We could do with some luck in our search, too. But, as the famous French bacteriologist Louis Pasteur remarked, "Luck favours the prepared mind." So, in our quest, we must be familiar with the patterns of potential theory, but not overly so, as too much complexity may obscure the prize for which we are searching.

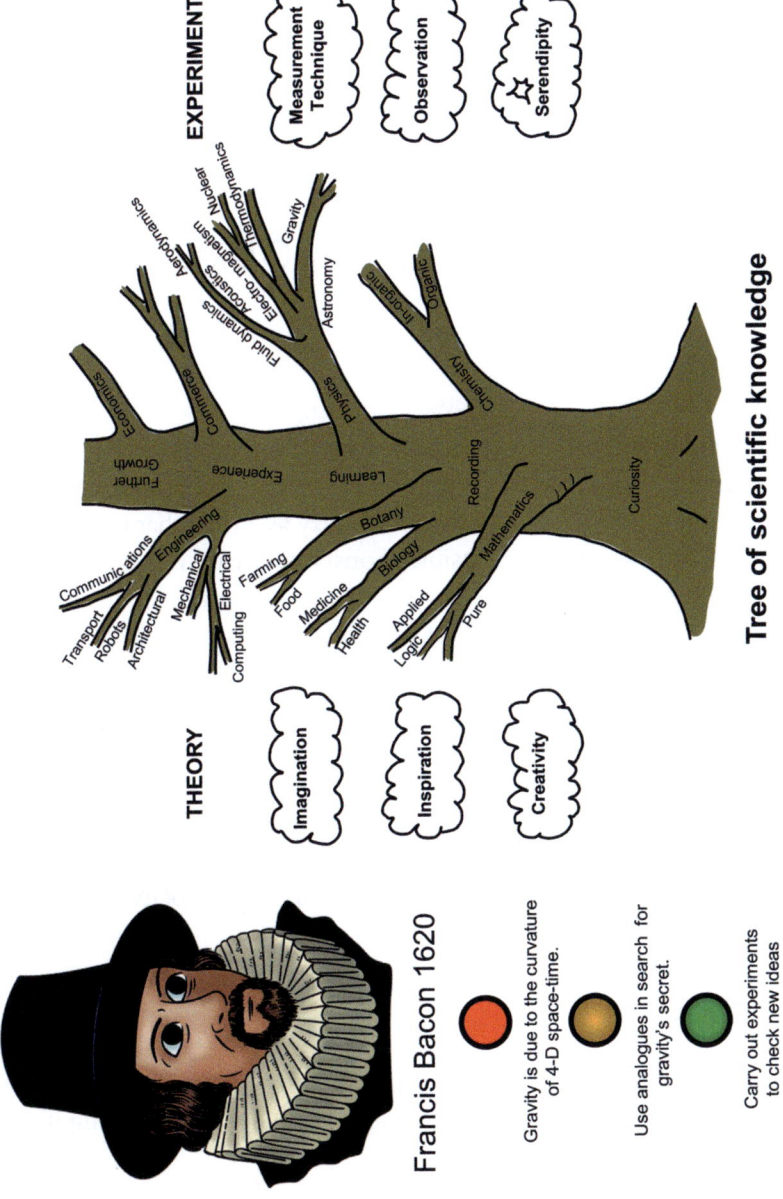

## CHAPTER 16

# FLUID DYNAMICS WITH SOURCES AND SINKS

The first thing mathematicians do, when confronted with a phenomenon to model, is to make some very simple assumptions about the phenomenon. So, for fluid dynamics they assumed that they had an ideal fluid which was incompressible (constant density $\rho$) and when moving was free of friction (viscosity $\eta = 0$). In doing this, it meant that they couldn't model some well-known phenomena of fluids (including air, water and treacle), such as sound waves and shock waves (both of which depend on compressibility) and vortices (which depend on viscosity). With these simple assumptions, the basic equation of motion for an ideal fluid, based on Newton's 2$^{nd}$ law of motion, was derived by the Swiss mathematician Leonhard Euler in 1755. From this equation, the Dutch mathematician Daniel Bernoulli, a close colleague of Euler's, derived the Bernoulli equation, which links pressure p with velocity v at any point in the flow.

$$\frac{p}{\rho} + \frac{1}{2}v^2 = \text{constant}$$

When the fluid speed v increases, the pressure p drops, and vice versa.

This explains why a door, just ajar, slams shut with a bang if there is a sudden draught through the gap.

The idea of point sources and point sinks of fluids was introduced by the Scottish engineer William Rankine in the mid 19$^{th}$ century. It is difficult to imagine a 3-D source emitting fluid from a point in space and spreading out in all directions. However, we can use the kitchen tap to see a 2-D source (So). When the tap is turned on and the jet of water hits the bottom of the sink, the water diverges outwards in a radial direction. We have created a fluid velocity field. Streamlines indicate the directions of the velocity at points in the fluid flow. Introducing ink at various points into the flow enables the radial streamline pattern to be seen. In theory, the tangent at any point on a streamline points in the direction of the fluid velocity there.

In reality, the flow starts out as a laminar (thin, flat and smooth) layer and ends with a circular hydraulic jump (the analogue in water of a shock wave in air) as the water backs up due to the side of the kitchen sink. Downstream of the jump, the water layer is thicker, slower and slightly turbulent.

Let's consider the ideal 2-D case where the radial flow, of unit thickness, has constant density $\rho$ and there is no hydraulic jump. Suppose the jet of water leaves the tap at a rate of Q kg per second. The source (So) is said to have strength Q. The point where the jet hits the floor and the flow becomes radial is the centre point of our 2-D source. Consider a circle of radius r surrounding the source centre. Since the flow of water is continuous, Q kg of water must cross the surface area $2\pi r$ (that is, the perimeter times unit thickness) every second. So at any point on the perimeter the velocity $\mathbf{v}$ of the water flowing radially outwards across it is given by

$$\mathbf{v} = \frac{Q}{\rho 2\pi r}\hat{\mathbf{r}} \text{ m/s where } \hat{\mathbf{r}} \text{ is the unit vector pointing in the radial direction}$$

For a 2-D sink (Si) of strength -Q, the ideal fluid flows inwards to the centre point.

For the 3-D source, Q kg of fluid per second must cross a spherical surface of radius r, with surface area $4\pi r^2$. At any point on this surface the fluid flows in a radial direction outwards with a velocity $\mathbf{v}$ given by

$$\mathbf{v} = \frac{Q}{\rho 4\pi r^2}\hat{\mathbf{r}} \text{ m/s}$$

We have defined a 3-D velocity field. The velocity $\mathbf{v}$ obeys an inverse square law of distance from the point source centre. The streamlines fan out, or diverge, from the source. We interpret the density $\rho$ as the permeability of the fluid to movement. If $\rho$ is high then movement is sluggish (low velocity), while if ρ is very low then movement is hardly impeded (high velocity).

The expression for velocity **v** is analogous to that for the gravity field of a point mass M, given by

$$\mathbf{g} = \frac{M}{\gamma \, 4\pi r^2}(-\hat{\mathbf{r}}) \quad m/s^2$$

γ is the gravitational permittivity of the ether of space. Bernhard Riemann speculated that a point mass was a 3-D ether sink.

Just as we can write force **F** = m**g**, where m is a mass in a gravity field **g**, so we can write force **F** = − q**v**, where q is the strength of a fluid source (-q for a sink) in a fluid velocity field **v**.

If two free fluid sources are situated fairly closely together then they will appear to attract each other. With all the fluid pouring out of each source, you might have thought that they would be forced apart. But in the region between the two sources the two approaching flows are diverted away, speeding up in the process. From Bernoulli's equation, we know that when a fluid speeds up the pressure drops, so the two sources are sucked together. This effect can be observed visually with streamlines. Knowing how the streamlines of a fluid diverge, or converge, is a useful characteristic of a flow. Where streamlines begin to bunch close together (converge), the flow is speeding up and there is a drop in pressure. When streamlines start to spread apart (diverge), the flow is slowing down and there is a rise in pressure. So, two 3-D fluid sources obey the inverse distance squared law of attraction, just like two masses attract each other.

The velocity field of a 3-D sink obeys the same inverse distance squared law but sucks the fluid in. The streamlines converge towards the sink hole. Two sinks attract each other. But a source repels a sink and vice-versa.

Introducing a source q into the velocity field of a source Q and holding it fixed means that q has some potential energy to do work.

A source of strength q in a uniform stream flowing with velocity U will experience a force upstream of qU, as though attracted by an upstream source at minus infinity.

Suppose we have a fluid flowing with velocity **v**. We consider an area A in the fluid. If the normal to A does not correspond with the direction of the fluid velocity **v**, then we use a directed area **A**. (A directed area has size and direction, so is a vector shown as a bold character.) We define the velocity flux of the fluid through the area A as **v.A**.

# FLUID DYNAMICS WITH SOURCES AND SINKS

Now consider a short length ds of fluid, which passes through A. The direction and magnitude of the velocity vector **v** may change slightly over the short length ds. The parameter v/ds is called the divergence of the velocity. Mathematicians introduce $\nabla \cdot \mathbf{v} = v/ds$.

The volume V occupied by the short length of fluid is given by $V = A \cdot ds$. This allows us to rewrite the velocity flux as

$$\mathbf{v} \cdot \mathbf{A} = v \cdot V/ds = (\nabla \cdot \mathbf{v})V.$$

The simplest example is to consider a source Q surrounded by a spherical surface A of radius r.

Since velocity $v = \dfrac{Q}{\rho \, 4\pi r^2}$ and $A = 4\pi r^2$ then $\mathbf{v} \cdot \mathbf{A} = (\nabla \bullet \mathbf{v})V = \dfrac{Q}{\rho}$

Finally, we can write

$$\nabla \bullet \mathbf{v} = \dfrac{\sigma}{\rho} \text{ where } \sigma = \dfrac{Q}{V} \text{ is the source density.}$$

This result is called the divergence theorem and its derivation is attributed to the German mathematician Karl Gauss. The divergence theorem has an interesting history in that in an analogous form for electrostatics, it was first derived by George Green, a Nottinghamshire miller, in his leisure time.

$$\nabla \bullet \mathbf{E} = \dfrac{\rho}{\varepsilon} \text{ where } \rho = \dfrac{Q}{V} \text{ is the charge density and } \varepsilon \text{ is the electrical perimittivity.}$$

Green lived at the turn of the 18th into the 19th century and he privately published his result in 1828. Green's windmill, at Sneinton in the city of Nottingham, is now a mathematics centre.

We can extend the flux analogue to gravitostatics and write

$$\nabla \bullet \mathbf{g} = -\dfrac{\rho}{\gamma} \text{ where } \rho = \dfrac{M}{V} \text{ is the mass (source of gravity) density.}$$

From the fluid analogue, the gravitational form of the Bernoulli equation is

$$\dfrac{p}{\gamma} + \dfrac{1}{2}g^2 = \text{constant} = \dfrac{1}{2}\Lambda c^4$$

The medium is the ether of space (for flat space-time $g_{pq} = 1$), p is the pressure, γ is the gravitational permittivity, g is the gravitational field strength and Λ is the cosmological constant. The gravitational field lines in the ether are analogous to streamlines in fluids. We can rewrite the equation as

$$p = -\frac{1}{2}\gamma g^2 + \frac{c^4}{8\pi G}\Lambda$$

The term ½γg² is the local gravitational energy density, which is modified by the cosmological constant term.

Dynamically, as galaxies swirl about and gravity fields change, the above equation shows that the ether undergoes changes in pressure which influence the movement of gravitational sources, namely astronomical masses. The result is an analogue of incompressible fluid flow. The analogue of compressible fluid flow suggests that for compressible ether, local changes in γ can occur.

The fact that a source and a sink both exist (at least in our imagination) for fluids, raises the question of why we only know of one source of gravity; namely mass. Where is the negative source, representing negative mass?

# Fluid dynamics with sources and sinks

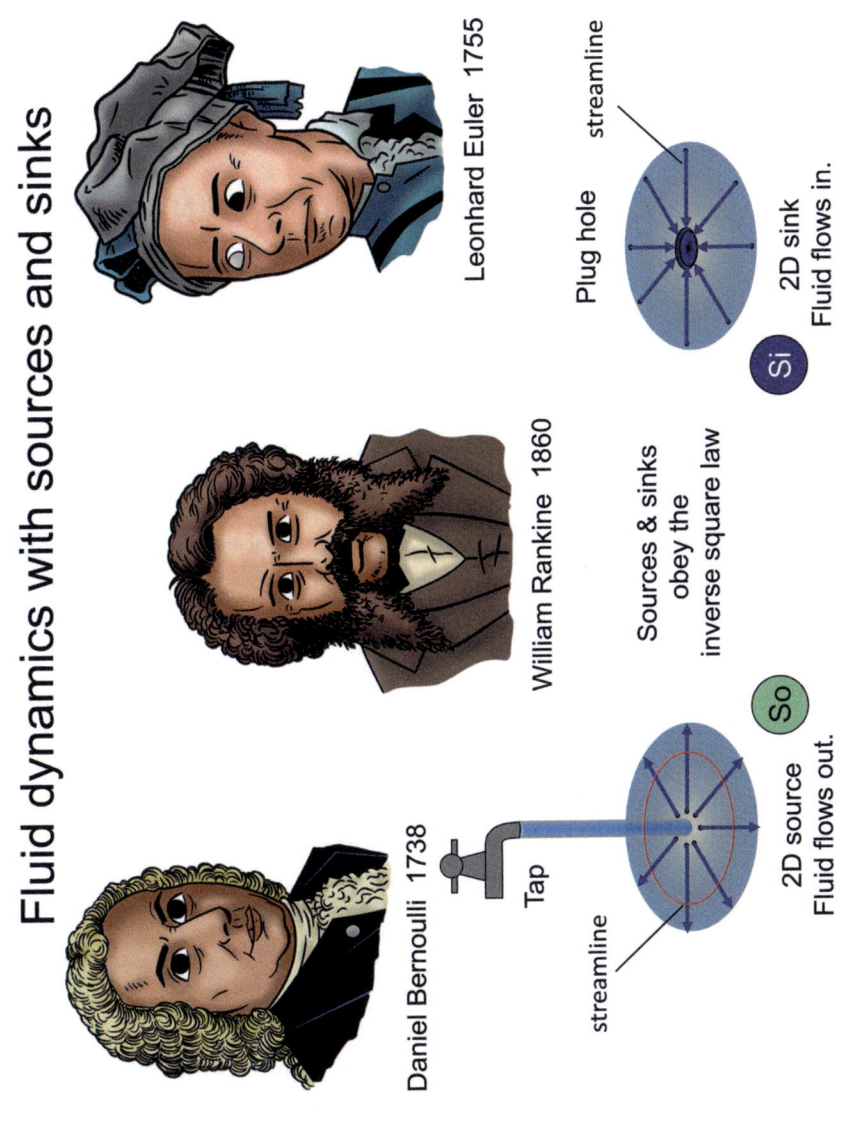

## CHAPTER 17

# THE SOURCE–SINK COMBINATION

Suppose a source and a sink of equal, but opposite, strength are fixed close to one another. In 2-D, the steady streamline pattern has mirror image symmetry, either side of the line joining the fixed source–sink combination. In the special case where the source and sink are coincident, then a dipole (sometimes called a doublet) is formed.

If the source and sink are free to move in a stationary external ideal fluid then, since they repel each other, they will move apart. However, if the source and sink are firmly coupled together, to form a source–sink combination, then the combination can't move.

In a real fluid, where some fluid friction (viscosity) is present, a fixed source–sink combination will cause the external fluid to move in the direction from the source to the sink. This process is called entrainment. If the source–sink combination is free then diffuse viscous entrainment will cause it to move. The nearest that we can get to modelling entrainment with an ideal fluid is to include a free stream with the source–sink combination.

We can demonstrate the ideal 2-D flow pattern of a fixed source–sink combination in a uniform stream with a water table. The Hele-Shaw apparatus is a flat metal surface covered with a glass sheet between which a thin layer of water is forced to flow. A number of ink ports are arranged in the metal plate upstream across the flow to mark out parallel streamlines for the channelled flow. Two small holes in the metal surface near the centre of the water table allow water to be fed in from one (to create a source) and sucked out from the other (to create a sink). The fluid from the upstream source and the downstream sink is contained in an oval-shaped self-contained region of circulating flow, separated from the main stream.

The 3-D streamline pattern for the 3-D source–sink combination is rotationally symmetrical about the line joining the source and sink. If a 3-D uniform stream is introduced then an ellipsoidal bubble of fluid separates the source–sink combination from the stream.

If the ideal source–sink combination is free to rotate about its centre then, if it is placed in a uniform stream, it will line up with the uniform stream of fluid, with the source upstream and the sink downstream, just like a magnetic compass lines up with an external magnetic field. In this condition, the ellipsoidal bubble is drag-free in an ideal fluid so the source–sink combination remains stationary in the uniform stream. Equally, the result applies to an ellipsoidal bubble moving through a stationary fluid. This result is contrary to real conditions, where an object moving through a fluid experiences drag due to viscous effects. The ideal, but unreal, prediction is called d'Alembert's paradox, named after the famous 18th century French scientist.

We must be careful not to dismiss ideas as impossible, just because theory says so. Theoretically, a propeller cannot work in an ideal fluid. But, in a real fluid, we know that it does. The missing ingredient is viscosity, which gives rise to the vortex phenomenon.

# The source-sink combination

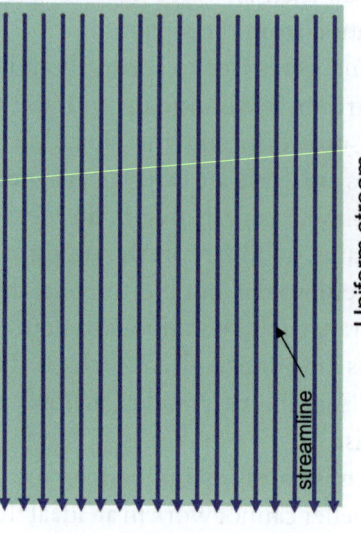

Uniform stream

streamline

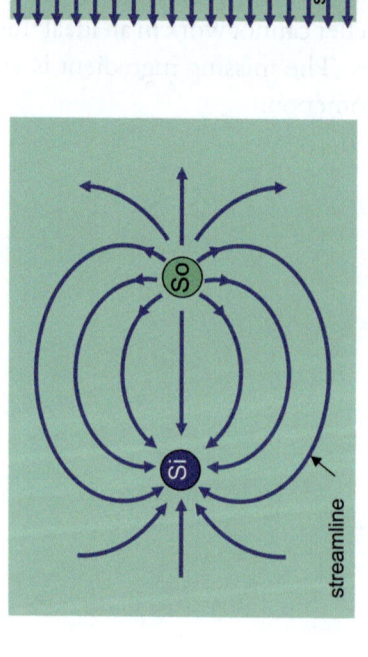

Dipole

streamline

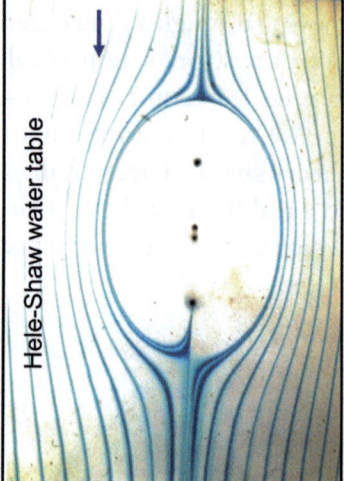

Hele-Shaw water table

CHAPTER 18

# THE LINE VORTEX AND ROTATING FLOWS

Two important types of fluid motion have been identified in terms of what happens to individual fluid particles. If the particles rotate as they move along then their motion is said to be rotational; otherwise, their motion is irrotational.

For circular flow, a good analogy to distinguish both types of motion is as follows. Place a pencil stub, representing a fluid particle, on a turntable. As the turntable rotates, the compass direction in which the pencil points changes. This is rotational motion. Replace the pencil with a magnetic compass. Now, as the turntable rotates, the compass needle always points north and doesn't change direction. This is irrotational motion.

We can use a ping-pong ball with red and white octant markings as a detector. On dropping the marked ping-pong ball into a stream of water, if it starts to rotate as it moves along then the flow is rotational.

Rotating flows occur everywhere, from spiral galaxies and giant swirling dust clouds in space; to hurricanes on planets with atmospheres; to tornadoes and their tiny counterparts, the street-corner dust eddies; to large tidal whirlpools, down to the rows of tiny vortices created as water swirls past rocks. These are further examples of nature's patterns being repeated; in this case, scale being a factor. Rotating galaxies in space are associated with gravity, while the vortices that we experience on Earth are associated with rotating fluids. Whenever a fluid rotates then viscosity $\eta$, or fluid friction, will create a vortex.

Earth's atmosphere is an example of a heat engine, with air and water vapour forming the driving medium and the Sun providing the heat source. As we will discuss in a later chapter, the thermodynamic cycle of a heat engine involves heat being transported by the medium from a high-temperature region to a low-temperature region. For the Earth, high temperature occurs at ground level, while low temperature occurs high up in the troposphere. Gravity also plays a part. Over the sea, the heated surface water evaporates and, being less dense, rises in Earth's gravity field forming an upward convection current of water-laden air. The geography of Earth's surface forms centres for these upward currents of air. High above us, the water vapour forms clouds. In time, the clouds become saturated with water vapour, and water condensation (increase in density) occurs, resulting in rainfall. On the surface of the Earth, where the air currents rush towards a convection current centre, the increase in air speed results in a region of low air pressure at the centre. The inrushing air forms a giant sink. Due to the Earth's rotation, the inrushing air also starts to rotate and a giant sink-vortex occurs. Satellite photos of clouds show these huge sink-vortices, which we know as hurricanes and cyclones.

Most of us are familiar with the combined sink and line vortex created in the bathwater after the plug is removed. The vortex core starts as a dimple at the water surface and extends downwards to the plug hole and beyond, with the bulk of the water rotating around the core. For a very strong vortex, the core may become a hollow tube with a funnel at the water surface. To examine the sink-vortex flow more carefully, scientists use a large circular tank of water, with a diameter of several metres, with a plug hole at the centre of the tank floor. The water is allowed to settle until it is completely still and then the plug is removed and the water begins to drain away. Initially, near the floor of the tank, the water moves radially towards the plug hole in a sink motion. However, for a wide tank of water, the local vertical component of the Earth's rotation $\Omega'_E$ and the radial velocity **v** of the escaping water combine to start the water rotating and a line vortex appears at the centre of the tank, with the core extending down to the plug hole.

Rotating flow patterns were originally investigated by Gaspard-Gustave de Coriolis, a French professor of mathematics, in the early 19$^{th}$ century. For those with a mathematical inclination, if m is the mass of a particle of water flowing with speed **v** towards the plug hole and the bulk of the water is made to rotate with angular velocity $\Omega$, then the force **F** experienced by the water particle is given by the following vector product

$$\mathbf{F} = m(\mathbf{v} \times 2\mathbf{\Omega})$$

The force **F**, called the Coriolis force, is perpendicular to the plane

# THE LINE VORTEX AND ROTATING FLOWS

containing **v** and **Ω**. In other words, the water particle is forced to circle around the plug hole.

In the Northern Hemisphere the clouds (water vapour and air) rotate in an anti-clockwise direction around a convection centre, while in the Southern Hemisphere they rotate in a clockwise direction. The same is true for the large tanks of water with a tiny plug hole used in science experiments. For the bath, where the plug hole is at one end, the shape of the bath wall and wall friction are the dominating factors determining the direction of the swirling water.

For our ideal (non-viscous and constant density) model of fluid flow, we can include a vortex and assume that it always existed. However, if we want to model the generation of a vortex then we must introduce viscosity, and the flow is no longer ideal.

For a closed path, or circuit, in a fluid, we introduce the parameter $\Gamma$ (Gamma is the Greek letter G) called the circulation. The value of $\Gamma$ is equal to the sum of the component fluid velocities in the direction of the path for all points along the path times the path length. Those flows where the circulation is always zero ($\Gamma = 0$), no matter what path is taken to complete the circuit (beginning and ending at the same point), we call irrotational, while those flows where the circulation is non-zero ($\Gamma \neq 0$) around some complete circuits are called rotational. In the case of a circuit enclosing a vortex core, we always find that $\Gamma \neq 0$.

A test source is a point source which, when placed in a larger unchanging (static) flow, leaves the larger flow pattern undisturbed. When a test source moves around a complete circuit in an irrotational fluid then its potential energy at the start and finish are the same. We call the irrotational velocity field a conservative force field. For fluid motions which are conservative, we find that we can write the fluid velocity vector, **v**, as the negative gradient of a scalar function $\phi$ = constant. In mathematical terms, we write $\mathbf{v} = -\nabla\phi$. This is analogous to the gravity field, where $\mathbf{g} = -\nabla\phi$. The field patterns are the same.

A vortex core filament is made up from a connected line of fluid particles with spin. The amount of spin accorded to each fluid particle in the vortex core is called its vorticity $\zeta$ (zeta is the Greek letter z), which is equal to twice the angular velocity ($2\Omega$) of each particle. A thick vortex core can be made from a bundle of vortex core filaments.

If you could grip a vortex core with your right hand and curl your fingers in the direction of the streamlines then your thumb will point in the direction of the vorticity $\zeta$. Since $\zeta$ has a magnitude (angular velocity) and a direction (axis of rotation), it is a vector. The Scottish theoretical physicist James Clerk Maxwell defined $\zeta$ by the mathematical expression $\nabla \times \mathbf{v} = \zeta$, which he called the curl of the velocity $\mathbf{v}$.

The laws obeyed by a vortex core in an ideal fluid were given by the German physicist Hermann von Helmholtz in 1858:

1. Either a fluid particle always has spin or it never has spin.
2. A fluid particle in a vortex core always remains in the core.
3. The cross-sectional area of a thick vortex core times its angular velocity $\Omega$ is constant over the whole length of the core.

Law 1 tells us that in an ideal fluid a vortex cannot be created. The flow either contains a vortex and the flow is rotational or it doesn't and the flow is irrotational.

Suppose we add two vortices with equal but opposite vorticity to an ideal irrotational flow. The circulation around a circuit containing both vortices is zero since the enclosed vorticity is zero, so the flow outside the circuit remains irrotational.

Law 2 tells us that a vortex is carried along by the flow. If a ping-pong ball is dropped onto the top of a vortex core, where the dimple is, it will stay on top of the vortex and be swept along by the flow.

Laws 2 and 3 tell us that if the vortex core is stretched then its rotation speeds up. This is what happens to the bath vortex. As the water flows away through the plug hole into the pipe below, the vortex is stretched and the vortex rotation speeds up. However, if the vortex core is compressed, or shortened, then it gets fatter and its rotation slows down.

There is an unwritten vortex law that says the ends of a vortex core either finish at a surface (fluid or solid boundary) or join together to form a ring.

# The line vortex and rotating flows

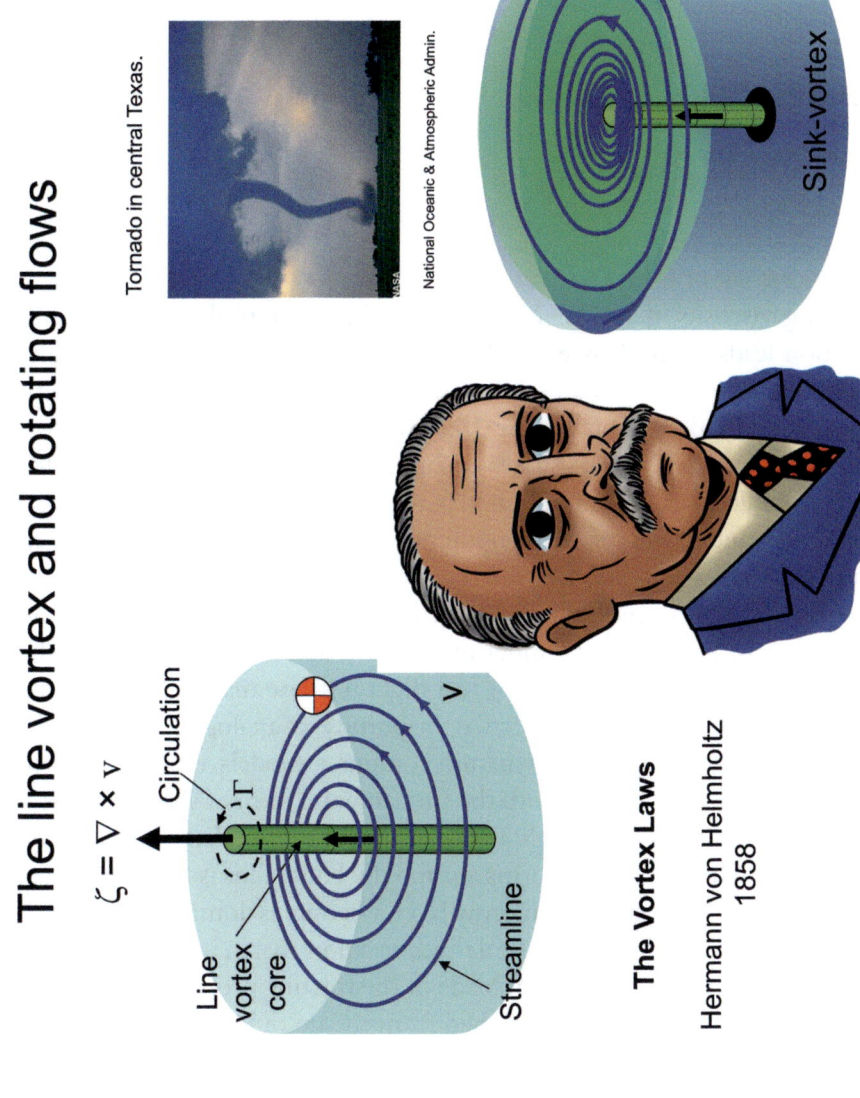

## CHAPTER 19

# VORTEX RINGS

Adding viscosity $\eta$ (units: $N.s/m^2$) into the theoretical model of fluid motion leads to the Navier–Stokes equation. The equation was initially derived by the French mathematician Claude-Louis Navier in 1822, with an improved derivation by the Irish mathematician George Stokes in 1845. Solving the Navier–Stokes equation analytically is only possible for a few simple cases. For complicated bodies it is necessary to use numerical methods and high-speed computers. Aerodynamicists, interested in modelling flow conditions around aircraft, have developed a number of computer programmes, which can predict the pressure distribution on a surface in flight. Wind tunnel tests on models are used to check the predictions and, when necessary, real flight tests are made to confirm the veracity of modelling. However, for the purpose of analogue 'read-across', the complexity of the aerodynamic computer models rather rules out their use. We need to stick with the fundamental form of the equations.

As a body moves through a real fluid, a thin boundary layer of fluid in contact with the surface forms around the body. This is fluid friction at work. Within this extremely narrow layer, viscosity is dominant. Although a propeller cannot work in an ideal fluid, we know that in a real fluid it does. The boundary layer along each blade of the rotating propeller generates a vortex around it, which is then shed from the tip of the blade. Generally, the blade vortex is not seen, unless smoke is introduced at some point along the blade. The circular plane area containing the rotating propeller blades is called the actuator disc. Suppose the propeller position is held fixed. The propeller draws fluid in from in front and propels it backwards. So the propeller forms an axially confined sink–source combination. The fluid flowing towards the sink at the front of the actuator disc converges and speeds up, so the pressure drops ahead of the disc. The fluid flowing out from the source behind the actuator disc expands and slows down, resulting in a pressure rise in the rear of the disc. If the propeller is free to move, the pressure difference across the actuator disc would cause it to progress forward into the fluid. If the propeller position is fixed, the vortices shed from the tips combine to form a vortex ring around the periphery of the actuator disc. If the propeller is allowed to move forward, the blade vortices spiral behind and disperse. In analogy with electromagnetism we can think of the vortex spirals forming a vortex solenoid in the fluid behind the actuator disc. A ducted propeller is much more efficient. This is a very powerful device, forming the basis of the jet engine.

In the reverse process, when fluid flows through the actuator disc, the blades are caused to turn. This is seen in the wind causing the windmill sails to turn and the spinning of the falling sycamore seed. In the hydroelectric scheme, water pours past turbine blades causing them to rotate, and the rotation is converted into electrical power.

An impulsive flow of fluid through a circular orifice into a stationary fluid also creates a vortex ring. A common example is the smoke ring. The cylindrical surface of the plug of ejected air forms a boundary layer between the stationary and the moving air. Since the plug surface is free at each end, the boundary layer rolls up to form a vortex ring which travels along with the plug of air. The $2^{nd}$ vortex law states that the fluid trapped in the vortex core stays in the core.

The initial impulse causes the vortex ring to move linearly through the stationary fluid. Each segment of the core induces a rotational velocity in the rest of the core and from symmetry the ring moves forward with constant velocity v in a direction perpendicular to the face of the ring.

To get a steady streamline pattern, we can impose a uniform stream v in the opposite direction. This is akin to observing the flow while travelling with the vortex ring. In other words, we use a Galilean transform to change our frame of reference. From this viewpoint, the ejected plug of fluid forms an ellipsoidal- shaped body around the vortex ring and separates it from the mainstream uniform flow.

For an ideal fluid, we can treat the vortex ring as a narrow circular line (with vorticity $\zeta$), so that the rest of the fluid, outside of the core, is irrotational. The vortex ring appears to have a virtual, axially confined source–sink combination but, unlike the ideal source and sink combination (which is unconfined and doesn't move), it moves through the stationary fluid. So, viscosity, confined to the vortex core, provides some form of traction. The streamline pattern associated with the vortex ring moving into a stationary external fluid with constant velocity v is unsteady.

There is a duality between vortex types:

1. A straight vortex core has circular streamlines around it.
2. A circular vortex core has a straight streamline through it.

For ideal flow, the information about flow conditions is transmitted instantaneously throughout the fluid. Introducing viscosity means that vorticity diffuses into the fluid from boundaries, in the same way that heat diffuses out of a solid body from the boundaries.

The Navier–Stokes equation was developed at about the same time that Faraday discovered electromagnetic induction. With hindsight, we know that the generation of a vortex ring created by an impulse of fluid through a sharp-edged circular orifice is very similar in pattern to that of the current induced in a conducting ring as a magnetic impulse passes through the ring. The situation is not quite the same between the two cases, since the vortex ring moves, while the conducting ring is stationary. But, if we travel with the vortex ring, we see that an analogue exists between the vortex core $\zeta$ and an induced current of electric charge. This was first mentioned by Helmholtz in 1858. By the same token, streamlines **v** are analogous to magnetic field lines **H**, and fluid sources and sinks are analogous to magnetic monopoles. In terms of 'read-across', we know that positive and negative charges exist, while vortons, the fictitious carriers of vorticity, only occur in pairs (Vortex law 1).

Equally, we could take the fluid velocity **v** as the analogue of the electric field **E** and the movement of positive charge. The vorticity vector $\zeta$ would then be the analogue of the magnetic field **H** and both are dipole phenomena. With this analogue, linear fluid flow is surrounded by concentric circular lines of vorticity, suggesting a link with turbulence. But an analogue is just that; how it is used for 'read-across' is the important consideration. But we are running ahead of ourselves, because we haven't looked at electromagnetism yet. It's as though in reading our guide book on analogies, we've skipped a few pages and having read on, we have to go back to read what we've missed.

# Vortex rings

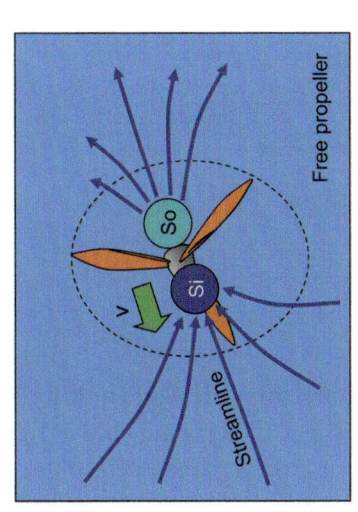
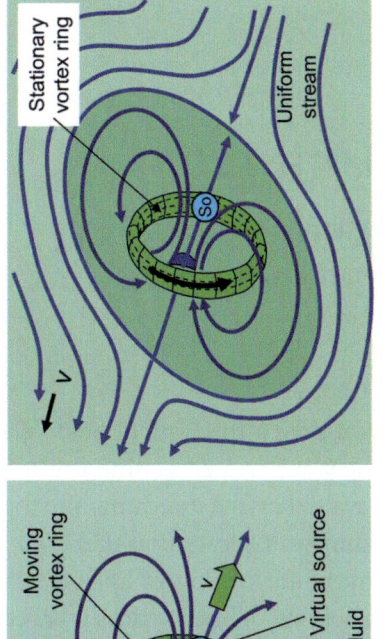
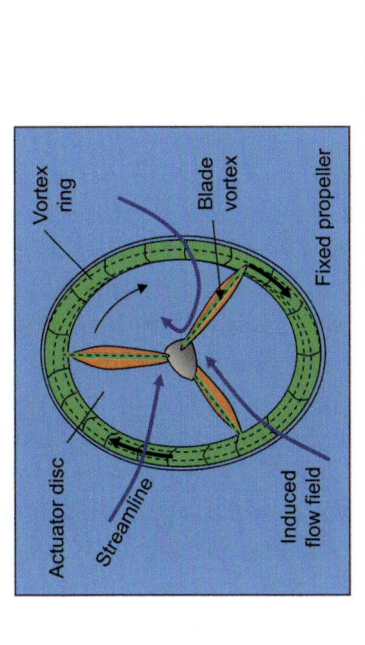
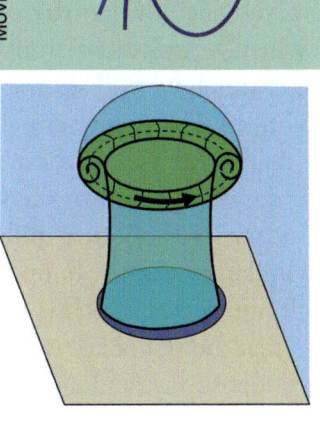

IMPULSE

UNSTEADY

STEADY

CHAPTER 20

# FIXED LINE VORTEX IN A UNIFORM STREAM

If you drag a thick stick, held vertically, through water and look closely at the wake created behind the stick you will see a number of dimples in the water surface indicating the tops of little vertical line vortices. Careful examination shows that there are two parallel rows of line vortices shed from the stick; a row of line vortices shed from each side of the stick. This parallel row of line vortices is called a vortex street. Leonardo da Vinci was fascinated by this pattern in nature and sketched a picture of the vortices shed from a small rock in a stream.

For an experimental study, we need a channel of freely flowing water. Suppose the flow is from left to right. We can use an upstream row of ink ports across the flow to mark out a set of parallel streamlines in the water, seen when we look down from above. If we place a cylinder in the middle of the channel, we will see the little vortices being shed into the wake of the cylinder. The water very close to the cylinder surface acts more like syrup than water. We talk about a 'no slip' condition on the cylinder surface where the water sticks to the surface. This viscous effect is confined to a boundary layer, about a millimetre thick. On the downstream side of the cylinder, part of the boundary layer breaks away and rolls up to form a small line vortex which is swept away into the stream. Close scrutiny shows that the boundary layer breaks away alternately, from one side of the cylinder surface and then the other, forming the wake of small vortices that we call a vortex street.

When a vortex breaks away, the cylinder receives an impulse perpendicular to the mainstream flow, followed by an impulse in the opposite direction when the next vortex is shed from the opposite side. If the cylinder has any freedom of movement, it will begin to vibrate in a direction perpendicular to the mainstream flow. For long, thin cylinders, like telephone wires, as the wind blows across them, they will vibrate like violin strings. So, the eerie whistle that you hear from the telephone wires on a windy day is caused by vortices being shed.

By rotating the cylinder very quickly, we can enforce the 'no slip' condition and keep the boundary layer attached to the surface of the cylinder. No little vortices are shed into the wake, and a concentrated vortex of strength $\zeta$ forms around the cylinder.

Suppose the vortex rotates in a clockwise direction when viewed from above, so that the vorticity vector $\zeta$ points downwards. Over the top of the cylinder, the water in the channel speeds up, as shown by the closeness of the streamlines, with a resulting drop in pressure. This is the Bernoulli effect in action. Around the bottom of the cylinder, the water slows down and the pressure increases. The overall result is for the cylinder to be sucked and pushed upwards. This force on the rotating cylinder is called the Magnus effect. It is named after the German scientist Heinrich Magnus, of the Humboldt University in Berlin. Although Newton had described the effect much earlier, it wasn't until 1852 that Magnus explained why the force occurred.

A simple demonstration of the Magnus effect in air can be done with a cardboard tube from a finished roll of kitchen towels. A long length of tape (1cm wide) is wound around the tube and the tube is then placed on the end of a table. A good tug on the tape, which must come free of the tube, will project the tube forward with backspin. The horizontally spinning cardboard cylinder will lift and fly. With some practice, you should even be able to make the tube do a loop-the-loop. The vortex effect is stronger if cardboard end-plates are fitted to the tube.

Birds can soar with extended wings in a strong wind. Even a simple kite, formed from a flat surface, can be made to lift in the air. Somehow, the shape of the surface of a wing or a kite creates a vortical flow around it. If you drag the bowl shape of a spoon sideways through the surface of a cup of coffee, you can see the small vortex shed from the spoon. But in generating the shed vortex, an equal and opposite (rotating in the opposite direction) unseen vortex is created around the spoon.

Using the Hele-Shaw apparatus, we can look at the flow pattern around a 2-D wing section in water with the streamlines marked out by ink introduced upstream of the wing. In this particular experiment, the flow around the wing is ideal (non-viscous), so no boundary layer is shed from the trailing edge. Consequently, no vortex with opposite rotation forms around the wing and no wing lift occurs. For a 2-D wing moving through the air, the streamline pattern is almost the same as that seen in the Hele-Shaw experiment, except that vorticity is shed from the trailing edge. The boundary layers (about 1 millimetre thick) around the top and bottom surfaces of the 2-D wing move at different speeds so that when they come together at the trailing edge and are shed, they form a thin viscous layer which breaks up into a series of mini-line-vortices. This leaves the wing with opposite vorticity, like the coffee spoon, and like the rotating cylinder, the wing gets lift.

For a 3-D wing, a vortex layer, or sheet, is shed from the trailing edge. Since the sheet is free at each wing end, it rolls up to form a pair of trailing vortices, with equal and opposite rotation. For aircraft the size of a Boeing 747, or an Airbus A380, these rolled up vortices are large and powerful. They can last for quite a while (several minutes) after an aircraft has passed by, so they can be a danger to smaller aircraft, particularly at airfields immediately after take-off, if time is not left for them to disperse.

# Fixed vortex in a uniform stream

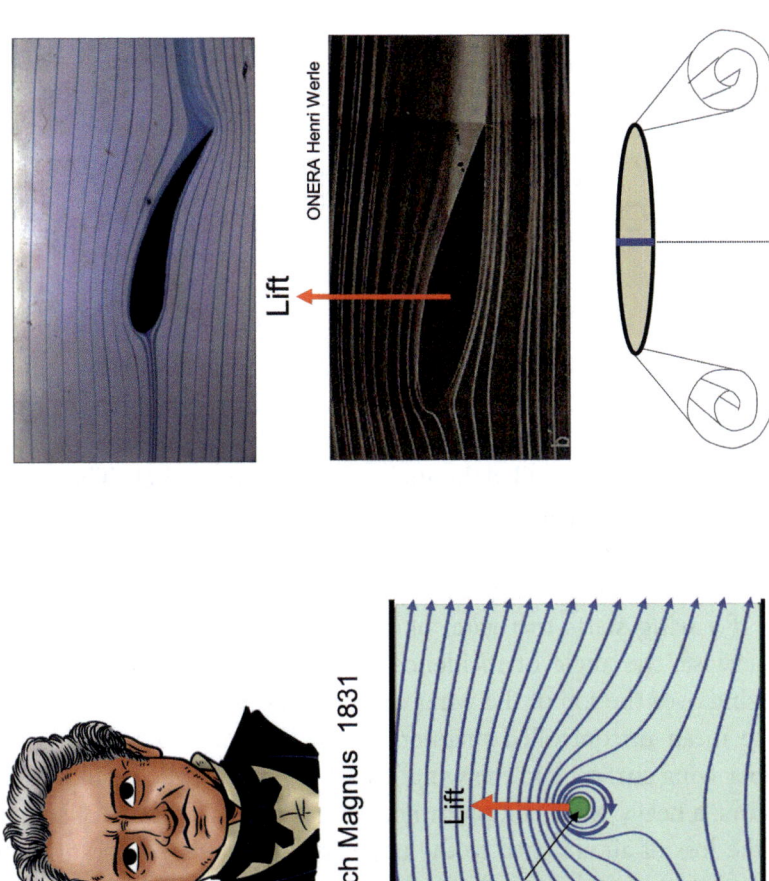

# CHAPTER 21

# AERODYNAMIC FLIGHT REQUIRES VORTEX CONTROL

Due to the Earth's gravity field, an aircraft has weight. For level flight, the weight is balanced by the wing lift. The angle the wing chord (straight line drawn from the leading edge to the trailing edge of the wing) makes with the wind direction is called the angle of incidence, or pitch. The oncoming wind will try to rotate the wing and increase the angle of incidence. You can try this out for yourself very easily by dropping a postcard edge on. It will spin as it falls to the ground.

If a wing starts to rotate, or pitch up, then it must be controlled; otherwise, the angle of incidence will increase until eventually the air passing over the top of the wing will break away from the wing's leading edge in an uncontrolled manner. A low pressure void forms over the upper wing surface and the wing loses lift. This is called stall. When this occurs, a heavy wing will plummet downwards. The cause of stall is due to the loss of air attached to the upper wing surface, so that no vorticity is shed from the trailing edge and no opposite sense vortex forms around the wing to create lift. For attached flow, the speed of the air over the upper surface is greater than that over the lower surface, so that from the Bernoulli effect (increase in speed means decrease in pressure), wing lift occurs. As demonstrated with the lifting cylinder, the difference in flow speed above and below the wing can be attributed to the presence of the vortex forming around the wing. This is called a bound vortex.

The way of preventing pitch up or down becoming too great is to use two wings spaced longitudinally apart; a primary plane which provides the lift, and a small secondary plane able to counteract unwanted pitching.

# AERODYNAMIC FLIGHT REQUIRES VORTEX CONTROL

The Wright Brothers' Flyer I, of 1903, was based on a box-kite with top and bottom surfaces, or planes, for lift. This biplane configuration was used in many early aircraft designs. For pitch control the Wright's used a foreplane ahead of the biplane. More conventionally, especially with monoplanes, the secondary plane is fitted downstream of the main wing, forming a tailplane. Over the past 100 years the tailplane has served aviation well with all passenger and cargo airliners, the majority of private aeroplanes and most military aircraft continuing to employ it successfully.

To turn an aircraft in flight requires banking, or rotating the body of the aircraft about a longitudinal axis. This requires wing vortex control, where the lift on one side of the main wing is increased (stronger bound vortex) and on the other it is decreased (weaker bound vortex). To obtain wing rotation, or roll, the Wright's employed wing warping, where the whole biplane wing (both sides) was twisted. For a monoplane, wing flaps, called ailerons, are fitted to the main wing trailing edges, moving in opposite directions, up or down, to create wing roll. As the whole wing rotates about a longitudinal axis, the lifting force is no longer vertical, and coupled with its weight the aircraft follows a circular path.

Aircraft climb or dive is achieved with flaps attached to the planes of the tailplane called elevators, which work up or down in unison.

For delta wings moving at low speed, high incidence is sometimes used to deliberately create controlled air-flow separation from the sharp leading edge. The result is a pair of tightly rolled up vortices above the wing creating suction suitable for low-speed landings. Concorde made use of this technique.

For fighter aircraft requiring supersonic performance, the leading edge of the wing is swept back to alleviate shock waves. Recent designs have seen the secondary plane placed at the front of the aircraft as a foreplane. This increases the overall lift, which is better for landing, and improves agility during subsonic flight, particularly relevant for air combat. With this so-called canard configuration, the main wing and the foreplane are arranged close together so that the air flow between them is highly interactive. This introduces stability problems.

Placing a canard configured model aircraft in a water channel and injecting coloured ink into the flow from the wing leading edges, we can make the vortices shed from the foreplane and the closely coupled main wing visible and see the interactions.

With the huge advances made in computer data storage, very sophisticated computational fluid dynamic models have been developed which solve the Navier–Stokes equation using numerical methods. With these models, the whole flow field around aircraft in flight can be predicted.

Numerical modelling coupled with wind tunnel experiments have shown that aircraft with a foreplane perform best when an optimum proportion of the total lift is borne by the foreplane, and this means that such aircraft are unstable in flight. It is not possible for the pilot of a high-speed, canard-configured aircraft to continually make the small adjustments to the foreplane and the main wing incidences to achieve stable flight. Instead, although the pilot still flies the aircraft, he has hidden assistance from an on-board computer system which electronically monitors the flight conditions and makes the incremental changes needed for stability.

In the British Aerospace Experimental Aircraft Programme (EAP), the demonstrator aircraft featured a closely coupled canard configuration, requiring active control technology to make it flyable. The aircraft made its successful maiden flight from Warton Airfield, near Blackpool in Lancashire, on Friday 8th August 1986. The EAP demonstrator was the forerunner of the Typhoon, the European combat fighter aircraft.

We can now control the vortex and use it to master flight in the atmosphere. We know that the vortex phenomenon occurs in electromagnetism, too, but in a different guise. When we can control the vortex in gravitation, we will become the masters of space flight.

# Aerodynamic flight requires vortex control

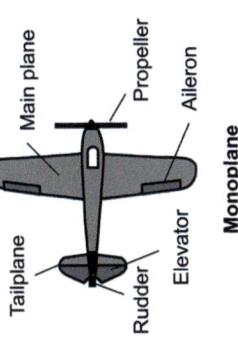

**Monoplane**

Orville & Wilbur Wright 1903

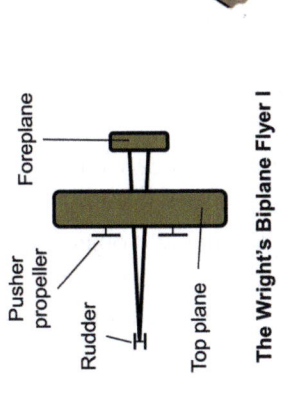

**The Wright's Biplane Flyer I**

Copyright BAE Systems / EAP / Geoff Lee

Copyright BAE Systems

CHAPTER 22

# MAGNETISM AND ELECTRICITY

The property of a certain type of stone to attract, or repel, other stones of the same type was known about in China at least 4,000 years ago. Each stone had two different ends, or poles, which attracted or repelled other stones. The ancient Chinese noticed that when suspended with their poles horizontal, these magical stones always pointed in the same direction. It is thought that they were originally used in the ancient mystical art of *feng shui* (wind water); the arrangement of objects to harmonise with their surroundings so as to provide a stress-free atmosphere. The Chinese also used these suspended stones for direction-finding, leading to the development of the crude compass. Today, we refer to these stones as magnets, since the ancient Greeks found their own supply of these stones from a site in Magnesia, in Asia Minor (modern-day Turkey), in about 600 BC. The ancient Greek philosopher, Thales of Miletus, referred to the curious attractive property of these stones.

The ancient Greeks also discovered that when they rubbed amber, a fossilised resin used in jewellery, it had a similar property to that of the magnetic stones, in that it would attract some lightweight objects and repel others. We now refer to this phenomenon as electrostatics. For more than 2,000 years, electrostatics and magnets remained separate phenomena until about 200 years ago when their intimate relationship was discovered.

In 1600, William Gilbert, court physician to Queen Elizabeth I of England, published his book *De Magnete*, which collected together all the previous knowledge about magnetism and electrostatics. It contained the observations that like magnetic poles repel while unlike poles attract. His book also contained the results of experiments. He magnetised a large iron sphere, so that it had a north and a south pole, and showed that the angle-of-dip of a magnetic needle on the surface changed with its position relative to the poles. Thus, he surmised, the known angle-of-dip with latitude indicated that the Earth contained a giant bar magnet at its centre, roughly in line with the axis of rotation.

Gilbert introduced the word electricity into the English language. During his study of electrostatics, he found that, as well as amber, other substances could be electrified by friction, such a rubbing an ebony rod with fur, or a glass rod with silk. But, apparently, metal could not be electrified.

Otto von Guericke, a German scientist, was interested in Gilbert's book, particularly the parts dealing with frictionally produced electricity. In 1663, he used a rotating sphere of sulphur to form a continuous rubbing machine which produced small sparks. Henry Oldenburg, a German scientist and the first Secretary of the Royal Society, demonstrated von Guericke's machine to the members of the Royal Society. Francis Hauksbee developed von Guericke's idea. He used a rotating sphere of glass containing a vacuum which glowed in the dark while rubbing it with wool. Isaac Newton, made President of the Royal Society in 1703, greatly encouraged such demonstrations, as they fascinated the Society members and stimulated further research into electrical phenomena.

Magnetism, by contrast, seemed to be an unchanging phenomenon, so little work was done on studying it. Nevertheless, during the mid-1740s, the English scientist Gowin Knight used iron filings to map out the magnetic patterns around magnets. These patterns are now referred to as magnetic fields and given the symbol **H**.

Between 1720 and 1740, Stephen Gray, a Fellow of the Royal Society, carried out experiments with frictionally generated electricity to show that electrical effects could be transmitted along some substances like brass wire and wet string (conductors), but not others like silk and wool (insulators).

Charles du Fay, in France, carried out electrical experiments for his own amusement, like many other well-off gentlemen at the time. He repeated Gray's experiments on transmitting electrical effects along brass wire and wet string. He also discovered that metal substances could be charged, providing they were insulated and held in an amber, or glass, handle.

In experiments carried out in 1733, du Fay found that charged substances of the same type, such as glass rods (charged by rubbing with silk), would repel each other but that they would attract some other charged substances, such as an ebony rod (charged by rubbing with fur). He concluded that there were two types of electrical fluid, which he called vitreous and resinous. We would say that like charges repel, whereas unlike charges attract. So, electric charges are analogous to magnetic poles.

In 1748, Benjamin Franklin, who had already made his fortune in printing in the British colony of Pennsylvania, began his interest in science, particularly in the new subject of electricity. Franklin read du Fay's work and was convinced that there was only one type of electrical fluid and that its transfer left substances either positively (vitreous) or negatively (resinous) charged. He guessed that the current of electrical fluid was the flow of positive electrical charges. Although his idea was right, he guessed wrong. About 150 years later, in 1897, the English scientist Joseph John Thomson demonstrated that an electric current is the flow of negative electric charges, called electrons. But an electron current flows in the opposite direction to Franklin's current, or what we now call the conventional current.

Franklin represented the British colony of Pennsylvania in Parliament, so he spent a lot of time in London and while there demonstrated his electrical experiments at the Royal Society. Franklin's parents were English (as a colonial, Franklin was British), and he enjoyed searching for his family's roots in England. In 1758, he attended a meeting of the famous Birmingham 'Lunar Society' and was feted by Matthew Boulton, a founder member. Franklin, who was deeply interested in electrical phenomena, encouraged Joseph Priestly, a member of the Lunar Society, to write a history of electricity. This Priestly did, publishing his *History and Present State of Electricity* in 1767. Based on the analogue with gravity, Priestly predicted that static electric charge would obey an inverse square law.

In 1785, using a very sensitive torsion balance housed in a glass case, the French scientist Charles Coulomb was the first to demonstrate that the force between electric charges did obey an inverse square law. The force **F** between two charges q and Q is given by

$$\mathbf{F} = \frac{qQ}{\varepsilon 4\pi r^2}\hat{\mathbf{r}} \quad \text{Newtons}$$

where $\varepsilon$ is the electric permittivity of the medium and the distance between the charges is r. The SI unit of charge Q is the Coulomb.

## MAGNETISM AND ELECTRICITY

For free space $\varepsilon_0 = 8.854 \times 10^{-12}$ Farads/m.

The force on charge q in the electric field (with symbol **E**) created by charge Q is written as **F** = q**E**, where

$$\mathbf{E} = \frac{Q}{\varepsilon 4\pi r^2}\hat{\mathbf{r}} \quad \text{Volts/m}$$

Following the pattern set by gravity, we introduce the electrical potential φ for a point charge Q. It is a scalar quantity given by

$$\phi = \frac{Q}{\varepsilon 4\pi r} \quad \text{Volts}$$

On the spherical surface of radius r, surrounding the charge Q, the potential φ is a constant. In analogy with gravity, the potential energy E of a point charge q in the field of a point charge Q is given by E = q φ.

Again, following the pattern set by gravity allows us to express the electric field **E** in terms of the negative gradient in electrical potential φ. In mathematical symbols, we write **E** = – ∇ φ.

Magnetic poles always come in pairs with the same strength but opposite polarity, say +m (north pole) and –m (south pole). For a very long and narrow magnet, the poles approximate to isolated monopoles. The unit for a magnetic monopole is the Weber. Experiment shows that the force **F** between two different magnetic monopoles, m and M, obeys the inverse square law and that like poles repel.

$$\mathbf{F} = \frac{mM}{\mu 4\pi r^2}\hat{\mathbf{r}} \quad \text{Newtons}$$

The parameter μ is the magnetic permeability of the medium. For free space $\mu_0 = 4\pi \times 10^{-7}$ Henries/metre.

From the analogy with the electric field, the force **F** on a magnetic monopole m in the magnetic field **H** surrounding a monopole M is written as **F** = m**H**, where

$$\mathbf{H} = \frac{M}{\mu 4\pi r^2}\hat{\mathbf{r}} \quad \text{Amps/metre}$$

Since molecules contain positive and negative charges, when some materials are placed in an electric field E, the molecular charges dissociate, or polarise, stretching the molecule in opposite directions, forming tiny electric dipoles. Such materials are called dielectrics. The displacement of molecular electric charges is characterised by the permittivity ε of the material. The ratio of the permittivity ε of the medium to the permittivity $\varepsilon_0$ of free space is called the relative permittivity $\varepsilon_r$.

$$\varepsilon_r = \frac{\varepsilon}{\varepsilon_0}$$

Apart from the type of material, $\varepsilon_r$ may be a function of electric field strength **E**, or pressure. Dielectric materials exhibit different electrical properties depending on their $\varepsilon_r$ value, from $\varepsilon_r = 2$ for basic materials like rubber and plastic through to $\varepsilon_r = 23,000$ for the exotic material barium zirconium titanate (BZT).

Studies of the atom during the early 20$^{th}$ century showed that the ultimate source of magnetism is the electron within the atom. There are two parts to magnetism which arise from the electron. Firstly, an orbiting electron forms a tiny ring current which creates an atomic magnetic dipole. As an electron shell fills up, electrons pair together so that the magnetic fields of their dipoles cancel out. It is only when the outer atomic shell contains an unpaired electron that the atom has a magnetic property. Secondly, the electron has spin which also contributes to the magnetic property of an atom. Materials with magnetic properties are composed of aggregates of atomic and molecular magnetic dipoles. They are characterised by their relative magnetic permeability $\mu_r$.

$$\mu_r = \frac{\mu}{\mu_0}$$

Magnetic materials exhibit vastly different magnetic properties depending on their relative permeability from superconductors ($\mu_r = 0$) to ferrites ($\mu_r \sim 5,000$). The $\mu_r$ are affected by temperature and high pressure.

The concept of storing electric charge was accidently discovered in 1745 by the German scientist Ewald von Kleist, an ex-student of the Dutch University of Leyden. Since water was known to be a conductor and glass was known to be an insulator, he was trying to store charge in water in a glass jar. The jar had a cork stopper through which a nail was pushed so that it just dipped into the water. While holding the jar in his hand, Kleist rubbed an ebony rod with fur and transferred the frictionally produced charges to the water by touching the protruding head of the nail with the rod. Because like charges repel, only a small amount of charge could be transferred at a time. So, the process was repeated a number of times. Kleist assumed that the charges accumulated in the water. He got a severe shock when he dipped the finger of his other hand into the water, as he provided a conducting path to Earth. The details of the discovery were passed to Pieter van Musschenbroek, the Professor of Physics at Leyden (now Leiden) University, in 1746, who publicised the discovery as the Leyden jar. Later, it was realised that the water inside the jar and the moist hand outside the jar were conductors. Opposite type charges resided on the conductors either side of the glass dielectric. Once it was known how the device worked, the water was discarded. The inside and the outside of the jar were covered in metal foil and the nail was connected to the inside metal lining with a brass chain. The Leyden jar was the world's first capacitor; a device for storing electrical charge.

The most simple capacitor is formed from two parallel flat metallic plates, each of area A, spaced a distance d apart. Suppose that charge $+q$ is stored on one plate and charge $-q$ is stored on the other plate. The difference in electrical potential between the plates is called the voltage V and the magnitude of the electric field **E** created between the plates is $E = V/d$. The amount of charge stored in the capacitor is called its capacitance C. By definition, capacitance $C = q/V$. The SI unit for C is the Farad. If the medium, or dielectric, between the plates has permittivity $\varepsilon$, then the capacitance C of the simple flat plate capacitor is given by $C = \varepsilon A/d$.

Energy is the capacity to do work. And work is equal to the force applied at a point times the distance moved. The electric field **E** stretches the capacitor dielectric, so that work is done and energy resides in the medium. It is analogous to the energy residing in a stretched elastic band. In effect, each capacitor plate pushes on the other plate with a force qE. Averaging the effect, the work done in charging the capacitor is $½(qE)d$, which is equal to the stored energy. Since $E = V/d$, the stored energy is $½qV$. If we introduce the capacitance C, the stored energy $= ½CV^2 = ½q^2/C$. The volume of the capacitor is Ad. Therefore, the energy density of the capacitor is $½\varepsilon E^2$.

The idea that lightning might be an electrical phenomenon was suggested, in 1708, by Dr William Wall, an English scientist. But it was not until 1752 that Wall's idea was confirmed by Benjamin Franklin. In a dangerous experiment, Franklin flew a kite up into the clouds of a thunderstorm. To the end of the string, he tied an iron key and, very wisely, he held the string with a length of insulating silk cord. Charge flowed down the wet string to the key, from which it was transferred to a Leyden jar. Subsequent experiments with the stored 'cloud charge' showed that it had the same properties as frictionally produced electric charge.

Having proved that lightning was an electrical discharge and knowing that lightning usually struck the highest, and sharpest, point on the Earth in the vicinity of the storm, Franklin advocated that tall buildings should be fitted with lightning conductors. This was a metal strip that extended above the top of the building, ran down the side of the building and was, then, properly earthed, since Franklin understood that the Earth was an electrical conductor.

This wasn't only a problem for buildings. The masts of sailing ships at sea were, usually, the highest point around for miles and were very vulnerable to lightning strikes. Wooden warships, with large gunpowder stores, were particularly at risk. Lightning had been known to travel down a wet mast, enter the gunpowder store, ignite it and cause the ship to be blown apart. As Franklin's idea become known, the British Admiralty arranged for the masts of all Royal Navy warships to have a strip of copper running down them, connecting (sometimes via chains) with the outside copper lining of the hull.

The skins of aircraft get highly charged through the frictional effects of the air flow. Usually, wicks are fitted to allow these charges to stream off into the wake. If the skin is metal, it acts as a Faraday cage and, ideally, no charges should penetrate through to the inside. The fact that charge resides on the outside of a hollow metal conductor was demonstrated by Faraday in his famous 'ice pail' experiment carried out in 1843.

Although aircraft in flight are not earthed, they can still be struck by lightning. An all-metal skinned aircraft offers the safest protection, but modern aircraft are made up from a variety of materials. Also, they are crammed full of avionic systems, the malfunctioning of some of which may affect the safety of the aircraft. Special electromagnetic compatibility measures (EMC) have to be introduced just to allow avionic systems to be sat side by side. On top of that, aircraft are full of fuel, and some military aircraft carry highly explosive stores. So, the threat of a lightning strike has to be taken very seriously.

For lightning protection alone, an enormous amount of work goes into designing special conducting paths to guide any lightning strike through an aircraft, away from sensitive areas. Ground test clearance of production Typhoon aircraft requires them passing a simulated lightning strike. A thick conducting rod protruding from a Marx electrical generator feeds a massive current of 200 kA into the aircraft through the radome, which is then drawn out through the wings and engine nozzles.

By the end of the 18$^{th}$ century, frictionally produced short discharge electrical currents were used in many experiments. The next major step forward was to find the means of producing continuous electric currents. The Italian physicist Alessandro Volta showed the way.

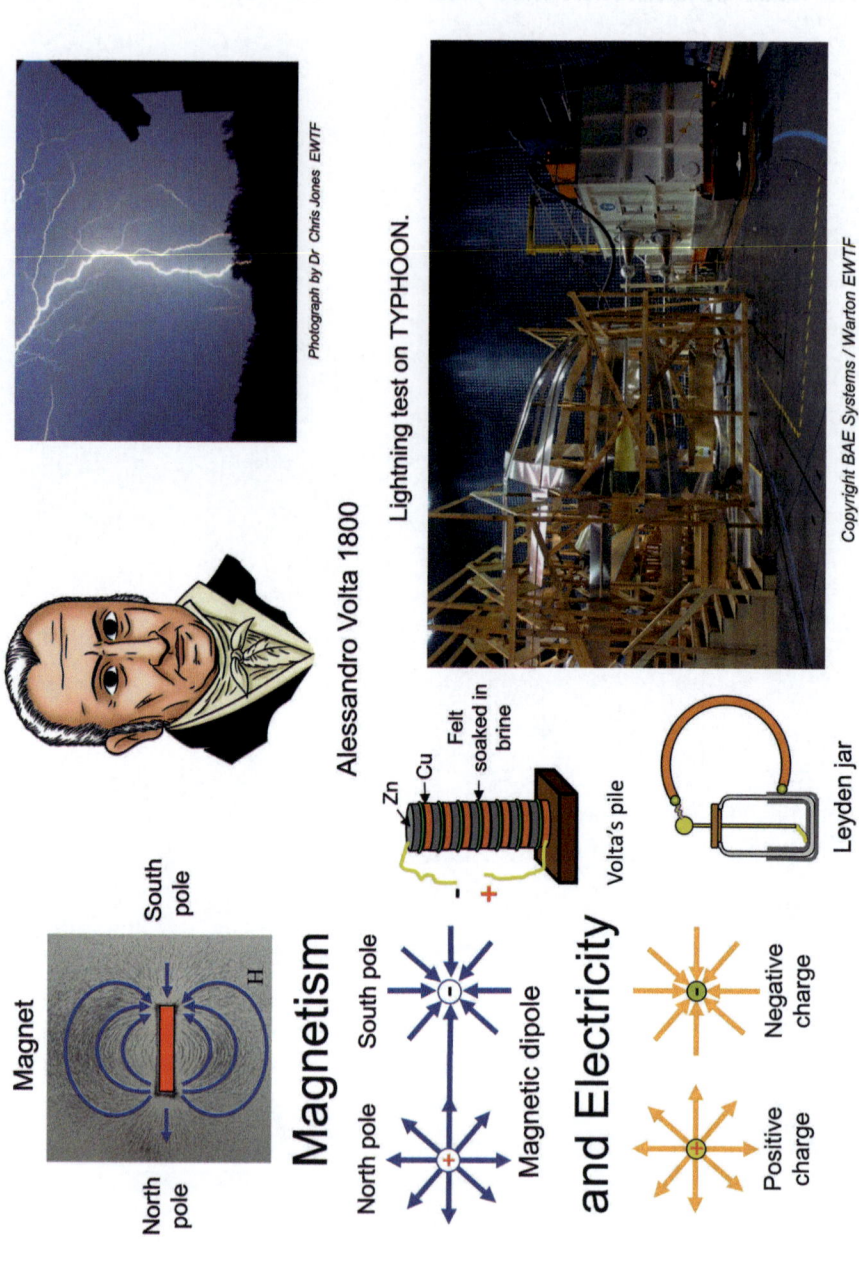

CHAPTER 23

# THE ELECTRIC VORTEX AND ELECTROMAGNETISM

Experiments by Stephen Gray and others in the 1720s had revealed that electrical discharges to a body caused the muscles to convulse if the body was in contact with the ground. Such electric shocks were treated as fun but could be very dangerous. In 1791, Luigi Galvani, an Italian professor of Anatomy, carried out experiments on frogs' legs with the aim of locating the source of 'animal electricity', which he thought was responsible for making muscles function. He hung a pair of dissected frog's legs from a copper hook over an iron tabletop and used short electrical discharges to make the frog's legs twitch. During further experiments, he noticed that the frog's legs twitched of their own accord whenever the frog's feet touched the iron tabletop. Examining the experiment more closely, he decided that the nerve exposed at the top of the frog's leg must be the source of 'animal electricity' responsible for the twitching.

Alessandro Volta, an Italian physicist, investigated Galvani's finding but concluded that the frog's legs were only acting as a conductor of electricity. The essential fact was that dissimilar metals, separated by a membrane of some kind, resulted in an electric current when one end was connected to the ground. With this knowledge, in 1799, Volta built his 'voltaic pile'; a column consisting of alternate layers of zinc and copper plates interspersed with wetted pasteboard (felt soaked in brine is better) acting as a membrane. Each sandwich of zinc, pasteboard and copper plate formed an individual cell. He stuck his wet fingers to the top and bottom of a small column of cells, thereby completing a circuit, and felt the tingle of electricity passing through his body. By analogy with fluid flow, the tingling sensation was thought of as an electric current. Volta's pile was the first means of generating a continuous electric current. In 1800, Volta wrote a letter to Sir Joseph Banks, the President of the Royal Society, in which he revealed details of his sensational device.

In 1801, Volta was summoned to Paris by Napoléon Bonaparte, the 1st French Consul and later Emperor, to demonstrate his voltaic pile. Bonaparte, a mathematical member of the Institute of France, was always keen to promote advances in science. He ordered 600 voltaic cells to be laid out horizontally and connected together in an array, thus forming the world's first battery. Developments led on to the accumulator and, later, the car battery. The dry cells that we use in our torches and mobile phones are based on the Bunsen cell, a development of Volta's original idea made by the German scientist Robert Bunsen. (Yes, it's the same Bunsen of Bunsen burner fame.) Today, new batteries are being developed to power electric cars. And the race is on to find a way of battery recharging, similar to filling up with a tank of petrol. And to think that the original breakthrough had come from studying twitching frog's legs! It seems too ridiculous to be true.

In SI units, the electrical potential difference $\Delta\phi$, or voltage V, created by the battery, is measured in Volts and the rate of flow of charge Q, called the current I, is measured in Amps. The electric field $\mathbf{E}$, measured in Volts/m, points in the direction of the current.

Once the secret of how to make a voltaic pile had been revealed, it was fairly simple for other scientists to copy the idea. In England, Sir Humphry Davy, at the Royal Institution in London, was very quick in following up Volta's discovery. Not to be outdone by the massive voltaic pile in Paris, the Royal Institution in London built one with twice the number of cells.

While experimenting with an early voltaic pile, in 1801, Sir Anthony Carlisle noticed that if wires from each end of the voltaic pile were dipped in water, then bubbles appeared around them. By chance, Carlisle had discovered electrolysis, whereby water is decomposed, giving off bubbles of oxygen at one wire and bubbles of hydrogen at the other. Working with William Nicholson, the process of electrolysis was rapidly developed. Sir Humphry Davy used electrolysis to decompose chemical solutions and discovered some new elements in the process. Michael Faraday, Davy's laboratory assistant, was involved in the work, too, and after many years of research formulated his laws of electrolysis in 1834. This period saw the beginning of the electroplating industry. Amazingly, from archaeological finds made in Baghdad in the 1930s, we now know that the ancient Babylonians used simple electrical cells, but their knowledge was lost for more than 2,000 years.

In 1810, Hans Christian Oersted, a Danish scientist, wrote, *It is my firm conviction that a fundamental unity permeates all Nature.* In 1812, he predicted that electricity and magnetism were connected. But it wasn't until 1820 that he finally discovered the link for himself. It was, as he admitted, a chance observation made during a laboratory demonstration. While using a Volta battery to create a current along a wire, he noticed that a nearby magnetic compass needle began to swing. Carrying out further experiments, he discovered that there was a circular magnetic field **H** around the wire. Oersted had discovered the electric vortex. If you grasp the wire in your right hand, so that your thumb points in the direction of the current, then your fingers will curl in the direction of the magnetic field **H**. If **v** is the velocity of the current of positive charges in the wire and $\rho$ is the density of the charges, then $\rho \mathbf{v} = \nabla \times \mathbf{H}$. That is, $\rho \mathbf{v}$ equals the curl of the magnetic field **H**.

Interestingly, Benjamin Franklin had reported earlier that a spark from a discharging Leyden jar disturbed a magnetic compass needle, but neither he nor any other scientists had followed up the observation. A breakthrough can sometimes be made by a scientist making the effort to investigate a quirky result, rather than just ignoring it.

Oersted published the details of his discovery and this set off an explosion of interest in the subject of electromagnetism. Within a few months of Oersted's discovery, Johann Schweigger, a professor of Mathematics at Halle University in Germany, showed that using many loops of wire to form a coil multiplied the magnetic effect. The French scientist François Arago then showed that the magnetic effect of the coil could be increased still further if an iron bar was placed along the axis of the coil. Later, in 1823, William Sturgeon, a little-known English inventor of electrical devices, demonstrated the first practical electromagnet. He coiled wire around an iron bar in the form of a horseshoe. When the current was switched on, the two exposed ends of the horseshoe became powerful poles of an electromagnet. This device has been used in many other electrical devices since.

Davy took his copy of Oersted's paper downstairs to the laboratory at the Royal Institution and repeated the experiments in order to familiarise himself with the subject and to verify the results. In assisting Davy with his experiments, Faraday quickly absorbed the new field of study. Then

Faraday, on his own initiative, used iron filings to make the magnetic fields visible around straight and circular wires carrying currents. He noticed the duality between the electric fields **E** (in the direction of the current I) and the magnetic fields **H**.

1. A straight current-carrying conductor produced a circular magnetic field **H** around it.
2. A circular current-carrying conductor produced a straight magnetic field **H** through its centre.

Although it was not realised at the time, the circular current is the electromagnetic analogue of the vortex ring, with virtual magnetic poles of +m (north pole) and –m (south pole). The strength of the current I is the analogue of the circulation $\Gamma$ around a vortex. However, the analogy is not exact, because the steady current in a conducting ring is not a propulsive device.

Soon after Oersted announced his discovery, Georg Ohm, a German scientist, began to investigate the electrical properties of different types of wire. Ohm based his idea of electric current flow on an analogy with the flow of heat, or calorific, as described in the book, *The Analytical Theory of Heat*, published by the French scientist Jean-Baptiste Fourier in 1822. From experiments based on his analogy with heat flow, Ohm discovered his famous law, $V = IR$, which he published in 1827. Initially, Ohm used Seebeck thermoelectricity (discovered in 1822 by the German physicist Thomas Seebeck) to create a steady voltage V (an electric pressure, or electric field **E**) along wire samples of length $\ell$ and cross-sectional area A. Later, he used Volta cells. The voltage V produced a current I. In today's terms, as the current flows through the wire, the moving free electric charges have to negotiate their way through the obstruction of the wire's atomic lattice, which is where the resistance R arises. When Ohm first proposed his law, it was mostly ignored. However, after Ohm was made a foreign member of the Royal Society, in 1842, scientists began to realise the importance of his empirical law.

The SI unit of electrical resistance R is the Ohm. The inverse of the resistance is called the conductance, with the SI unit of the Siemen. The electrical conductivity $\sigma$ is defined by $\sigma = \ell/AR$ with the SI unit of the Siemen/m. When the resistance to current flow is infinite, the material is non-conducting and the electrical conductivity is zero. The analogue of the electrical conductivity $\sigma$ is the thermal conductivity k. But beware!

Electrical conductivity σ relates to moving electrical sources (charges), while thermal conductivity k relates to the permeability of heat flow, not to the movement of heat sources.

Ohm's law can be written in a form analogous to Fourier's law of heat flow. The electric field E across a length $\ell$ of wire is given by $E = -V/\ell$, where V is the voltage, or potential difference, across the length $\ell$. The current $I = dQ/dt$ and the resistance $R = \ell/\sigma A$. So, Ohm's law can be written as

$$E\ell = \frac{dQ}{dt}\frac{\ell}{\sigma A}$$

The amount of charge Q in the volume of wire of length x and cross-section A is $Q = \rho A x$, where ρ is the charge density. Thus, $I = dQ/dt = \rho A v$, where $v = dx/dt$. Hence, Ohm's law can be rewritten in the form

$$\rho v = \sigma E$$

With our modern understanding of an electric current I in a wire, we know that an electric field **E** creates a potential difference, or voltage V (analogous to pressure), along the wire, forcing free positive charges to flow along the wire. In fact, we now know that the free charges are electrons with negative charge, and we talk about an electron current. But the original model, using positive free charges, continues to be used and we refer to the current I as the conventional current.

From the Lorentz transforms, we can show that the appearance of the magnetic field around the wire is due to a 'special relativity' effect, which arises because the electrons in the wire are moving relative to an external observer. Surprisingly, the averaged velocity of the current electrons is only about 1 m/hr and yet it is this small velocity that leads to the creation of the magnetic field, amplified in size by the presence of billions of free electrons all drifting along in the wire. By contrast, the electrical influence associated with the charges passes through the wire at nearly the speed of light.

# The electric vortex and electromagnetism

Hans Oersted 1820

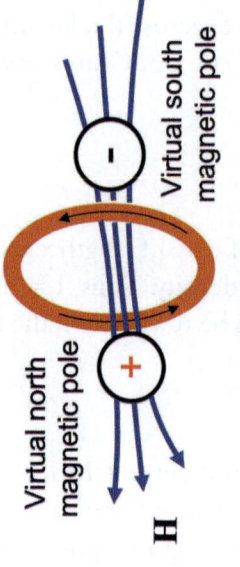

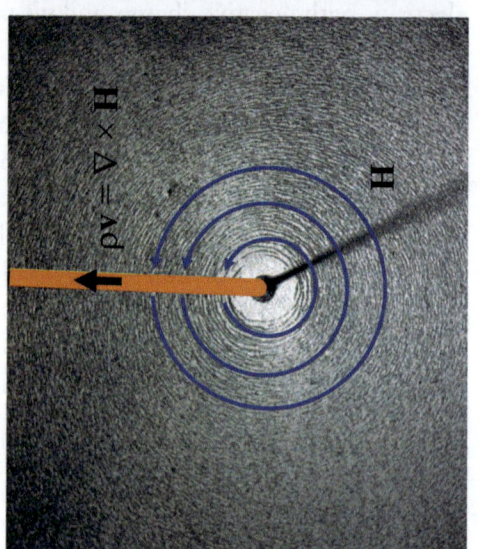

Electromagnet.

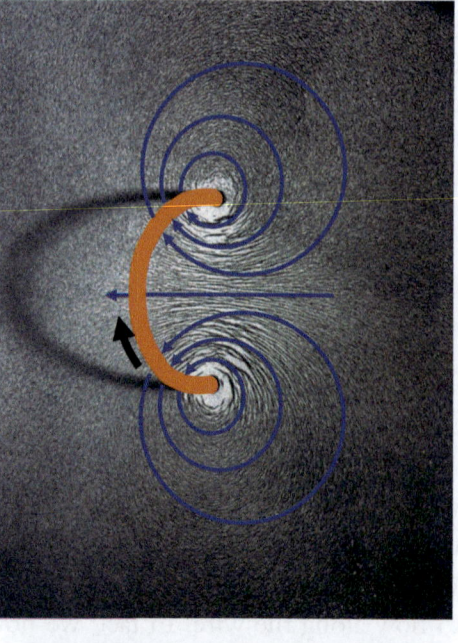

CHAPTER 24

# THE ELECTRIC VORTEX IN A MAGNETIC FIELD

Following Oersted's discovery of electromagnetism, Faraday dabbled on his own with electric currents and magnets. In 1821, he made a novel discovery which, eventually, led to the invention of the electric motor. He had a small basin of mercury and, using candle wax, he stuck a short magnet upright in the centre, so that the north pole was level with the surface. He then dangled a live (current-carrying) wire in the surface of the mercury, close by the top pole, and watched the wire end rotate around the pole. He visualised the rotation in terms of interacting magnetic fields. In Faraday's case, the top pole created a radial magnetic field near the surface of the mercury. This simple demonstration made Faraday's name and led to his being elected a fellow of the Royal Society in 1824.

To understand why the rotation occurred in Faraday's experiment it is easier to see what happens when a straight live wire is placed in a uniform magnetic field. Suppose we create a uniform magnetic field between north and south magnetic poles. Let us now place a wire between the poles so that it is perpendicular to the magnetic field. We know that when we pass a current through the wire it will create a circular magnetic field which will interact with the uniform magnetic field and cause the wire to move.

Professor Sir John Ambrose Fleming of Imperial College (the inventor of the thermionic valve, or diode) introduced Fleming's 'left hand rule', as a simple way of determining the direction of the force on the wire. With the thumb, first and second fingers of the left hand all at right angles, the rule states that if the **F**irst finger points in the direction of the external magnetic **F**ield, and the se**C**ond finger points in the direction of the **C**urrent along the wire, then the direction of **M**otion of the wire is given by the thu**M**b.

Suppose a charge Q inside the wire moves along with speed **v**, then the force **F** that it experiences, due to the interaction of its own circular magnetic field, with the external magnetic field **H** is

$$\mathbf{F} = Q(\mathbf{v} \times \mu\mathbf{H})$$

The term **v** × μ**H** is called a vector product and the force **F** on the wire is perpendicular to **v** and **H**.

The resulting magnetic field pattern is similar to the streamline pattern that arises when a line-vortex is generated in a uniform stream. This is the Magnus effect. The wire and the vortex both experience a sideways force, but they are in opposite directions. The two analogues illustrate Helmholtz's observation, that vorticity $\zeta$ is analogous to the electric field **E** and that streamlines **v** are analogous to magnetic field lines **H**. However, there is another difference in 'reading-across'. Electric sources Q moving with speed **v** exist, but moving vortex sources (vortons) don't. We always have to be careful when we 'read-across' between analogues.

Control of the electric vortex led to the invention of the first useable electric motor by William Sturgeon in 1832. Modern electric motors, although vastly improved, can be traced back to Sturgeon's original design. A flat rotatable coil was placed between the poles of a powerful magnet. When the current was switched on, the coil became an electromagnet and interacted with the external magnetic field. The coil would rotate and try to line up with opposite poles facing each other, but just as that condition was approached, the direction of the current was switched and the interaction became one of repulsion, which kept the coil turning. Sturgeon invented the commutator to allow this switching. The rotating axle of the coil could be used to power machinery. For some reason, Faraday chose to ignore Sturgeon's contributions to electrical science. Perhaps it was because Faraday considered himself to be a natural philosopher; a discoverer of new physical truths, not a mere inventor linked with trade.

The world's climate seems to be changing for the worse, and the link has been made with the pollution of the atmosphere by hydrocarbon particulates which arise in the burning of fossil fuels. One of the major sources is thought to be vehicles driven by combustion engines, so there is a move towards electrically driven vehicles. Thus, the electric motor will play an increased role in the future if we are to clean up the atmosphere.

Electric motors work by interacting magnetic fields, and these interactions were first observed 4,000 years ago by the ancient Chinese.

Adding the force on a charge moving in a magnetic field to the electrostatic force, the total force experienced by a charge in combined static electric **E** and magnetic **H** fields is called the Lorentz force and is given by

**F** = Q**E** + Q(**v** × μ**H**)

As we have seen, it is the second part of the Lorentz force (the vector product) which is responsible for the motion of a current-carrying wire in a magnetic field, which led on to the electric motor. An electric current is taken to be the flow of positively charged particles. A single particle, with charge Q, on crossing a magnetic field, will follow a spiral path, rotating one way or the other depending on the sign of its charge. The spiralling pattern is the same as the rolling-up of a vortex, yet another example of repeated patterns in nature. Nuclear physicists examine the spiralling paths of charged particles in magnetic fields in their attempts to characterise them.

From symmetry, it seemed reasonable to suppose that if an electric current could create a magnetic field then, somehow, it ought to be possible to reverse the effect and convert magnetism into electricity. The leading scientists of Faraday's time all believed this to be true. But no one knew how to do it!

# The electric vortex in a magnetic field

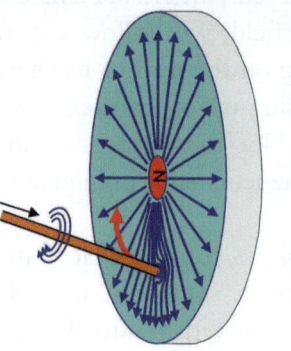

Fleming's left-hand rule.

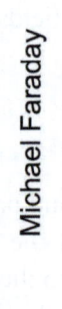

Michael Faraday

William Sturgeon

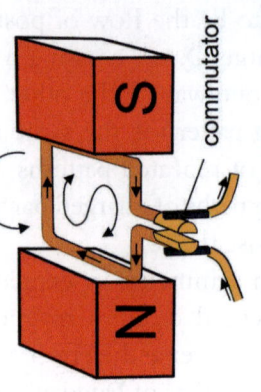

Faraday's rotating wire 1821

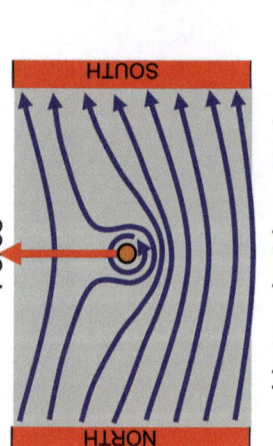

Sturgeon's electric motor 1832

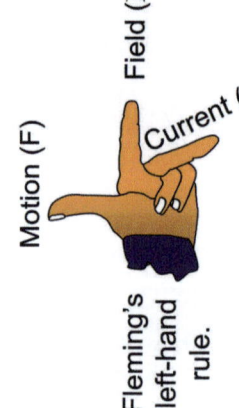

Linear electric current in a magnetic field

CHAPTER 25

# FARADAY AND THE CONTROL OF ELECTROMAGNETISM

Following Oersted's discovery of electromagnetism, there began a race by scientists in Europe and North America to be the first to discover how to turn magnetism into electricity. Oersted had shown that a magnetic compass could be used to detect an electric current, via its magnetic field, but it wasn't a very sensitive device. The most important instrument developed at that time for electrical research was the galvanometer, capable of detecting small electric currents. Schweigger developed the first crude galvanometer by placing a compass needle in the middle of a short coil. Any current through the coil produced a magnetic field which overcame Earth's magnetic field and caused the compass needle to react. In 1826, Johann Poggendorff, a German scientist at the Humboldt University in Berlin, increased the sensitivity of the galvanometer by horizontally suspending the magnetised needle on a vertical thread, to which a mirror was attached. A light beam shone on the mirror was reflected onto a horizontal scale. Any tiny rotation of the needle was then amplified by the movement of the light beam across the scale. Improved galvanometers of this type are now referred to as tangent galvanometers. Much later, around 1882, long after the breakthrough in understanding of the phenomenon of electromagnetism had been made, a more sensitive galvanometer was invented. This was the moving coil galvanometer. At its heart was a wire coil suspended between the poles of a magnet so that its two open ends faced the poles. The galvanometer coil was included in the circuit for which an electric current was sought. Any current flow through the coil created a magnetic field, and the coil would twist in the external magnetic field causing a rotation of the pointer attached to the suspension, thus detecting the presence of a current.

In 1824, the French scientist Dominique Arago discovered that when a magnetic compass was centrally placed on a glass plate held just above a rotating copper disc then the compass needle was deflected. When the copper disc had a high angular velocity, the needle rotated continuously in the same direction as the disc. Where did the rotating magnetic field come from that overcame the Earth's magnetic field? We don't know what led Arago to try such an odd experiment.

Faraday investigated the strange phenomenon of Arago's disc. In 1825, he noted in Davy's laboratory diary that he thought the explanation for the rotation of the disc was a link between magnetism and electricity. But his experiments came to nothing. Other scientists also investigated the phenomenon. Most famously, Charles Babbage, the Lucasian professor at Cambridge University, carried out experiments with Arago's disc, but he, too, was unable to explain the motion of the compass needle. It was Babbage who designed and built the difference engine and partially built the analytical engine, a mechanical computer using punched cards. Arago's disc remained a puzzle.

Faraday then tried a completely different set of experiments. He connected a wire to a simple galvanometer to form a complete circuit and then placed a live wire with a steady current I alongside it to see whether the magnetic field **H** induced an electric field **E** in it and, therefore, an electric current. But the galvanometer showed no induced current. He tried coiling the two wires together to increase the magnetic influence but, again, he found nothing. Reluctantly, Faraday put the research aside.

In 1831, Faraday worked with Charles Wheatstone studying the patterns made in sand on vibrating metal plates, first investigated by the German acoustic scientist Ernst Friedrich, in 1809. Faraday noticed that one plate set vibrating would, via acoustical vibrations through the air, disturb the sand on a nearby plate and set it vibrating, too. Faraday wondered whether the magnetic field around a live wire might be some form of vibration of the ether which could cause a vibration in a nearby wire and, hence, create a current of electricity. He decided to have another try at turning magnetism into electricity.

He made a circular wooden frame and coiled the live wire and the other wire connected to the galvanometer about it. While steady current was flowing there was definitely no effect induced in the other wire. But he noticed that on switching the current off, or back on, the magnetised galvanometer needle twitched momentarily. By chance, he had stumbled on the secret that many other scientists were searching for. As the magnetic field changed with time, $\partial \mathbf{H}/\partial t$, growing, or collapsing, inside the live coil, it induced a current in the coil connected to the galvanometer. But once a steady current was reached, the magnetic field $\mathbf{H}$ it created also became steady and the galvanometer needle settled back to give a zero reading. He had made the breakthrough in understanding! The magnetic field $\mathbf{H}$ had to change with time to induce an electric field $\mathbf{E}$ in a nearby wire, resulting in an induced current.

After various experiments, he made a paper tube and coiled a wire around it and connected the ends to a galvanometer. Then he plunged a strong magnet into the paper tube, into the centre of the coil. Sure enough, as the magnetic field changed with time, the galvanometer needle momentarily flicked into motion across the scale. He pulled the magnet out and the needle momentarily moved in the opposite direction. Changing the magnetic flux though the coil created a voltage in the wire of the coil. The voltage induced a current to flow around the coil, which created a changing magnetic field exactly in opposition to the moving external magnetic field.

At the same time, in the USA, Joseph Henry, whose parents were Scottish, was carrying out similar experiments to Faraday. Henry discovered the secret of electromagnetic induction, independently, in 1832. This is an example of the synchronicity in scientific discoveries. But Faraday published his results first and claimed the credit for the discovery. Henry visited Faraday at the Royal Institution in 1837 and the two scientists got on well together. Henry was also a professor of Mathematics and later became the first Director of the Smithsonian Institute in Washington, founded in 1846.

Interestingly, the Smithsonian was funded with money willed by the Englishman James Smithson, the natural son of the Duke of Norfolk, while the founding of the Royal Institution of Great Britain, in London in 1800, was due in part to funding from the former North American colonial Benjamin Thompson (Count Rumford).

In 1834, the Russian scientist Heinrich Lenz, a professor of Physics at the University of St Petersburg, explained the meaning of Faraday's discovery. It was all to do with satisfying Newton's 3rd Law of action and reaction. As Lenz stated, "The direction of the induced current is such that it opposes the motion producing it." If the north pole of a magnet approaches the coil, the induced current creates a north pole in the end of the coil facing the approaching magnet, in order to oppose it, since like poles repel. When it's the north pole of a magnet being withdrawn from the coil, then the induced current creates a south pole, which attracts the magnet and opposes its motion.

Now a coil possesses a characteristic of its own, called self-inductance. A steady current I through the coil creates a steady magnetic field. But, if the coil current changes then the magnetic field that it creates also changes and, in so doing, it induces a voltage in the coil itself to oppose the current change. This self-induced voltage V is directly proportional to the change in current dI/dt. This property of the coil is expressed as

$$V = L\frac{dI}{dt}$$

Where L is called the self-inductance of the coil. In SI units, L is measured in Henrys.

The magnetic energy E of the coil is given by

$$E = \frac{1}{2}LI_0^2$$

Where $I_0$ is the final steady current. There is a gravitational analogue with a spinning ring mass.

With hindsight, we can now see why the magnetic needle rotated in Arago's disc experiment. Imagine the copper disc, containing free charges (actually electrons with negative charge), rotating in a clockwise direction when viewed from above. As a small region of the disc approaches beneath the north pole of the magnetic needle, it is subjected to an increasing magnetic field. This causes an anti-clockwise electrical eddy current (analogous to a vortex ring) to arise in the disc surface around the region, thereby generating a virtual north pole. Since like poles repel, the virtual north pole pushes the real north pole away, causing the needle to rotate clockwise. A region of the copper disc moving away from the north pole of the magnetic needle experiences a decrease in magnetic field, giving rise to clockwise eddy currents which generate a virtual south pole in the disc surface. Since unlike poles attract, the virtual south pole drags the real north pole after it, also causing the needle to move in a clockwise direction.

A similar story explains what happens in the copper disc in the vicinity of the south pole of the compass needle. Overall, the compass needle is induced to rotate in the same direction as the copper disc.

## FARADAY AND THE CONTROL OF ELECTROMAGNETISM

Faraday turned his attention to a variation of Arago's disc experiment. He arranged for the outer periphery of a revolving copper disc to pass between the poles of a very strong horseshoe magnet. Conducting strips in contact with the perimeter of the disc and the axle of the disc were connected to either side of a galvanometer, thus forming a closed circuit. As the disc rotated, the galvanometer needle showed a constant deflection, indicating the generation of a constant current. Rotary mechanical motion had been converted into the continuous flow of electricity. This was the world's first dynamo generating continuous direct electric current, and it stemmed from Arago's odd experiment. But how did it work? Free charges Q in the disc on passing vertically through the magnetic field **H** experienced a force $\mathbf{F} = Q(\mathbf{v} \times \mu\mathbf{H})$, which resulted in an inward radial current I from the outer disc edge to the axle. Professor Fleming's 'right hand rule' gives the current direction, where the thumb and first two fingers of the right hand are perpendicular to each other. If the thu**M**b points in the direction of motion (**v**) of the free charge and the **F**irst finger points in the direction of the magnetic field (**H**), then the se**C**ond finger points in the direction of the induced current (I).

Very quickly, following Faraday's and Henry's breakthrough, several practical electric generators were built. The first was made by the French scientist Hippolyte Pixii, in 1832. It was capable of generating alternating current (ac) and direct current (dc), the latter requiring commutators; an idea devised by Sturgeon. The steam engine was ideally suited to provide the power needed for rotary motion, and the electrical power generating industry began to grow.

Until the development of the dynamo and the widespread distribution of electrical power, the way that we lived wasn't much different from the way people lived in ancient times. The steam age had begun the process of change but the age of electric power accelerated the change. Now our houses are full of electrically powered labour-saving devices. Electric lighting has extended the use that we can make of each day. In the factories, electric motors first powered drilling tools. In the US, Thomas Davenport was the first to patent such a device in 1837. In 1885, Nikola Tesla invented his brilliant induction motor with its rotating magnetic field, used in many of today's electric devices. For the last 100 years, we have been able to communicate with people nationwide and across the world using electromagnetic waves. The advent of the electronic computer has increased the rate of change. The increasing amount of leisure time that we gain is being filled with electrically powered gadgets to entertain us. Soon electrically propelled vehicles, some driverless, will be the norm for all forms of transport. In our factories, fixed, electrically powered robotic machines are producing more and more sophisticated devices to assist and amuse us. In the future, it seems that we will live with mobile robots with enormous intelligence, capable of carrying out all forms of dexterous tasks. The development of the electric motor and the dynamo has led to a huge change in the way that we live. However, it is important that humans retain the key to turn the power off.

# Faraday and the control of electromagnetism

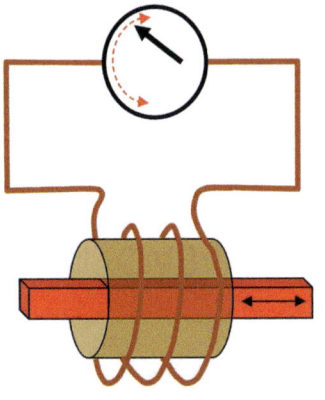

The Discovery.

The relative movement of a magnet inside an electrically conducting coil generated a current in a closed circuit.

Motion (F)

Field (H)

Induced current (I)

Fleming's right-hand rule.

Michael Faraday 1831

Joseph Henry 1830

The World's first Dynamo, generating continuous direct electrical current

CHAPTER 26

# MAXWELL, HERTZ AND THE REVOLUTION IN COMMUNICATIONS

In 1855, the young Scottish mathematician James Clerk Maxwell had just graduated from Cambridge University. He was fascinated by Faraday's discoveries in electromagnetism made 20 years earlier, but there was still no complete mathematical model to describe the phenomenon. Faraday's use of iron filings had revealed the pattern of the magnetic field circling around a current-carrying wire. In 1858, the German scientist Hermann von Helmholtz pointed out that the current-carrying wire was like a vortex core and the circling magnetic field was analogous to circling streamlines. Maxwell sought to make use of the analogy.

From Oersted's discovery, we noted that for the steady flow of current I we have $\rho \mathbf{v} = \nabla \times \mathbf{H}$. As the charges move along the wire with speed v, a circular magnetic field $\mathbf{H}$ is created around the wire. If A is the cross-sectional area of the wire then the current $I = \rho vA$, where $\rho$ is the charge density. Compare this with the vortex case, where we had $\zeta = \nabla \times \mathbf{v}$ and the fluid analogue of electric current is $\Gamma = \zeta A$, where $\Gamma$ is called the circulation.

Electricity $\mathbf{E}$ and magnetism $\mathbf{H}$ are analogues of each other as well as being coupled together. One may be taken as the dual of the other. So, from duality, we might guess that for steady conditions $\rho' \mathbf{u} = \nabla \times \mathbf{E}$, where $\mathbf{u}$ is the velocity of magnetic monopoles and $\rho'$ the density of their distribution. But magnetic poles always appear together as positive and negative poles, forming dipoles, so the density $\rho'$ of magnetic monopoles is always zero. Thus, for steady conditions, $\nabla \times \mathbf{E} = 0$, which explains why Faraday found no effect when the magnetic field was steady.

But for the case where the magnetic field changes with time, we have

$$-\mu \frac{\partial \mathbf{H}}{\partial t} = \nabla \times \mathbf{E}$$

This is the mathematical model of the law of electric field induction, discovered experimentally by Faraday in 1831. The minus sign indicates that the effect of the induced $\mathbf{E}$ field is to oppose the changing magnetic field. The parameter $\mu$ is the magnetic permeability of the ether.

Maxwell then turned the duality between $\mathbf{E}$ and $\mathbf{H}$ round the other way and suggested that, even if there was no current flow of charge ($\rho \mathbf{v} = 0$), a time-changing electric field might induce a magnetic field given by

$$\varepsilon \frac{\partial \mathbf{E}}{\partial t} = \nabla \times \mathbf{H}$$

The conduction current is represented by the $\rho \mathbf{v}$ term, so the $\varepsilon\, \partial \mathbf{E}/\partial t$ term must represent some other form of current that was time-dependent. Maxwell noted that some materials became slightly strained when they were placed in an electric field. These materials were called dielectrics. Although dielectric materials were neutrally charged, when they were placed in an electric field pairs of equal, but opposite, charges within the material were forced to move in slightly opposite directions, resulting in the material being stretched. The slight displacement of the charge pairs meant that tiny currents occurred during the stretching process. Maxwell called these tiny currents displacement currents and identified them with the stretching caused by the changing electric field $\partial \mathbf{E}/\partial t$.

Today, we are familiar with the model of an atom as the analogue of a mini Solar System with a positively charged nucleus at the centre, around which move negatively charged electrons in circular orbits. With this model, we can see why some materials, when placed in an electric field, become stretched, because the atomic nuclei move very slightly in one direction while the circular orbits of the atomic electrons become elliptically distorted in the opposite direction. However, proof of the existence of atoms did not occur during Maxwell's lifetime.

Charge separation strains the dielectric (stretches it like a spring) and explains where the stored electrical energy resides in a capacitor. The first capacitor was the Leyden jar, and the stored electrical energy resides in the stressed glass walls of the jar, not, as was originally thought, in the water.

Maxwell assumed that the vacuum of space was filled with ether, an incompressible type of fluid, which could be stressed by electric and magnetic fields. His mathematical model wasn't quite the same as that of fluid dynamics, but it was fairly close. With just a few changes it had provided a set of mathematical equations that unified electricity and magnetism. In 1873, Maxwell published his theory in his *Treatise on Electricity and Magnetism*. He used Cartesian components (the x, y and z components of a vector) and quaternions, a rather obscure mathematical form particularly useful for dealing with rotations, as well as his curl notation. Oliver Heaviside, a self-taught English mathematical genius, converted Maxwell's equations into the simpler vector form, retaining the curl term. And it's this form of Maxwell's equations that engineers and scientists have used ever since.

Maxwell's four famous equations describing electromagnetism are:

$$\nabla \cdot \mathbf{E} = \rho_e/\varepsilon$$
$$\nabla \cdot \mathbf{H} = 0$$
$$\nabla \times \mathbf{E} = -\mu \, \partial \mathbf{H}/\partial t$$
$$\nabla \times \mathbf{H} = \rho_e \mathbf{v} + \varepsilon \, \partial \mathbf{E}/\partial t$$

The first equation describes steady conditions for the electric field $\mathbf{E}$. Here, $\rho_e$ is the charge density (the suffix e is usually omitted) and $\varepsilon$ is the electric permittivity of the ether. The equation is called the divergence of $\mathbf{E}$ and arises because a point charge has a diverging radial $\mathbf{E}$ field. The equation is the analogue of the gravity case, where $\nabla \cdot \mathbf{g} = -\rho/\gamma$.

The second equation reflects the fact that monopoles always occur in plus and minus pairs, so that the density of magnetic poles is zero.

The third equation describes the electric $\mathbf{E}$ field which is induced by a change in magnetic field $\mathbf{H}$. Since the density of magnetic poles is zero, this term is missing from the equation.

The fourth equation describes the magnetic field $\mathbf{H}$ created in two ways. Firstly, by the motion of electric charges, with density $\rho$ and velocity $\mathbf{v}$, and, secondly, due to the time-changing electric field $\mathbf{E}$ resulting from displacement currents.

Maxwell thought of the ether as a di-electric medium. In his mind's eye, he saw that the disturbance of the ether by an electric field would briefly create a displacement current in the ether. Taking it a step further, he thought that a varying electric field might create an electromagnetic wave in the ether, radiating away from the disturbance centre. Maxwell called his version of the ether the luminiferous ether. In today's quantum dynamical view of the ether, the electromagnetic ether is envisaged as a seething ocean of virtual electromagnetic particles, called photons, with a plethora of frequencies and each only existing for a brief moment. By comparison, Maxwell's electromagnetic wave may be seen as the propagation of real photons, with a single frequency, passing through the ocean of virtual photons.

From his two curl equations, with $\sigma \mathbf{E}$ replaced by $\rho \mathbf{v}$ (see Ohm's law), Maxwell derived an equation for the electric field $\mathbf{E}$, now known as the equation of telegraphy. In the 1-D x-direction, it is

$$\frac{\partial^2 \mathbf{E}}{\partial x^2} = \sigma\mu \frac{\partial \mathbf{E}}{\partial t} + \varepsilon\mu \frac{\partial^2 \mathbf{E}}{\partial t^2}$$

The varying magnetic field $\mathbf{H}$ satisfies the same equation in the 1-D y-direction.

Maxwell assumed that the ether was a non-conducting medium (unlike the Earth's atmosphere, which supports lightning strikes) with conductivity $\sigma = 0$ and resistance $R = \infty$. The resulting equations are

$$\frac{\partial^2 \mathbf{E}}{\partial x^2} = \varepsilon\mu \frac{\partial^2 \mathbf{E}}{\partial t^2} \quad \text{and} \quad \frac{\partial^2 \mathbf{H}}{\partial y^2} = \varepsilon\mu \frac{\partial^2 \mathbf{H}}{\partial t^2}$$

These equations describe the motion of electromagnetic $\mathbf{E}$ and $\mathbf{H}$ sinusoidal waves moving together with speed $c = 1/\sqrt{\varepsilon\mu}$ in the z-direction, with the $\mathbf{E}$-field vibrating in the xz-plane and the $\mathbf{H}$-field vibrating in the yz-plane.

Such waves had never been seen. But when Maxwell calculated the value of the wave speed c, he found that it was equal to the known speed of light, so he concluded that light itself was an electromagnetic $\mathbf{E}$-$\mathbf{H}$ wave. This was a brilliant piece of speculative mathematical physics, but did it model reality? Were there other forms of electromagnetic waves, apart from light? Maxwell included his ideas about electromagnetic waves in his book, *Treatise on Electricity and Magnetism*.

Hermann von Helmholtz was very interested in the phenomenon of electromagnetism. He thought about the electrical discharge from a Leyden jar. This could be done with a conducting arm touching the outside metal surface and the other end close to (forming a spark gap) the ball on the conducting stalk attached to the inner metal surface. On discharging, the current would first surge one way and then the other, so that the inside and outside metal coatings of the glass jar would keep reversing in charge polarity (positive or negative), until eventually the oscillation died away. In 1847, Helmholtz suggested that the electrical charges crossing the spark gap were probably performing simple harmonic motion. Later, in 1853, the British scientist William Thomson (Lord Kelvin) provided a mathematical model supporting Helmholtz's idea. Clearly, the Leyden jar had capacitance C, and the conducting arm used to discharge the jar introduced induction L and resistance R. When oscillations of charge are involved, the joint effect of the L and C properties is called reactance, which combined with R is called impedance. Thomson predicted that the resonant frequency $f_{Res}$ of the oscillating electrical discharge would be

$$f_{Res} = \frac{1}{2\pi\sqrt{LC}} \text{ Hz}$$

When Helmholtz eventually read about the idea of electromagnetic waves existing in the ether, in Maxwell's *Treatise on Electricity and Magnetism* published in 1873, his interest was re-aroused. In 1879, he asked his PhD student, Heinrich Hertz, whether he would be willing to carry out some experiments to search for the existence of the electromagnetic waves predicted by Maxwell's theory. However, Hertz felt that the task was far too risky for a doctoral dissertation and he opted, instead, for a purely theoretical study. Nevertheless, Helmholtz had planted the seed in Hertz's mind, and after completing his PhD, Hertz turned his attention to Maxwell's theory of electromagnetism, corresponding with scientists in Great Britain who were particularly interested in the possibility of electromagnetic waves; notably Oliver Heaviside, Oliver Lodge and George Fitzgerald. Hertz could see that a spark might well create an electromagnetic disturbance in the ether; probably an oscillation that would quickly decay, but how could it be detected?

Hertz considered sympathetic resonance and an analogue with tuning forks. Suppose two tuning forks have the same resonant frequency. If the first fork is struck and held near to the second, then the second will begin to vibrate in sympathy with the first, as it is subjected to the incident acoustic waves from the first. Thus, the second fork can be used to detect acoustic waves from the first fork. Faraday and Wheatstone had noticed a similar effect with neighbouring Chladni plates, where one vibrating plate of sand induced vibrations of the sand on a nearby plate. Sympathetic resonance doesn't arise if the frequency of the forks and vibrations of the plates are not of the same frequency (not in tune), nor if there is no medium, namely air, for the waves to travel through. This latter point was important with regard to the supposed existence of the ether.

Hertz considered Helmholtz's idea that the discharge from a Leyden jar would be oscillatory. If the conductor used to provide the discharge path contained a spark gap, then the Leyden jar could be charged up until eventually voltage breakdown occurred across the gap and a spark appeared. While the spark formed, the current would surge backwards and forwards around the circuit.

The problem with the capacitive discharge from a Leyden jar was that the current oscillation rapidly decayed and the sparking stopped. Instead, Hertz decided to use an induction coil to differentially charge up a pair of capacitor plates separated by a spark gap. In this way, regular oscillatory sparking could be maintained. He then planned to try to detect any oscillatory disturbance of the ether using sympathetic resonance.

The large capacitor plates were connected by wire to two small polished brass knobs, which were arranged fairly close together to form a spark gap. Opposite electric charge supplied by the induction coil gradually built up on each plate, increasing the magnitude of the electric E-field between them. Suddenly voltage breakdown occurred and a spark jumped across the gap. During this discharge, a large current flowed along the wires between the plates. As the current surged along the wire and across the gap, it was accompanied by a circular magnetic H-field.

When the voltage across the spark gap dropped, so that no spark could cross over, the charges began to increase on the capacitor plates with the opposite polarity until, eventually, the E-field across the gap was great enough to cause a spark to jump across in the opposite direction, generating an H-field in the process. Thus, the oscillatory sparking continued, like pushing a swing, with the E and H fields reversing directions, until the induction coil was switched off.

Providing the oscillatory discharge is great enough, the E and H fields are radiated away out into space, forming electromagnetic waves of many frequencies. The important thing about Hertz's apparatus was that it created a regular electromagnetic disturbance of the ether, allowing time for its presence to be searched for.

At this stage, let us look at a model of the type of electromagnetic waves transmitted through the ether with Hertz's device. For one particular frequency, the outward radiation is composed of two transverse waves. By transverse, it is meant that the waves oscillate in a direction perpendicular to that in which the radiation is moving. There is a vertical electric sine wave and a horizontal magnetic sine wave and they oscillate in phase (increasing and decreasing together). Each wave is called a progressive wave, because the wave shape moves outwards, away from the generating device.

To detect the electromagnetic wave, Hertz used a conducting ring with a small spark gap, rather like a copper bracelet used by some people to alleviate arthritis. This was his analogue of the tuning fork. The circumference of the ring was important for tuning in to a particular electromagnetic wave and allowing induced oscillatory currents to resonate in sympathy with it. The waves were detected, in a darkened room, by observing the appearance of a tiny spark across the gap. The gap size was made adjustable to allow some fine tuning. Using this crude method, Hertz, by then a professor at the University of Karlsruhe, detected the presence of electromagnetic waves, in 1888. It had taken 10 years to sort out the experiment after the idea of searching for electromagnetic waves had first been suggested to him by Helmholtz.

Hertz also reflected the transmitted waves back from a plane metal sheet, which was positioned so as to cause the formation of electromagnetic standing waves. At any point, the outward wave causes an oscillation and the return wave also causes an oscillation. For the right conditions to cause a standing wave there are some points where the combined oscillations of the two waves cancel out, so that there is no E-field. Such points are called nodes. Exactly between two nodes there is a point, called an antinode, where the combined oscillations always vibrate and cycle through a maximum and minimum E-field. Moving his detecting ring backwards and forwards, Hertz was able to find out the length of the transmitted and reflected waves that he was tuned into by finding the nodes (no spark) and the antinodes (large spark). The distance between nodes, or between antinodes, was equal to half a wavelength ($\lambda/2$).

In 1889, just a few months after Hertz had detected electromagnetic waves, Oliver Lodge (then a professor at Liverpool University) gave a demonstration at the Royal Institution of his method of syntonic tuning into an electromagnetic wave using a pair of Leyden jars. One Leyden jar had its inner and outer metal surfaces connected with a wire circuit (in a vertical plane) containing a spark gap. This jar was connected to a device (a Wimshurst machine) to create continuous sparking across the gap. A few feet away stood an identical Leyden jar with a wire circuit (also in a vertical plane) connecting its inner and outer metal surfaces. In addition, a metal strip was connected to the inner metal surface and looped over the edge of the jar with its free end about 1mm from the outer metal surface. The wire circuit of the second jar could be varied (made longer or shorter) to tune into the electromagnetic wave generated by the spark of the first Leyden jar. In other words, the length of the circuit of the second jar was altered until resonance was obtained with the oscillations in the first jar. Resonance was observed when a tiny spark appeared at the free end of the metal strip, nearly touching the outer metal surface of the second jar.

In 1890, the French physicist Edouard Branly made an electromagnetic wave detector, based on a device containing a ground copper sandwich. When an electromagnetic wave passed across the device, there was a sudden drop in resistance across the sandwich, caused by the ground copper particles cohering together. A sharp tap of the device freed the copper particles, making it ready to detect an electromagnetic wave again. The device was called a coherer.

Oliver Lodge greatly improved Branly's coherer in 1894. Even Ernest Rutherford (he who split the atom!), while at the Cavendish Laboratory at Cambridge University, developed a coherer based on a sandwich of magnetic needles. In 1895, Rutherford was able to transmit an electromagnetic wave in the laboratory and detect it in his Cambridge lodgings, half a mile away.

Less than 20 years after Hertz's discovery, Guglielmo Marconi, a wealthy Italian-Irish enthusiastic radio entrepreneur, had developed Hertz's apparatus and combined it with detector technology to create a wireless communication system. In 1896, he demonstrated his equipment to British military personnel on Salisbury Plain. The Royal Navy were impressed with the technology and by 1899 it had three ships fitted with wireless apparatus. In the same year, Marconi established wireless communication between England and France and, by 1902, between England and America. Initially, Marconi's transmitters and receivers were untuned, but after 1900, tuning was included. For his development of wireless radio, Marconi was awarded the Nobel Prize in Physics in 1909.

We now know that there is a huge spectrum of electromagnetic waves, starting with enormously long radio waves, kilometres long, then the shorter waves for TV and mobile telephones, followed by radar waves of the order of centimetres (as discovered by Hertz), then the rainbow of light waves with lengths of about half a micron ($0.5 \times 10^{-6}$m), then into the broad range of X-rays (Röntgen rays) with incredibly small wavelengths starting at the nanometre ($10^{-9}$m) scale and reducing to pico-metres ($10^{-12}$m) and, finally, γ-rays (gamma rays). Apart from light, everything else was totally unknown and unsuspected 130 years ago.

# Maxwell, Hertz and the revolution in communications

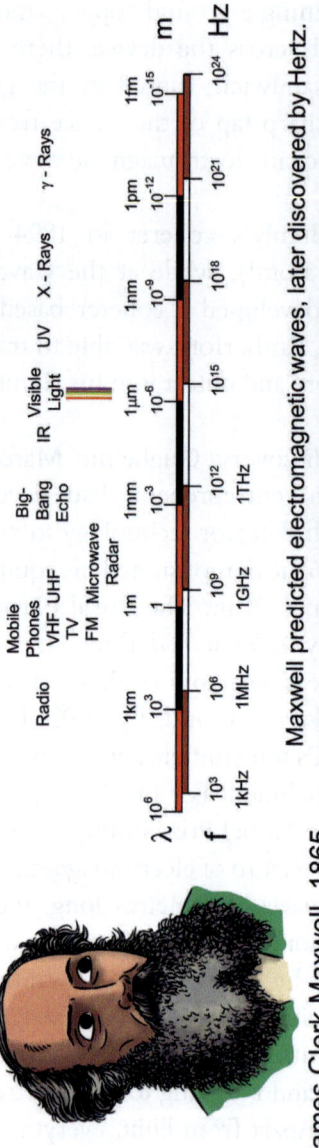

James Clerk Maxwell 1865

Heinrich Hertz 1888

Maxwell predicted electromagnetic waves, later discovered by Hertz.

CHAPTER 27

# HEAT AND THERMODYNAMICS

Hero of Alexandria (circa 50 AD) is reckoned to be the last and the greatest of the ancient Greek experimenters. He was a mathematical engineer who taught at the museum, part of the University of Alexandria, in Egypt. He collected together work done by earlier mathematical engineers, most notably Ctesibius (circa 200 BC) and Archimedes (circa 200 BC), and developed their ideas. Hero wrote a number of works on plane and solid geometry (*Metrica*), on surveying (*Dioptrics*), on the mechanical advantage of levers and pulleys and the use of gears (*Mechanica*), on optics (*Catoptrics*) and on hydraulics and pneumatics (*Pneumatica*). Much of his writing is lost, but that which remains gives details of some of his amazing inventions (*Automata*), such as a mechanism to open a temple door, involving steam pressure, siphoned water to a bucket suspended from a pulley attached to ropes wrapped round an axle connected to the door, a vending machine worked by water pressure which dispensed a small amount of holy water after the insertion of a coin, several devices worked by air and water pressure to give apparent life to automata and other extraordinary ideas, including the piston-operated syringe. On a more mundane level, the area A of a triangle with a half-perimeter length $s = \frac{1}{2}(a+b+c)$, where a, b and c are the lengths of the sides, is given in *Metrica* as $A = \sqrt{s(s-a)(s-b)(s-c)}$ and is known as Hero's formula.

The ancient Greeks conceived the idea of temperature $\theta$ as the degree of hotness or coldness of matter. A common observation is that heated bodies expand. In Hero's work *Pneumatica,* he described a simple experiment, attributed to Philon of Byzantium (circa 200 BC), which demonstrated that heated air expanded. To a hollow sphere he attached a tightly fitted tube and placed the open end under the surface of water in a jug. When the Sun shone on the sphere, the air expanded and bubbles of air were expelled into the water. When the sphere was shielded from the Sun's heat, the air contracted and water was sucked up the tube.

In 1589, Galileo obtained a copy of *Pneumatica* and read about Philon's device. He then made his own version using a glass tube fitted with an inverted corked flask at the upper end with the open lower end of the tube placed in a bowl of water. He called the device a thermoscope. The height of the water in the tube gave an indication of the temperature of the air in the flask. It was the forerunner of the thermometer, but its failing was that it was subject to change in air pressure.

The first recognisable thermometer, with liquid in a bulb connected to a glass tube allowing for liquid expansion and sealed at the tube's other end, was invented by Ferdinand II, the Grand Duke of Tuscany, in 1654. The height of the liquid column in the tube was related to the temperature of the liquid in the bulb. During the 1660s, Robert Hooke, the curator of experiments at the Royal Society, adopted the single fixed reference point for temperature as that at which water begins to freeze. The first sealed mercury thermometer was made by the German instrument maker Gabriel Fahrenheit in 1714. He was, also, the first to introduce a graded scale on the thermometer tube, where the height of the mercury column was related to the temperature of the bulb. Water froze at 32°F and boiled at 212°F on his scale. Later, in 1742, the Swedish physicist Anders Celsius introduced the centigrade scale of 100 divisions, with water freezing at 0°C and boiling at 100°C.

The first vacuum pump was built by the German scientist Otto von Guericke. In 1657, he demonstrated the immense force that the weight of the atmosphere has at ground level (approximately 14.7 lbs/in$^2$ or $1 \times 10^5$ N/m$^2$). He fitted two copper hemispheres together to form a sphere, of diameter 1m, and then using his vacuum pump evacuated it. Two teams of six horses could not pull the sphere apart, due to the pressure of the atmosphere outside it.

Before the formation of the Royal Society, Hooke worked as an assistant to Robert Boyle and built a version of von Guericke's vacuum pump for him. In 1660, Boyle published his famous law which states that at a fixed temperature the gas pressure multiplied by the gas volume is constant. Over a century later, in 1787, the French experimental physicist Jacques Charles made a curious discovery about confined gases. He found that a gas maintained at a constant pressure increased its volume by 1/273 of its volume at 0°C for every 1°C rise in temperature. Charles is better known for his earlier exploit of flying in a hydrogen balloon. He flew above Paris in 1783 just ten days after the Montgolfier brothers made their historic first flight in a hot air balloon.

During the second half of the 19th century, the Scottish theoretical physicist James Maxwell and the Austrian theoretical physicist Ludwig Boltzmann developed the kinetic theory of gases, based on the idea that gas temperature θ is linked with the speed v of gas molecules. In their model, the kinetic energy of a gas molecule is equal to $3/2\, k_B\theta$, where $k_B$ is Boltzmann's constant. So, the square root of gas temperature θ is proportional to the averaged velocity v of the gas molecules. Pressure on the container walls is due to the impact of gas molecules. The higher the temperature of a gas in a container of fixed volume, the higher the molecular velocities and the greater the pressure.

The Irish scientist William Thomson (later Lord Kelvin) thought about Charles' law and wondered about reducing the gas temperature as far as possible. Extending the graph of Charles' law backwards, he showed that absolute zero temperature, which he defined as 0 K, corresponded to -273°C. Reducing the temperature of a gas reduced the speed of its gas molecules, so Kelvin reasoned that the motion of the molecules must tend to zero as they approached 0 K. Although the process of liquefaction and solidification did not lead to zero volume, Kelvin thought the energy content of a body might be zero at 0 K. Kelvin's supposition was supported by experimental evidence, although even at 0 K, a body contains some infinitesimal molecular movement, giving rise to zero-point energy.

The study of the flow of heat through matter is called thermodynamics. But what actually is heat and how does it flow? Since the time of the ancient Greeks, heat has been associated with motion, or vibration. Through rubbing, objects get frictionally heated. Sir Francis Bacon was one of the first English scientists to investigate the properties of heat. In his book, *Novum Organum Scientiarum*, published in 1620, Bacon declared that, *Heat itself is motion and nothing else.* Newton held the same view.

In 1665, Robert Hooke wrote that *Heat is a property of a body arising from the motion or agitation of its parts; and therefore whatever body is thereby touched must necessarily receive some part of that motion whereby its parts will be shaken.*

Much has been discovered about the constituents of matter since the 17th century. Taking into account the development of atomic theory, by Rutherford and Bohr during the first part of the 20th century, has allowed us to gain a stronger insight into the phenomenon of heat. Vibration of solid matter at the atomic scale is associated with the elasticity of the lattice supporting the atomic masses and with acceleration and, via equivalence, with gravity. Although atomic, and molecular, masses are generally neutrally charged, we know that they contain equal amounts of positive and negative charge. The vibration of the atomic, and molecular, masses causes the charges to vibrate at the same frequency, resulting in electromagnetic radiation either internally (between atoms) or into the surrounding space from surface atoms, or molecules. Thus, heat is an electromagnetic radiation. So, we can think of an atom as a point heat source with a specific frequency. Collections of atoms, and molecules, will form a more complex heat source exhibiting a range of vibrational frequencies. If heat is fed into matter, it will absorb the electromagnetic radiation and cause the atoms to vibrate more and vice versa.

There are three forms of heat flow: heat conduction through a solid, where internal electromagnetic radiation is exchanged over atomic distances; heat convection through a liquid, or a gas, where molecules are free to move and exchange electromagnetic radiation; heat transport across the vacuum via pure electromagnetic radiation. In this chapter, we will mostly be concerned with heat conduction, but we will look briefly at the other two forms of heat transport.

Since all heat transfer mechanisms may be viewed as electromagnetic, any link between heat and gravity is really a link between electromagnetism and gravity; the much sought-after unification of the two forces. But that's not much help. We already know that electric charge is linked with gravity, since charge has mass.

Long before electromagnetic radiation was known about, Newton developed an empirical law of cooling for the convection of heat through air from a hot surface. Once heat radiation was recognised as being associated with electromagnetic waves, scientists began to examine the frequency and energy content of radiated heat from surfaces at various temperatures. Most of these experiments were carried out by German and Austrian scientists at the end of the 19th century. They used an oven which could be heated to a precise temperature $\theta$ and allowed a tiny light ray of heat radiation to escape from the black interior of the oven through a tiny hole in the door. Examining the light ray led to the discovery of the black body electromagnetic frequency spectrum, where the distribution of heat energy for different frequencies was dependent on the internal temperature of the oven, but independent of what the internal oven surface was made of.

In 1893, the German scientist Wilhelm Wien discovered his experimental 'displacement law', which stated that as the temperature $\theta$ of a black body increased, the wavelength $\lambda_{Max}$ for the maximum heat emitted decreased such that their product was a constant.

$$\lambda_{Max}\theta = \text{a constant} = 2.9 \times 10^{-3}\,\text{m.K}$$

In other words, the maximum heat emitted by a black body at a temperature $\theta$ coincided with the maximum frequency $f_{Max}$ of the vibration of the emitter.

In his attempt, in 1900, to derive a mathematical formula for the black-body spectrum, the German theoretical physicist Max Planck was led to conclude that the energy content of radiated heat came in discrete quantities he called quanta. Planck postulated that each quanta of radiation had energy E equal to its frequency f times an incredibly tiny constant h, called Planck's constant. The constant h has dimensions of angular momentum, and Planck was unaware of any natural phenomenon associated with h, so he was rather apprehensive about his idea.

Another way of writing the energy quanta is as $E = h/\tau$, where $\tau$ is the wave period. In this form, the equation indicates that the angular momentum h is generated during a wave cycle. Although the rate of change of angular momentum varies for quanta, the angular momentum h generated is always the same for each quanta of energy.

Quanta are now known as photons, each having discrete energy $E = hf$. Photons have a temperature, depending on their frequency.

At this point, it is important to establish the difference between the temperature $\theta$ of a heated body and the radiation of heat energy. Temperature $\theta$ is a classical concept and can change in a continuous manner. The radiation of heat energy from a body at a particular temperature is a process containing discrete energy quanta, or photons, with a wide range of frequencies. Based on Wien's law, we can associate the temperature $\theta$ with the maximum frequency $f_{Max}$ of vibration of the heated body.

In 1905, Einstein provided support for Planck's speculation about the discrete energy form of radiation, through his theoretical explanation of the photo-electric effect. In an experiment carried out in 1888, Wilhelm Hallwachs had negatively charged a zinc metal plate and noticed that the charge was lost when the zinc surface was bathed in low level ultraviolet light, but not with much brighter light of less frequency. In his model, to explain the photo-electric effect, Einstein used photons to represent electromagnetic waves. A high-intensity (bright, with many photons) but low-frequency electromagnetic wave could not dislodge negative charge (electrons) from the zinc surface. However, a low-intensity (not so bright, not so many photons) but high-frequency electromagnetic wave could. It was all down to the magnitude of the discrete energy ($E = hf$) of a photon striking a negative charge on the zinc surface, not how many photons were hitting an individual negative charge. Einstein was awarded the Nobel Prize in Physics in 1921 for his explanation of photo-electricity.

Planck's speculation about the discrete nature of radiation, with Einstein's support, led on to the important development of the theory of quantum mechanics. Planck's idea of the existence of a minimum value of angular momentum led others to consider uncertainty principles for particles with angular momentum made up from separate values, such as momentum times distance (Heisenberg) and energy times time (Einstein), since their combined value must always be greater than Planck's constant h. Quantum mechanics plays a major role in today's micro-electronics industry.

In 1911, the experimental physicist Ernest Rutherford, from New Zealand, suggested that the atom was analogous to a miniature Solar System, with the nucleus as the Sun and the orbiting electrons as planets. In 1912, the Danish physicist Niels Bohr improved on Rutherford's atom with the idea that the electrons orbited in discrete energy levels around the nucleus. Later, following the spectral analysis of heated gas flames, Bohr showed from his model that as gas molecules increased in temperature, the outer orbiting electrons of an atom, on absorbing a photon of discrete energy, jumped to a higher orbit. With the Solar System model of the atom, the photo-electric effect becomes clearer. Above a certain frequency, an electron in the outer orbit of a surface zinc atom absorbs enough energy from an incident photon to escape from its orbital confine and fly away from the surface. But remember, the theoretical model is based on an analogue which, to be of any use, must fit in with the results of experiments.

There were two competing models for heat conduction. The first, and oldest, idea was that heat is passed on by the vibration of contiguous particles of matter. The second, and relatively new, idea was based on an analogy with fluid flow and assumed that heat Q was associated with a substance, called caloric, which existed in a solid body and could flow through it.

The competition between the two heat models reached a zenith in the early part of the 19th century, with much of the terminology of caloric surviving but the earlier idea of heat flow being due to the vibrations of contiguous particles eventually winning the scientific argument. But, for a while, the caloric fluid model achieved some success.

During the 18$^{th}$ century, the Scottish chemist Joseph Black was one of the first scientists to carry out proper controlled heat experiments using thermometers. He was the discoverer of latent heat, associated with a body's change of state, from solid to liquid, or liquid to gas. During the change of state, heat is added without a change in temperature occurring.

At the time, the view of a body of matter was that internally it was made up of fundamental particles of mass that attracted each other. In his model of heat, Black assumed that permeating the interstices between the fundamental particles of mass was a substance called caloric, made up of discrete self-repellent mass-less particles. Although this was just prior to the advent of atomic theory, it may be helpful to think of the body of fundamental mass particles being held in a lattice. As the temperature of matter increased, Black assumed that more caloric particles were freed from the fundamental mass particles of the body, thereby increasing the amount of caloric substance in the region between the fundamental mass particles. In particular, it was thought that frictional wear released caloric particles from the fundamental mass particles at the surface of matter.

Thus, hotter bodies contained more free caloric particles than colder ones. And matter expansion was due to the repellent nature of the caloric particles. The hotter the body, the more free caloric particles there were, and the greater the expansion of the body.

In 1760, Black described his caloric model of heat flow as the flow of a substance made up of discrete self-repellent mass-less particles. Compare this with the modern view of an electric current, with its flow of self-repellent electric charges. The direction of the flow of heat was from the hotter part of a body to a colder part, due to the repellent nature of caloric.

The amount of caloric present in matter determined the thermal state of the matter and the amount of heat Q. Thus, the heat Q contained in a body is measured in calories. By definition, $Q = Mc\theta$, where M is the mass of the body, c is the specific heat (unit: cal/kg.K), determined for particular matter by experiment (assumed to be constant), and $\theta$ is the temperature (unit: K). Heat and temperature are often taken as averaged values for a distribution of matter, although we sometimes use point values in models. The amount of heat Q that can be stored in a mass M at a temperature $\theta$ is called the heat capacity $C_H$. Therefore, $C_H = Q/\theta$.

To develop a mathematical model of Black's flow of caloric substance, we can use a fluid analogue. In this way, we can form a potential-type model for thermodynamics. A 3-D point heat source of strength H emits H calories of caloric fluid per second. The heat sink of strength -H absorbs H calories of caloric fluid per second. The heat source density $\rho_H$ = H/V, where V is the volume occupied by the heat source ($\rho_H$ unit: (cal/s)/m³). We assume that caloric fluid is incompressible and that heat sources and heat sinks are situated in an infinite solid. The permeability of a solid to the flow of heat through it is called the thermal conductivity k (unit: cal/s.m.K), which must not be confused with Boltzmann's constant $k_B$. The introduction of more incompressible caloric fluid in a region will cause local expansion of the background solid. We can treat the classical temperature θ as a thermal potential where the temperature θ is given by

$$\theta = \frac{H}{k4\pi r} \quad \text{Kelvin (K)}$$

Although based on a fluid analogue, you may recognise temperature θ as the analogue of the gravitational potential φ. There is a difference between the two in that temperature θ is a positive quantity, while gravitational potential φ is a negative quantity. Suppose we have a background solid with no heat sources. That is, the background density of heat sources $\rho_H$ = 0. With no heat sources, the background temperature is zero (θ = 0 K). Suppose we introduce a heat source H into the background solid of zero temperature. Surrounding the point heat source there are spherical isothermal surfaces, reducing in positive temperature θ with distance r from the source centre. We can only introduce heat sources into the background solid of zero temperature, since heat sinks would introduce negative temperatures. For gravity, the situation is the same. For a background containing zero mass, the background mass density ρ = 0 and the background gravitational potential φ = 0. If we introduce a point mass M (a gravitational source) into the background, the potential φ everywhere remains negative.

However, in relative terms, for a solid background temperature θ greater than zero (say, on the isothermal surface surrounding a heat source), we have the possibility of relative heat sources and sinks. Heat dipoles also exist if the background temperature is greater than zero. Practically, a thermocouple, created by a thermo-electric Peltier effect, forms a heat dipole. Similarly, if the background mass density is greater than zero then we can have a relative positive mass (one with greater density) and a relative negative mass (one with lesser density), compared to the density of the background. This leads to the concept of buoyancy.

We see that a mass and a heat source share the same mathematical model. A point heat source harbours a point mass. Suppose that, somehow, we could manufacture a true negative mass (one with negative density). Would it contain a positive or a negative heat source? Or is the idea of a negative heat source with negative temperatures too strange to contemplate?

For gravity, the potential energy E of a point mass m on a potential surface φ (created by a mass M, say) is $E = m\phi$. The classical thermal analogue is that the energy E of a point heat source h on a surface of constant temperature θ (created by a heat source H, say) is given by

$$E = h\theta \quad \text{unit: cal.K/s}$$

We require energy E to be measured in Joules, but our model, based on the fluid analogue, has left us with (cal/s).K. This is an indication that the model variables need to be re-examined and their units modified.

We speculate that two heat sources, of magnitude H and h, in a solid, will obey the inverse square law of repulsion.

$$F = \frac{hH}{k4\pi r^2}\hat{r} \quad \text{unit: cal.K/s.m}$$

However, we require that the force unit cal.K/s.m is equal to the Newton (N). This is another sign that the model variables need to be redefined.

As we will see later, following experimental studies by the English scientist James Joule in the middle of the 19th century and a lot of persuasion thereafter, scientists now agree that heat Q is a form of energy. Thus, the modified unit for Q is the Joule and the modified unit for the strength of the heat source $H = Q/t$ is the Joule/s. From our analogue model we have $E = h\theta$, which means that the temperature θ must have a unit of time (s). So, an interpretation of the temperature at a point is the period τ of the maximum vibration there. Coldness is associated with longer vibrational periods and lower frequencies (hardly any movement), and vice versa for hotness. This model is more in line with the original vibrational notion of heat. Now our model units fit in with conventional units. But note that a true negative heat source would have negative temperatures, linked to negative time, which probably rules out the possibility of such a phenomenon.

We can stretch the analogy further by introducing the thermal field **T** analogous to the gravity field **g**. In terms of gradient of gravitational potential, we have $\mathbf{g} = -\nabla\phi$, and by analogy we have $\mathbf{T} = -\nabla\theta$, where $\nabla\theta$ is the gradient of temperature. In our modified units $\mathbf{T} = -\nabla\theta = -\nabla\tau$.

The concept of work W done on a body is that it is equal to the force **F** applied to the body times the distance **d** that it moves, so that $W = \mathbf{F}\cdot\mathbf{d}$. Work W is measured in units of Newton.metres (N.m).

Consider the case for gravity. If we introduce a test mass m into a gravity field **g** it will experience a force $\mathbf{F} = m\mathbf{g}$. On forcing the mass to move around a circuit in the gravity field (crossing potential surfaces) and coming back to the start, its energy at the start is the same as at the finish. The system is closed and gravity is said to be a conservative field. Suppose we let a test mass m fall a distance $d = \ell$ through a gravitational field **g** from potential $\phi_1$ to potential $\phi_2$. The work done by the gravitational field on the mass during the fall is $W = Fd = mg\ell = m(\nabla\phi)\ell = m(\Delta\phi/\ell)\ell = m(\phi_1 - \phi_2)$. This is the change in potential energy of the mass. The same amount of work is required to raise the mass back up again to its starting potential energy.

Similarly, in our simple thermodynamic model of a heated solid, the thermal field is conservative. In terms of analogues, gravitationally we have force $\mathbf{F} = m\mathbf{g}$ and thermally (using our redefined variable units) we have $\mathbf{F} = h\mathbf{T}$. Applying the analogue, the work done on a test heat source h in a heated solid falling from temperature $\theta_1$ to a temperature $\theta_2$ is $W = h\Delta\theta = h(\theta_1 - \theta_2)$. The same amount of work is required to raise the heat source back to its original temperature $\theta_1$ and energy $h\theta_1$. In general, heat sources don't move in a solid, so we are just establishing the theoretical pattern for the analogue.

In the special case of the heated solid being in temperature equilibrium then there is no temperature gradient ($\nabla\theta = \Delta\theta/\ell = 0$) and $\mathbf{T} = 0$, so the force **F** on the test heat source h is zero and no work can be done. As we will see, this is a form of the second law of thermodynamics.

In 1822, the French scientist Jean-Baptiste Fourier published his book, *The Analytical Theory of Heat*, basing the flow of heat on the caloric theory. We talk about the amount of heat Q and the heat flow current, or rate of heat flowing past a fixed point dQ/dt, measured in calories per second.

Experiments show that heat Q flows in the direction from high temperature $\theta_1$ to low temperature $\theta_2$ and that the rate of heat flow is proportional to the temperature gradient and thermal conductivity k. Suppose that the temperature gradient is in the x-direction, so that heat Q flows in the x-direction. The change in temperature between two points separated by a distance $\ell$ along the x-axis is $\Delta\theta = \theta_1 - \theta_2$. The negative temperature gradient is $\Delta\theta/\ell = -\nabla\theta$, where we have introduced the mathematical notation $\nabla\theta$ for the gradient of temperature.

Based on these experimental observations, Fourier derived a 1-D equation for the steady flow of heat, or caloric fluid, passing along a solid bar of length $\ell$ and cross-sectional area A.

$$\text{Rate of heat flow per unit area} = \left(\frac{dQ}{dt}\right)\frac{1}{A} = -k\nabla\theta \quad (\text{cal/s})/\text{m}^2$$

Many Physics students will know of the Searle's bar apparatus (invented by George Searle at the Cavendish Laboratory). This is a short bar of metal, usually copper, with a heat source at one end and a heat sink at the other. Searle's bar is a thermal capacitor, no less. The bar is lagged to prevent any heat loss, and a steady temperature gradient $\nabla\theta$ is established between its ends. Assuming the truth of Fourier's heat flow equation, Searle's bar is used to determine the thermal conductivity k of the bar.

The German scientist Georg Ohm assumed that an electric current in a wire was the flow of an electric fluid. He based his idea on the analogue with Fourier's model of heat flow in a solid. He did not know that the analogue of current was not exact, since electric sources (electrons) flow, whereas heat sources (in general) don't. But analogues only provide ideas for consideration. Their veracity must be tested by experiment. Ohm's experiments led him to the discovery of Ohm's law, V = IR, which he published in 1827. He equated the electrical potential difference, called the voltage V, between two points along a wire of length $\ell$ and cross-sectional area A (analogue of temperature difference between two points along a heated bar of length $\ell$ and cross-sectional area A) with the current strength I (analogue of heat flow) times the flow resistance R.

We can work the analogue in reverse and consider the thermal case as an analogue of the electrical case. We can rewrite Fourier's equation in the form

$$-\Delta\theta = \frac{dQ}{dt}\frac{\ell}{kA}$$

This is the thermal analogue of V = IR, where the temperature difference is the analogue of V (the potential difference), I is the heat current and $R_H = \ell/kA$ is the thermal resistance to the flow of heat sources. An electrical capacitor stores positive and negative charges ($\pm q$) on opposing sides of a pair of electrically conducting surfaces. An electrical potential gradient, or electric field **E**, is created across the intervening medium with electrical permittivity ε. This suggests the analogue of a thermal capacitor for storing positive and negative heat sources, or heat sources and heat sinks (±h), on either side of a medium with thermal conductivity k. A temperature gradient, or thermal field **T**, forms across the medium. This analogous view leads to the concept of heat source storage capacitance C. This parameter must not be confused with the heat capacity $C_H$. Based on the analogue with the electrical capacitor, in terms of energy storage E, we can write $E = \frac{1}{2}C\theta^2$ and $E = \frac{1}{2}h^2/C$.

The shared pattern provided by analogues can be used to look for patterns apparently missing in some phenomena. They are probably there but undetected. A method of detection must be found and an experiment must be performed. The success of an experiment is the final arbiter.

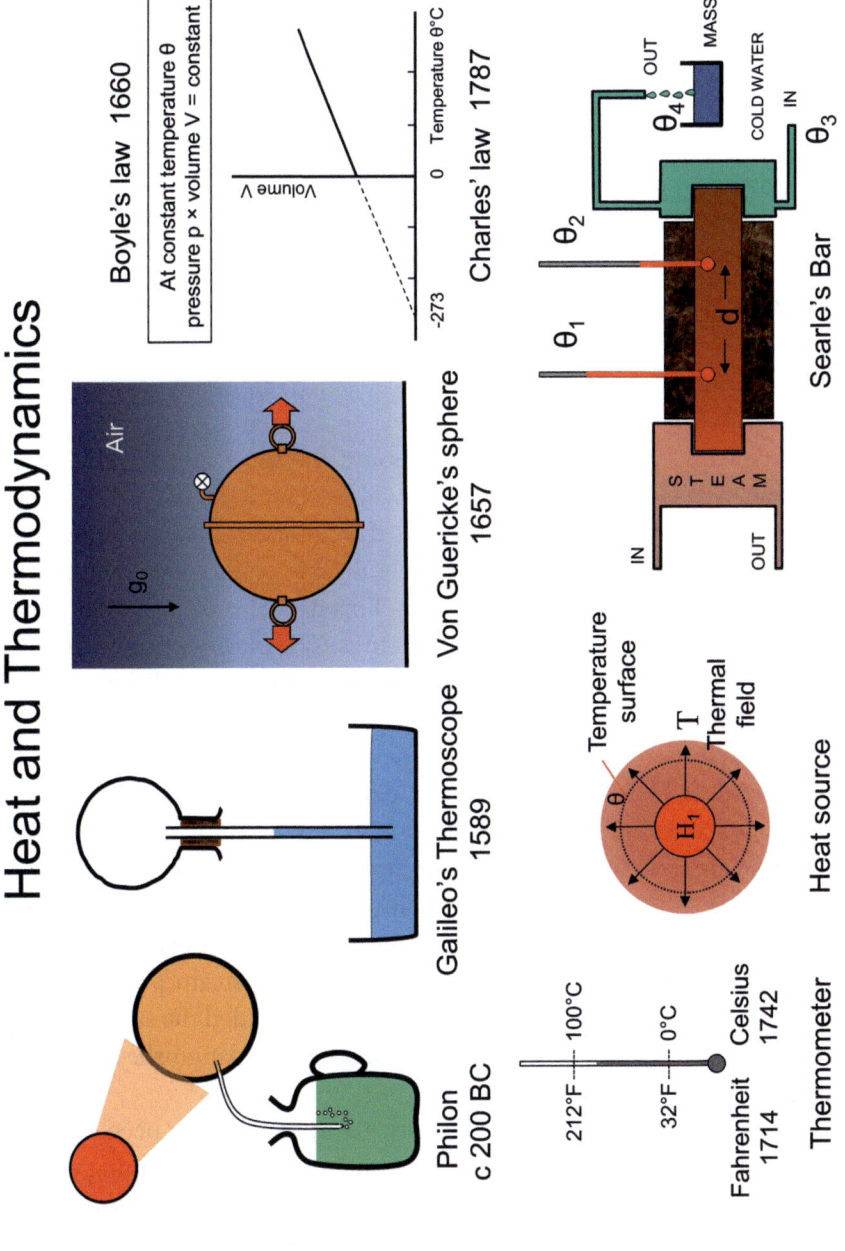

CHAPTER 28

# THE STEAM-DRIVEN INDUSTRIAL REVOLUTION

The twin concepts of atmospheric pressure and the vacuum were known to the ancient Greeks. It is not known who invented the first piston pump to suck up water. What is recorded is that in about 200 BC, Ctesibius, at the University of Alexandria, built a double piston-suction pump to provide a continuous stream of water. This device was used throughout the cities of the Roman Empire to fight fires. With various improvements, Ctesibius' pump continued to be used for firefighting well into the late Middle Ages in Europe.

The first use of steam to power a device was the Aeolipile, which was mentioned in Hero's *Pneumatica* and demonstrated by him in about 50 AD. It was a hollow copper sphere, supported on a horizontal axle, into which steam was fed. The steam escaped from two tubes on opposite sides of the sphere, aligned tangentially, so that the steam jets made the sphere rotate. The device was merely a toy and was not exploited further by Hero. The idea of a mechanical system, using steam pressure and a vacuum to cyclically drive a piston and do work, evaded the ancient Greek scientific minds. Hero's basic idea of steam power languished for many centuries.

In England during the 18th century, there was a great need to pump out floodwater from coal and tin mines but, often, hand pumping was not feasible. British engineers responded by building steam engines coupled to standard pumping mechanisms. They quickly became world leaders in building steam engines. Their engineering skill was based on 'rule-of-thumb' techniques gained from experience and experimentation.

# THE STEAM-DRIVEN INDUSTRIAL REVOLUTION

The French scientist Dionysius Papin is recognised as the pioneer of the steam engine. He moved around Europe but ended up in London. He worked with Christiaan Huygens and with Robert Boyle. In 1682, he was elected to become a fellow (FRS) of the Royal Society. He built a crude steam-driven piston engine but it was not a success. As an off-shoot of his work, he invented the 'pressure cooker', used to make food more digestible. The world's first successful steam engine was built in 1696 by Thomas Savery, an English inventor and engineer who was familiar with Papin's work. It was used to drive a water pump. In 1699, Savery demonstrated a model of his steam engine to the Royal Society. In 1703, Isaac Newton became the President of the Royal Society and Savery was elected a fellow in 1705.

The world's first practical steam engine was built in 1710 by Thomas Newcomen, an English engineer, using many features of Savery's design. In Newcomen's engine, steam forced a piston to rise in a cylinder and then water condensed the steam, creating a partial vacuum, enabling atmospheric pressure to push the piston back down to its starting position. The piston rod was connected by a chain to one end of a pivoted beam, so that as the piston rose and fell, the beam end rocked up and down. During the rocking motion, a chain attached to the other end of the beam worked a conventional suction pump to lift the floodwater out of the mine. This mechanical approach was successful in draining mines, overcoming the limitations of human pumping.

For 40 years, Newcomen's steam engines reigned supreme until their design was greatly improved by James Watt, a Scottish engineer who began his working life as an instrument maker at Glasgow University. Watt's important innovation was to have a separate cylinder to condense the steam. Watt's partnership with the Birmingham factory owner Matthew Boulton, in 1775, led to the mass production of efficient steam engines which were still, mostly, used to remove floodwater from mines. The Scottish engineer William Murdock deserves to be better known, as he developed coal-gas lighting on a commercial scale, with gasometers for gas storage. Murdock worked for the Boulton & Watt company, and it was he who devised the way to turn the up and down motion needed for pumping into rotational motion, similar to that already used in wind- and watermills to drive machinery.

Rotary output steam engines replaced the need for wind- and watermills. Instead, purpose-built factories relying on rotary steam power could be built almost anywhere, providing a source of coal and water was handy. This led to the start of the Industrial Revolution in Great Britain with factories springing up everywhere, many powered by Boulton & Watt engines. Mobile rotary output steam engines on metal tracks could be used to pull loads, leading to the introduction of railways. Over the following centuries, the Industrial Revolution gradually spread to the rest of the world.

Matthew Boulton, the far-sighted manufacturer and pioneer of new technologies, was a leading figure in the formation of the famous Lunar Society of Birmingham, in around 1775. The Society brought together scientists, engineers and medical men; including Erasmus Darwin (grandfather of Charles), Joseph Priestly, James Watt, Josiah Wedgewood and others, with the occasional visit by Benjamin Franklin during the latter's frequent trips to England. They were all fellows of the prestigious Royal Society, but the Lunar Society probably did more to stimulate the Industrial Revolution in Great Britain than did the Royal Society.

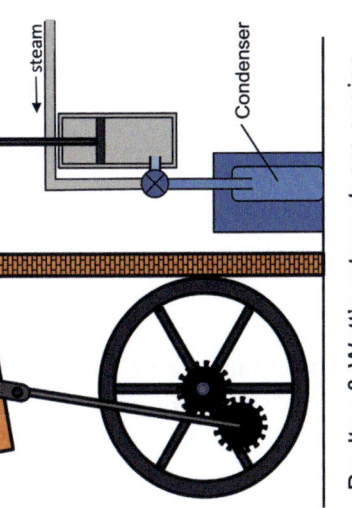

CHAPTER 29

# THE THERMODYNAMIC LAWS

Watt was fully aware of Boyle's law. When a gas expands at constant pressure, work is done equal to the pressure times the change in volume. Watt developed a theoretical model of the operation of his steam engine, with a pressure-volume indicator diagram for the steam cylinder from which the work done could be read. But, at the time, details of Watt's indicator diagram were shrouded in secrecy to avoid competitors learning about it.

The first theoretical model of an idealised steam engine was introduced, in 1824, by Sadi Carnot, a French army engineer. Sadi's father, Lazare Carnot, was also a French army engineer, well versed in mathematics. Carnot Senior published work on projective geometry and a book on engineering mechanics. After the French Revolution of 1789, Lazare Carnot entered politics and became an influential figure in the government of the French Republic. Carnot was one of the five members of the Directorate that ruled France, until it was swept away by Napoleon. Napoleon was actually a protégé of Carnot. After falling out with Napoleon, when he crowned himself Emperor, Lazare Carnot retired from politics and spent the time educating his children.

Sadi Carnot called his idealised steam engine a heat engine and published details of it in his book, *Reflections on the Motive Power of Fire*, greatly extending ideas started by his father. As Fourier had done, Sadi Carnot visualised heat flow as the flow of caloric fluid. He thought of his heat engine as analogous to a water wheel where, to turn the wheel and do work, water had to fall on the blades from a higher gravitational potential. To turn the wheel back to its starting position, the direction of the gravitational field had to be reversed. It was not possible for gravity, but the thermal field could easily be reversed just by reversing the position of the heat source and heat sink responsible for the temperature gradient.

Carnot's model describes an engine that uses the effect of heat on an ideal, or perfect, gas (one which obeys the gas laws) in a piston enclosed chamber to do external work on a load attached to the piston. Carnot introduced a comprehensive pressure-volume indicator diagram to describe the cycle of his heat engine and the underlying principles involved. If, instead, external work is used to drive the piston then the operating cycle is reversed and heat is pumped out of the engine. This is the process behind refrigeration.

Carnot's model engine came in five parts. The prime part was a chamber containing ideal gas enclosed by a piston connected to a load. Then there were four separate pieces which could be connected to the piston chamber, one used for each step in the operating cycle. In order of use, these parts were a heat source, a non-heat conducting block and a heat sink. Carnot assumed that each step was taken under equilibrium conditions so each step would take an exceedingly long time.

The next paragraph describes the operation of Carnot's heat engine in detail. You may want to skip over it for now and perhaps read about it later.

We start at point A (on the indicator diagram) with chamber conditions, pressure $p_1$, volume $V_1$ and temperature $\theta_1$, labelled as $A(p_1, V_1, \theta_1)$. In step 1, the heat source was attached to the base of the chamber, with the piston at its lowest position and the enclosed gas compressed. Heat $Q_1$ was absorbed from the heat source by the gas causing it to expand, thereby forcing the piston to rise, while keeping the temperature $\theta_1$ constant. This was an isothermal gas expansion taking us to point B, with conditions $B(p_2, V_2, \theta_1)$. In step 2, the heat source was replaced with the non-conducting block. The expanding gas continued to force the piston to rise to its maximum extent. Since no heat was allowed to enter the chamber there was a resulting drop in temperature to $\theta_2$. This was an adiabatic gas expansion taking us to point C, with conditions $C(p_3, V_3, \theta_2)$. Work was done by the gas on the piston in step 1 (A to B) and step 2 (B to C). In step 3, the heat sink was attached to the base of the chamber. As the piston began its descent, it forced the gas to become compressed. The heat $Q_2$ emitted in the process of compression was extracted by the heat sink, while keeping the temperature constant at $\theta_2$. This was an isothermal gas compression taking us to point D, with conditions $D(p_4, V_4, \theta_2)$. In step 4, the heat sink was replaced with the non-conducting block. As the piston continued to descend to its lowest position, the gas was further compressed, resulting in a rise of temperature back to $\theta_1$. This was an adiabatic gas compression taking us back to A, with the conditions A($p_1$, $V_1$, $\theta_1$). During step 3 (C to D) and step 4 (D to A), work was done by the piston on the gas. The piston had now completed its operating cycle and conditions were back to those at the start.

Carnot assumed that as his heat engine went through its cycle, it started with an input of $Q_1$ calories of heat into the piston chamber, the expanding gas doing work on the piston, and ended up with an output of $Q_2$ calories of heat, arising from the work done by the piston in compressing the gas in the chamber. Carnot defined the efficiency of his ideal heat engine as

$$\text{Efficiency} = \frac{Q_1 - Q_2}{Q_1} = \frac{\theta_1 - \theta_2}{\theta_1} = 1 - \frac{\theta_2}{\theta_1}$$

To get a maximum efficiency of 1, every bit of available heat must be used in the last stage, meaning $Q_2$ equals zero, which is not possible. To get the efficiency in terms of temperatures, we have used heat $Q = Mc\theta$. This is an oversimplification, because the specific heat c changes with pressure and volume, so we have to assume some averaged value of c.

According to Carnot, heat engines operating in forward or in reverse mode over the same temperature difference are equal in efficiency. So, in theory, it is possible for a heat engine working in forward mode to be continually replenished with caloric pumped from a heat engine working in reverse mode. But there is no excess caloric available to do any external work.

In 1830, Carnot realised that his model of an ideal reversible heat engine needed to be improved. A missing element was frictional heating due to moving matter, including the piston in contact with the chamber wall and the swirling cloud of steam. The consequence of internal frictional heating in a heat engine means that it is not possible to run a heat engine in forward mode supplied with caloric from an engine working in reverse mode, because caloric is being lost from both engines. Carnot, therefore, ruled out the possibility of perpetual motion using coupled heat engines working for each other, as an external supply of caloric would be needed.

From notes left by Carnot, discovered in 1870, it is clear that Carnot contemplated the idea that caloric heat was equivalent to mechanical energy. Unfortunately, Carnot died in a cholera epidemic in 1832, aged only thirty-six, before he was able to update his model.

The German theoretical physicist Rudolf Clausius wanted to get a better understanding of the Carnot cycle. In 1865, he introduced a quantity called the entropy S of an isolated system. For a perfectly lagged body of mass M (say, a solid block, or a box of gas) at a uniform temperature θ, the entropy of the body is

$$S = \frac{\text{Heat}}{\text{Temperature}} = \frac{Q}{\theta} = \frac{Mc\theta}{\theta} = Mc \quad \text{unit: cal/K}$$

From our simple model of thermodynamics we know the temperature θ at any point on an isothermal spherical surface, of radius r, surrounding a point heat source of strength H. So the rate of change of entropy of the body contained within the volume surrounding the heat source is

$$\frac{\Delta S}{\Delta t} = \frac{H}{\theta} = 4\pi k r \quad \text{unit: cal/s.K}$$

Clausius was more interested in determining whether a Carnot cycle was reversible, or not. This involved determining the change of entropy ΔS of the enclosed gas in Carnot's heat engine during each equilibrium step of the cycle, not the rate of change of entropy.

$$\text{Change in entropy of system} = \Delta S = \frac{\text{Change in heat}}{\text{Temperature}} = \frac{\Delta Q}{\theta} \quad \text{unit: cal/K}$$

During step 1 of the Carnot cycle, heat $Q_1$ was absorbed by the gas. So, the entropy change (assumed positive) is given by

$$\Delta S_{AB} = \frac{+Q_1}{\theta_1}$$

In step 2, there is no change in heat, so $\Delta S_{BC} = 0$. In step 3, heat $Q_2$ was emitted by the gas, so the change in entropy (assumed negative) is

$$\Delta S_{CD} = \frac{-Q_2}{\theta_2}$$

In step 4, there is no change in heat, so $\Delta S_{DA} = 0$.

During the adiabatic step 2 (B to C), no heat is added, so the entropy remains constant. Likewise applies for step 4 (D to A), where no heat is extracted. Since the expansion step 1 (A to B) and the compression step 3 (C to D) both take place between curves where the entropy is constant then the magnitude of the change in entropy $\Delta S_{AB}$ and $\Delta S_{CD}$ must be the same. So, for a reversible Carnot cycle, the total change in entropy $\Delta S_{ABCD} = \Delta S_{AB} + \Delta S_{CD} = 0$.

Although for a completely reversible Carnot cycle $\Delta S = 0$, in step 3 (C to D), where the gas cools, $\Delta S < 0$. From the kinetic theory of gases, this suggests lower molecular speeds and more order.

When friction is involved, extra heat is absorbed by the gas in all steps, so that $\Delta S_{ABCD} > 0$, and the Carnot cycle is not reversible.

Friction is the resistance to motion of bodies in contact. Contact is the important element, since moving bodies will interact via their gravitational fields, but for small bodies the effect across space is infinitesimal. In crude terms, friction is assumed to be due to the roughness of the surfaces of bodies in contact. Leonardo da Vinci, the famous 15th century Italian artist, inventor and scientist, was the first person to investigate the phenomenon of friction, but his findings did not incorporate the idea of force, provided later by Newton. It wasn't until 1699 that the French engineer Guillaume Amontons published the laws of friction that we recognise today. Nearly another century passed before the experimental laws were verified by the French scientist Charles Coulomb. Coulomb made the distinction between static friction, the resisting force before a body starts to slide, and kinetic friction, the resisting force once sliding is underway.

Recent research work suggests a quantum mechanical explanation for frictional resistance and frictional heating. In her *Scientific American* article, of October 1996, entitled *Friction at the Atomic Scale*, Jacqueline Krim noted that friction arises from atomic-lattice vibrations which occur when atoms close to one surface are set into motion by the sliding action of atoms in the opposing surface. These atomic-lattice vibrations set up sympathetic resonances in neighbouring surface atoms and extract energy from the motion. The work done by the motion in providing the vibrational atomic energy manifests itself as frictional resistance. In quantum terms, the atomic vibrational energy absorbed by the bodies is in the form of phonons, both thermal and acoustic. At the macroscopic scale, it is argued that friction is linked with quantum effects across surfaces in contact. At the microscopic scale, there is disagreement over whether quantum friction exists for nano-scale surfaces moving past each other, although not quite touching. This is important for forming something called a Casimir cavity, where interest is in extracting energy from the quantum vacuum.

Friction is often demonstrated as a classroom experiment. A sideways force is applied to a block of wood resting on a horizontal table and a resisting frictional force F is seen to exist, which prevents any movement. This is static friction in action. From Newton's 3$^{rd}$ law, the vertical weight W of the block is resisted with an equal and opposite force R by the table. The empirical law for static friction is that $F = \mu_s R$. As the sideways force on the block is increased, a maximum (limiting) static frictional force is reached after which sliding occurs. Sliding is kinetic friction in action and is represented by $F = \mu_k R$. The value of the coefficient $\mu$ depends on the two materials involved. The maximum value of $\mu_s$ is always more than the value of $\mu_k$. Experiments have shown that friction F is independent of the area of the surfaces in contact and that $\mu_k$ is almost independent of the sliding velocity.

If you push down on a copper coin on a wooden tabletop and apply a sideways force to make the coin slide, the coin becomes noticeably hot in the process. However, the laws of friction do not concern themselves with the production of frictional heat during sliding contact between bodies.

Benjamin Thompson was born, in 1753, in Massachusetts, one of the six British North American colonies that formed New England. At the beginning of the American War of Independence, in 1765, Thompson was a very young teenager with mixed views on the war, but with leanings more to the loyalist side. As he grew older, he became very interested in science. Maybe this helped him to block out the wartime situation and avoid making a decision on what side he supported. His much older compatriot, Benjamin Franklin, had already achieved fame as a scientist. But Franklin's views on the war were quite clear, as he signed the Declaration of Independence. In 1775, while the revolutionary war continued, Thompson sailed to England, where he became an under-secretary in the Colonial Office. While in London, he continued his interest in science and in 1779 became a fellow of the Royal Society, meeting many of the leading scientists of the time. In 1783, he briefly returned to New York to support, somewhat ambivalently, the independence movement. The war finally ended in 1785, but a year earlier, in 1784, Thompson had left North America to take a job in Munich, working for Karl Theodor, the Elector of Bavaria. In 1790, the Elector expressed his gratitude to Thompson for his outstanding work, much of it to do with social reforms leading to improved living conditions, by making him Count von Rumford of the Holy Roman Empire. (Rumford was a town in the British colony of New Hampshire where Thompson owned an estate. The town is now called Concord.) Count von Rumford became the Bavarian Minister of War, with responsibility for the military arsenal. While observing the boring of cannon guns in Munich, he was much taken by the colossal amount of frictional heat generated which had to be conducted away by spraying the bore site with cold water. The accepted opinion at that time was that as the metal was cut away it released heat in the form of caloric fluid. But Rumford noted that with a blunt borer, where hardly any metal was shed, an even greater amount of heat was generated by friction. Moreover, the supply of heat seemed to be inexhaustible. He concluded that the heat was generated by mechanical motion, heat itself being a form of vibratory motion which had nothing to do with the release of caloric fluid. With the start of the Napoleonic War, Rumford returned to England as the Bavarian Ambassador at the Court of St James, although, due to his British nationality, his credentials were never accepted. In 1798, he presented a paper to the Royal Society entitled, *Experimental Inquiry concerning the source of Heat excited by Friction,* which dismissed the caloric theory of heat flow. However, his view was largely ignored at the time.

In 1799, Count Rumford, Sir Thomas Bernard (whose father had been the governor of Massachusetts before the American war of Independence) and Sir Henry Cavendish (of Big G fame) provided the financial backing, along with membership subscriptions, to support the founding of the Royal Institution in London. The purpose of the Institution, which opened in 1800, was to improve the general public's understanding of science and to advance technology to benefit the living conditions of all.

The famous astronomer William Herschel investigated the use of red filters for his telescope to enable him to better observe sunspots. He thought that the red filters were quite hot so, in the vein of Isaac Newton, he used a prism to create a spectrum of sunlight and then used a thermometer to measure the temperature of each of the colours. As he expected, he found that the red end of the spectrum gave the highest reading. But, to his complete surprise, when he moved the thermometer just below the red colour (into a dark region), the thermometer registered an even greater temperature. This is the infra-red (IR) region of the spectrum, which is invisible to humans. Herschel reported his experimental discovery to the Royal Society in 1800. Most of the Sun's radiant heat is transported in the IR region. In 1801, Thomas Young, while the Professor of Natural Philosophy at the Royal Institution, established the wave theory of light. He also deciphered Egyptian hieroglyphics! Young concluded that radiant heat, like light, was a wave motion, suggesting that the source of heat in a solid was due to vibration.

Humphry Davy, an up-and-coming chemist, took an interest in the phenomenon of heat and agreed with Rumford's view that heat was generated by frictional motion and not by the release of caloric fluid. In Davy's view, caloric didn't exist; heat in solids was linked with vibration.

Count Rumford invited Humphry Davy to visit the Royal Institution. In 1801, on Rumford's advice, Davy was appointed as the Assistant Lecturer in Chemistry at the Royal Institution. It was a good move, as Davy was a brilliant presenter and made the Royal Institution famous. Later, he became the Professor of Chemistry and even more prestigiously, as Sir Humphry Davy, he became the President of the Royal Society.

Although Davy, and later Faraday, took an interest in the phenomenon of heat, they did not pursue Rumford's experimental work any further. Rumford's idea remained dormant for nearly half a century until it was taken up by James Prescott Joule. Joule was born, in 1818, in Salford, now part of Greater Manchester. At an early age, he inherited a successful brewing business from his father. With his financial security, Joule indulged his passion as an amateur scientist in his leisure time, with particular interest in heat experiments. Joule had the idea that the losses of various forms of energy were converted into heat.

With our modern understanding, we know that work done W is equal to the change in energy E. During the 1840s, Joule showed that many forms of work (particularly electrical and gravitational) degenerated into heat Q, all of which could be linked with the energy of motion via a conversion factor $J = 4.12$ Joules/calorie, termed the mechanical equivalent of heat, such that $E = JQ$. Since the increase in heat energy was directly linked with the loss of another form of energy, it suggested that energy was conserved. When Joule tried to explain his idea to scientists, with experimental results, there was little interest. At about the same time, the German scientists Julius Mayer and Hermann von Helmholtz published their theoretical papers on the conservation of force and the work done by force (energy), although, unlike Joule, they had no experimental evidence to back up their work.

Fortunately, for scientific progress, the Irish scientist William Thomson (Lord Kelvin) championed Joule's work and its correctness was eventually realised. The conservation of energy is now referred to as the 1$^{st}$ law of Thermodynamics. Today, the SI unit of work and energy, formerly the N.m, is called the Joule.

In Joule's famous calorimeter experiment, done in 1849, his apparatus consisted of a mass suspended at one end of a string, which was fed over a pulley, with the other end being wound around the vertical axle of a paddle wheel placed in a cylindrical container of water. Each blade of the paddle wheel was riddled with holes. On releasing the mass, the paddle wheel turned and water was forced through the tiny holes creating a multitude of tiny vortex rings, decaying into turbulence, until the mass stopped moving and the water came to rest. The process of rewinding the string, raising the mass and allowing it to fall again was repeated many times. A thermometer eventually showed that the water had heated up. Joule argued that the increased amount of heat Q (measured in calories) contained in the water was caused by the frictional heat generated by stirring the water, which was equal to the work done W (measured in Joules) by the falling mass, which was equal to its lost gravitational potential energy. To get the same units, Joule used his conversion factor J = 4.12 Joules/calorie. Thus, the work W done by the falling mass (resulting in a loss of gravitational energy) was equal to the increase in energy JQ of the heated water. Even so, Joule's work did not reveal how the heat originated, other than implying it was due to frictional resistance.

Using a quantum approach to thermodynamics, we can formulate a simple model to describe what is going on inside a steady-state heated body, and from the model obtain a crude estimate for J, the mechanical equivalent of heat. Suppose we have a body of mass M (kg) at a constant temperature $\theta$ (K), with specific heat c (cal/kg.K). Within the body, we assume that the fundamental particles (say, molecules) are vibrating and exchanging quanta of energy, without change in temperature. We assume that for a temperature $\theta$, the maximum frequency of each molecular vibration is $f_{Max}$, so that the averaged energy of a vibrating molecule is given by

$$E = hf_{Max} = \frac{hc_L}{\lambda_{Max}} \text{ where } c_L = f\lambda \text{ and } c_L \text{ is the speed of light.}$$

The number N of molecules in the mass M is given by

$$N = \frac{MN_A}{\text{Atomic Weight(A.W)}} \text{ where } N_A = \text{Avogadro's Number.}$$

Therefore, the total energy of all the molecules in the body is

$$E_{Total} = N\frac{hc_L}{\lambda_{Max}} = \left\{\frac{N_A}{A.W}\frac{1}{c}\frac{hc_L}{\lambda_{Max}\theta}\right\}Mc\theta = \{J\}Q$$

where $\lambda_{Max}$ is given by Wien's law.

Trying several materials, we get a value for J lying between 2 and 6. By experiment, Joule found that J was a constant given by J = 4.12 J/cal.

During his honeymoon in Switzerland, in August 1847, Joule took a very accurate thermometer with him and measured the temperatures at the top and bottom of several waterfalls with large vertical drops to see whether they confirmed his idea. However, the tiny difference in temperature would have made detection extremely difficult.

A modified version of Joule's waterfall experiment is the falling (flow) of lead shot in a glass cylinder of viscous liquid. With all the lead shot at the bottom of the cylinder, it is inverted and the lead shot falls through a gravity potential gradient to the bottom again. The lead shot starts and finishes with no kinetic energy. The loss of gravitational potential energy of the lead shot gives rise to an increase in temperature of the mixture. The process is repeated a number of times until a temperature change can be accurately measured. There is an analogy with electrons falling in an electrical potential gradient, so there is a gravitational analogue of Ohm's law, too. The lead shot falling through the viscous liquid experiences a resistance and so, by analogy, there will be a gravitational conductivity $\sigma_G$ (unit: $kg.s/m^3$) associated with the flow of lead shot.

Joule became a Fellow of the Royal Society in 1849 and published his paper on *The Mechanical Equivalent of Heat* in 1850. Joule's work demonstrated that different forms of energy could be converted into heat, showing that heat Q was a form of energy E = JQ.

So, the flow of heat along a bar was a flow of energy along the bar. With the unit for heat energy E in Joules, the unit for thermal conductivity k as Joules/s.m.K and the unit for the temperature gradient T as K/m, the energy form of Fourier's equation is

$$\text{Rate of energy flow per unit area} = \left(\frac{dE}{dt}\right)\frac{1}{A} = -kT \qquad \left(\frac{\text{Joules}}{s}\right)/m^2$$

Since our model for thermodynamics is based on fluids, we should expect an analogue of Bernoulli's equation. But, to see this, we must employ the modified units. The temperature $\theta$ is measured in time (s) equal to the period $\tau$ of vibration. The unit for temperature gradient T (where $\mathbf{T} = -\nabla\theta = -\nabla\tau$) is s/m, and the unit for thermal conductivity k is Joules/$s^2$.m. So, for steady thermal conditions, the thermal pressure p along a thermal field line T satisfies

$$\frac{p}{k} + \frac{1}{2}T^2 = \text{a constant}$$

This is the thermal analogue of Bernoulli's equation and it can be obtained from Fourier's equation. The pressure p = Force F/unit area A and the energy E gained equals the work done W = F.d, where d = distance moved. The volume V = Ad, so the energy density E/V of the thermal field is ½kT². The same result can be obtained from a 3-D extension of Fourier's modified equation, using the fact that Force **F** = H**T**, where H is the strength of a heat source and **T** is its thermal (temperature gradient) field.

Through a long series of experiments, Joule had shown that different forms of energy could be converted to heat. This being so, it was postulated during the 1840s, notably by Helmholtz and Mayer in Germany and by Joule and Kelvin in Britain, that one form of energy could be converted into another. This led to the 1$^{st}$ law of Thermodynamics, which is:

Although the form of energy may change, the magnitude of its content remains unchanged, provided the mass involved remains constant.

Based on experimental observations, a 2$^{nd}$ law of Thermodynamics was formulated during the early 1850s by Rudolf Clausius in Germany and Lord Kelvin in Britain. A rather long-winded form of the law is:

1. In nature, heat flows from a hot region (high temperature) of a body of matter to a cold region (low temperature).
2. Through matter expansion and contraction, heat does work.
3. When frictional heat is incurred, the direction of heat flow cannot be reversed without the aid of an external source of work.

The simpler Kelvin-Planck version of the law is:

It is impossible to extract heat from a body of matter at constant temperature and use it to do work.

The laws of Thermodynamics have been derived experimentally, based on the behaviour of heat flow within matter composed of huge numbers of contiguous particles. Whether the macroscopic rules of heat flow apply exactly at the atomic level is currently being examined. Quantum thermodynamics is an important new field of study (*Nature*, 2 November 2017, Vol. 551, p. 20).

In fact, using the term *law of Thermodynamics* is restrictive. It is really one form of the law of potential theory dynamics. In another form, we might introduce the 2$^{nd}$ law of Gravitodynamics as "It is impossible to extract gravitational energy from a body of matter at constant gravitational potential and use it to do work." It is interesting to contemplate the idea of gravitational energy flowing through a gravity field.

The 2$^{nd}$ law of Thermodynamics is usually linked with entropy S in the form S = energy/temperature. For an irreversible process of an isolated system, we have found that the entropy increases.

For a very simple model we assume that heat energy initially resides in a very small volume of matter with a uniform temperature and that matter outside the volume has no energy (zero temperature and motionless). As the heat spreads out, the volume of the heated matter increases and the temperature decreases, although the total dispersed energy remains the same. Therefore, the entropy of the increasing volume of heated matter increases. The increase in entropy is linked with an increase in volume of vibrating matter. This is often taken to mean that an increase in entropy of a system is associated with an increase in disorder, or chaos, within the system.

At the beginning of the 20$^{th}$ century, this simple model was applied to the Universe. It was assumed that eventually the whole Universe would reach an equilibrium condition, filled with matter at a very low uniform temperature, such that the entropy reached a maximum value. Since no work could be done, this condition was thought of as the final heat death of the Universe.

However, the model is far too simplistic. The presence of gravity, leading to supernova and black holes, means that equilibrium conditions are not reached. The formation of gigantic swirling galaxies in space is confirmation of this. The Earth's weather pattern is another example, where heat from the Sun disturbs the atmosphere and can lead to the formation of hurricanes, etc. On a small scale, a vortex street appears in a turbulent flow and hexagonal Bernard convection cells appear in a thin layer of boiling gravy in a saucepan. These irreversible patterns are known as 'self-organising' linear systems and they remain in existence, not far from equilibrium, while there is a steady throughput of energy. Other examples appear in chemistry, where spontaneous patterns appear during chemical reactions. In biology, self-organising systems probably led to the creation of life. So, it is now realised that the 2$^{nd}$ law of Thermodynamics only applies to closed linear systems close to equilibrium. Far from equilibrium non-linear systems have been studied by Professor Ilya Prigogine at the Free University of Brussels, winning him the Nobel Prize in Chemistry in 1977.

As far as we know, time flow is irreversible. That is, time is a positive quantity which always increases. So, we can't use negative time to make the rate of heat flow negative, namely $-(dQ/dt)$, thereby making heat flow from cold to hot. So, we can never get back to the original state where we started. But, as the example of the fridge shows, with an open system and an external energy supply, we can reverse the direction of heat flow. This has led to the thought that it might be possible to build a time-reversal machine. We know that gravity affects time and we have speculated that gravity is linked with heat flow. Heat flow is linked with temperature gradient and temperature is associated with vibrational time. So, perhaps, the area to search over in the quest to build a time machine is the combined phenomena of gravity and heat flow. But first we must learn how to manipulate gravity.

The 'Age of Steam' went into gradual decline following the development of the internal combustion engine by the German engineers Gottlieb Daimler and Wilhelm Maybach, towards the end of the 19$^{th}$ century. But thermodynamics is used to model the processes involved in the working of the combustion engine. Like much progress in science and engineering, practical devices come first, followed by theory, leading to a better understanding of how they work, followed by waves of refinements until something different, but better, arrives on the scene. Thus, the steam engine was eventually replaced with the combustion engine. What's next?

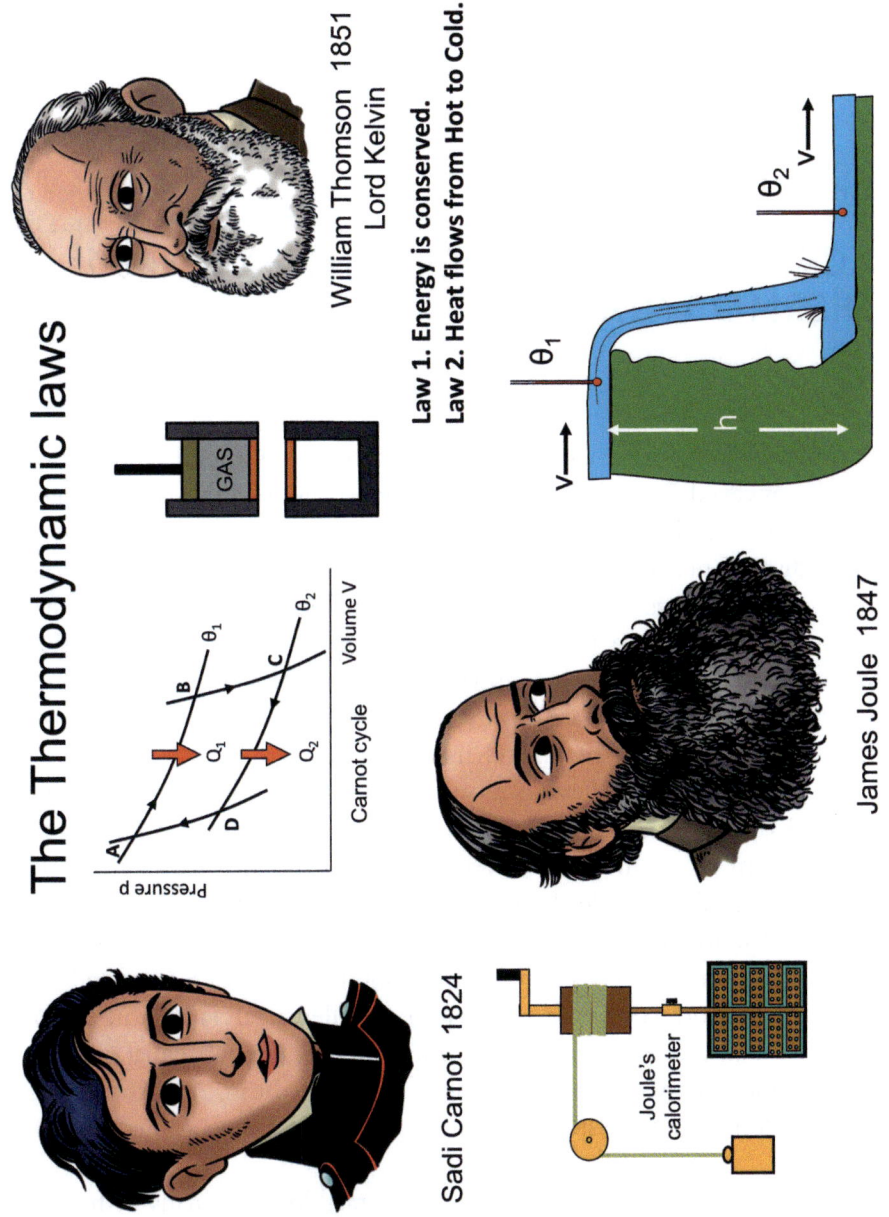

CHAPTER 30

# ACOUSTICS, DIFFUSION AND EXTENDED THERMODYNAMICS

Any disturbance within a mass medium, whether it is a solid, a liquid or a gas, is conveyed around by waves. The denser the medium is the greater is the wave speed.

A disturbance in the ether is conveyed around by waves moving at the speed of light, suggesting that the ether, apparently empty, is extremely dense. Whatever form mass takes it exists in the ether, suggesting that wave motions are always accompanied by, generally unobserved, ether waves. Such ether waves are gravity waves. Maybe electromagnetic waves in the ether only occur because charge has mass.

A common form of wave motion is the longitudinal compressible wave. A disturbance in a 1-D compressible medium causes all points of the medium to expand and contract as the disturbance passes through, setting up oscillations, leading to a train of waves. Although the waves move in the longitudinal direction, they are often depicted in books in a transverse form. The wave peak corresponds to a compression and the wave trough corresponds to an expansion (called a rarefaction in a gas). Each wave has kinetic energy due to its vibration and an equal amount of potential energy due to the elasticity of the medium. Acoustics is the study of longitudinal disturbances in media which move at the speed of sound $c_s$.

A long, solid bar can be made to vibrate longitudinally either by striking the bar or by stroking it. The disturbance moves through the bar at the speed of sound $c_s$. At any point within the bar, the disturbance causes matter there to be displaced by a tiny amount x. Elasticity provides the restoring force to try to bring the matter back to its original position, but over-shooting in the process, leading to a vibration. From thermodynamics we know that a perfect exchange between the kinetic energy and the potential energy of a wave is impossible, due to frictional heating during the expansions and contractions of the material, so the vibration will soon come to a stop.

Suppose a continuous source of vibration is applied to one end of the solid bar, creating a continuous series of internal 1-D waves moving through the bar. The speed of the waves is $c_s$, being that of sound. The wave speed $c_s = f\lambda$, where f is the frequency of vibration and $\lambda$ is the wavelength. The 1-D wave equation of the matter displacements in the solid bar is

$$\frac{\partial^2 \xi}{\partial x^2} = \frac{1}{c_s^2} \frac{\partial^2 \xi}{\partial t^2}$$

The wave speed

$$c_s = \sqrt{\frac{E}{\rho}}$$

where E = Young's modulus and $\rho$ = bar density.

If the bar is not of infinite length, then we must apply boundary conditions at the ends. Suppose that the internal waves are reflected backwards and forwards from each end of the bar. Then, for certain wavelengths equal to an integer number times $\lambda/2$, resonance ensues and standing waves form in the bar. At the antinodes there is a maximum region of vibration, while at the nodes there is almost no vibration. Vibrating matter is linked with the production of heat. A thermal camera view of the bar would show hot spots coinciding with the antinodes. This phenomenon is an exact analogue of a laser light beam shone into a bar of glass, where mirrors at each end reflect the beam backwards and forwards. Again, for the right conditions, standing waves are set up in the glass bar with bright spots for the antinodes and black regions around the nodes.

We can imagine the internal structure of a solid bar to be a 3-D lattice network with point masses representing atoms at the intersections. A sharp disturbance would cause the masses to move with the elasticity of the lattice trying to suppress the movement, resulting in mass vibration. Local gravitational changes would also lead to bar vibration. During the 1970s, Professor Joseph Weber of the University of Maryland in the USA used a cylindrical bar to search for vibrations caused by gravity waves from outer space striking the bar (see Chapter 36).

In his book *Principia*, Newton described how, from the echo of a hand clap, he used a simple pendulum to measure the speed of sound $c_s$ in air. In his experiment, he found that $c_s$ was 300 m/s. Modern measuring equipment shows $c_s$ = 333 m/s. Newton modelled the sound wave in air as a compression wave. This model gave the speed of sound in air as $c_s = \sqrt{(p/\rho)}$, where p is the air pressure and $\rho$ the air density. Newton's predicted value for $c_s$ was 15% too low. As was pointed out by the French scientist, Pierre Laplace, 150 years later, Newton hadn't allowed for the frictional heating of the air during compression and cooling of the air during rarefaction. When you pump up your bicycle tyres, the compressed air becomes hot. Allowing for this effect in the model of the sound wave, $c_s = \sqrt{(B/\rho)}$, where B is called the bulk modulus of the air. $B = \gamma_c p$, where $\gamma_c$ is a function of the specific heat of air.

The more compact the medium, the greater the speed of the disturbance through it. For water $c_s$ = 1400 m/s, while for a copper bar $c_s$ = 3800 m/s.

Another type of wave associated with vibration is the transverse wave, where the vibration is perpendicular to the direction in which the wave moves. For example, the waves formed on a vibrating stretched string are transverse. Another example of transverse wave vibration is that of surface water waves created by throwing a stone into a calm pond. The waves arise because of surface tension. The disturbed water level rises up and down in a perpendicular direction to that of the direction of the wave motion. The surface ripples travel at speed v away from the source of the disturbance, where the stone enters the water. Each ripple, or wave, has a peak followed by a trough. If the length of one wave is $\lambda$ then $v = f\lambda$, where f is the wave frequency.

The electromagnetic wave is a disturbance in the ether. If we accept that the ether can be squeezed and expanded, we might treat the electromagnetic wave as a longitudinal compression wave moving through the ether. But the electromagnetic wave is subject to polarisation, suggesting the existence of a transverse wave. A simple view of a polariser is to think of it as a length of railing. A skipping rope held between two vertical railings can vibrate in the vertical direction but not in the horizontal direction. A light polariser works in this way, where the polarising film acts like the railings, allowing light with electric field vibrations in a vertical plane to pass through, but blocking other planes of vibration.

In fact, the electromagnetic wave comprises an orthogonal pair of transverse waves, with an electric **E** wave and a magnetic **H** wave oscillating in planes perpendicular to each other and travelling in the direction of the beam. Such waves are transverse waves. The energy in the combined waves moves longitudinally at the speed of light c, where $c = 1/\sqrt{\varepsilon\mu}$.

$$\frac{\partial^2 \mathbf{E}}{\partial x^2} = \frac{1}{c^2}\frac{\partial^2 \mathbf{E}}{\partial t^2} \quad \text{and} \quad \frac{\partial^2 \mathbf{H}}{\partial y^2} = \frac{1}{c^2}\frac{\partial^2 \mathbf{H}}{\partial t^2} \quad \text{with the beam in the z-direction.}$$

A source of electromagnetic waves in the microwave region is the magnetron. This device can be viewed as the analogue of an acoustic siren. The siren produces sound waves in air, but does it also produce hitherto unobservable, ether waves? Detecting such waves would require more sensitive technology than that employed to detect gravity waves.

The maser amplifies weak microwave signals by combining them with a source of stimulated microwaves. It was the forerunner of the laser. The laser provides coherent (all in phase) light waves and now forms part of many electronic devices. The acoustic analogue of the laser is called a saser (sound amplification by the stimulated emission of radiation). During the late 1990s British Aerospace scientists at the BAe Sowerby Research Centre investigated work done earlier by Soviet scientists on developing a saser. Research work continues elsewhere, but a useful application for the saser has yet to be found.

What these analogues, or forms, show is how often nature's patterns repeat themselves. As Sir Francis Bacon realised, the studying of forms could provide clues of what effects to look for when exploring new phenomena. The possession of a clue about an effect might suggest the need for a new detection system to reveal the effect. This approach is very useful in developing new areas of research.

In the acoustic case, the wave in a mass medium is a 1-D compression wave, with a longitudinal direction of vibration, moving at speed $c_s$. It is assumed that it is not possible to polarise a sound wave, so no transverse planes of vibration are involved. This view might be challenged at some time.

In the thermal case, temperature fluctuations at one end of a solid bar result in temperature disturbances moving through the bar. But the disturbances are heavily suppressed, so that the speed $v_D$ at which the temperature changes move along the bar is much less than $c_s$.

Given an isolated solid which is heated and left to settle, the heat diffuses throughout the body until a uniform temperature is reached.

# GRAVITOMAGNETISM

Consider a 1-D solid bar of mass M, of length x and cross-sectional area A. Then M = ρAx, where ρ is the density, assumed constant. If the bar is at a uniform temperature θ, then the heat Q contained in the bar is Q = (ρAx)cθ, where c is the specific heat.

Heat diffusion is a process analogous to the diffusion that occurs between two different gases, or two liquids with different concentrations c (not to be confused with specific heat), which are in contact across a surface area A. When left to settle, Q (kg) amounts of gas, or liquid, diffuse, or move, across the area A until, eventually, the gas mixture, or the liquid concentration, on both sides of A reach the same value.

In 1856, the German scientist Adolf Fick based his diffusion law for liquids, during the settling period, on the analogue with Fourier's equation. For 1-D, where the gradient of concentration across the area A is dc/dx over a short length, his law is

$$\frac{dQ}{dt} = -kA\frac{dc}{dx}$$

Here, Q is the amount of liquid concentrate crossing the area A, analogous to the amount of heat crossing A. The diffusion coefficient k is analogous to the thermal conductivity k. Fick's law also applies to two different gases in contact across an area A.

Like Ohm's law, based on the analogue with Fourier's equation, the analogue is not exact. In the thermal case, energy is transported; in the electrical case, charge is transported and in the fluid diffusion case, mass is transported. But, at the end of the day, the analogues provide ideas for experiments. All three laws have been verified by experiment.

We will now look at an extension of Fourier's equation for heat flow, which is based on an analogy with electromagnetism. For those with a mathematical taste, an equation is like a picture. It can stimulate the imagination and say so much. However, I appreciate that, to some people, mathematical symbols are just like Egyptian hieroglyphs and are almost meaningless. For those people not mathematically inclined (which included Michael Faraday – a brilliant experimenter), I suggest that you skim-read, or skip, to the last paragraph of this chapter.

Let our 1-D body, or bar, in the x-direction, be subjected to a temperature gradient T. The heat Q reduces with distance x. For a short length dx, the mass dM = ρAdx. Assuming the temperature remains constant in this short length, the drop in heat dQ = – (ρAdx) cθ so that

$$\frac{dQ}{dx} = -\rho A c \theta$$

Differentiating Fourier's heat conduction equation with respect to x gives

$$\frac{d}{dt}\left(\frac{dQ}{dx}\right) = -kA\frac{d^2\theta}{dx^2}$$

Substituting for dQ/dx and rearranging the terms, we get

$$\frac{d^2\theta}{dx^2} = \frac{\rho c}{k}\frac{d\theta}{dt}$$

This is called the heat diffusion equation. It applies equally to gas and liquid diffusion and to the diffusion of vorticity in a gas and a liquid. We expect there is an analogue with gravity and electromagnetism, too.

For steady conditions

$$\frac{d\theta}{dt} = 0;$$

and we have

$$\frac{d^2\theta}{dx^2} = 0.$$

This is the thermal form of Laplace's equation for potential theory in 1-D. In vector operator form in 3-D it is $\nabla^2\theta = 0$. For vector operators the divergence of the gradient is always zero, so that $\nabla^2\theta \equiv \nabla \bullet \nabla\theta = 0$. This means that for steady thermal conditions we can introduce a field vector $\mathbf{T} = -\nabla\theta$ to represent the temperature gradient in 3-D.

Given initial starting conditions for temperature θ at a position x (usually x = 0) and time t (usually t = 0), the change of temperature θ at any point x along the bar for any time t can be calculated. The diffusion equation works perfectly well, but there is a flaw. If the temperature θ at some point is changed abruptly, the information about the change is broadcast instantaneously (in zero time) to all parts of the bar. This is not realistic and defies the idea of relativity. Newton's model of gravity suffers from the same problem. We expect a disturbance in a medium to be distributed in some form of wave motion, which takes time to travel.

Differentiating the heat diffusion equation with respect to x, we get the thermal field version, where T = dθ/dx. To show that T is a function of time and position, we can use the curly symbol ∂ for differentiation.

$$\frac{\partial^2 T}{\partial x^2} = \frac{\rho c}{k} \frac{\partial T}{\partial t}$$

The flaw in the heat diffusion equation interested James Maxwell. From his model of electromagnetism he had predicted that changes in the electric and magnetic fields in the ether were conveyed by waves (undetected at the time) moving at the finite speed of light. Maxwell was the master of analogues and he suspected that something similar to the electromagnetic case happened in the heat diffusion case, too.

Suppose that in the x-direction we have a bar of heat conducting material at a temperature θ. We can replace it in our imagination with a line of longitudinally vibrating atoms, each atom acting as a heat source. For a bar of uniform temperature, the strength H is the same for all sources. If we create a temperature gradient T = dθ/dx then in the higher temperature region of the bar, the atoms will be vibrating more quickly than those atoms in the lower temperature region.

With our mental picture, we can visualise that heat sources in a bar under a temperature gradient **T** would, if they were free, move in the x-direction with velocity **v**. Each heat source H experiences a force **F** = H**T**. For a bar at uniform temperature, the centres of the heat sources remain fixed in position but vibrate (expanding and contracting longitudinally), giving rise to a term $\partial \mathbf{T}/\partial t$. Of course, for each source there is the possibility of a combination of both movements.

We can think of the thermal gradient **T** as the analogue of the electric field **E**. With Maxwell to guide us, we speculate that there exists a phenomenon **R** (the analogue of magnetism **H**) such that for each source

$$\nabla \times \mathbf{R} = \rho_H \mathbf{v} + k \frac{\partial \mathbf{T}}{\partial t} \quad \text{where the unit for } \mathbf{R} \text{ is } \frac{\text{cal}}{\text{m.s}^2} \text{ or } \frac{\text{kg.m}}{\text{s}^4}.$$

The first term on the right is associated with a moving heat source (analogue of moving electric charge), and the second term is associated with a vibrating heat source (atomic vibration is analogue of displacement).

Based on its unit, the **R** field appears to be related to jounce, the rate of change of the rate of change of acceleration, hinting at a link with gravity. The temperature gradient is **T**, the thermal conductivity is k, the velocity of heat sources (and their inherent masses) is **v** and the density of heat sources is $\rho_H$.

For a constant thermal gradient **T**, the equation $\nabla \times \mathbf{R} = \rho_H \mathbf{v}$ is the thermal analogue of Oersted's discovery of the magnetic field around a wire carrying a steady current of electric charges. However, in the thermal case, heat sources are, in general, not free to move, so $\mathbf{v} = 0$ and the $\mathbf{R}$ field does not arise.

For divergence, the analogues for electricity and gravitation are

$$\nabla \bullet \mathbf{E} = \frac{\rho_e}{\varepsilon} \text{ and } \nabla \bullet \mathbf{g} = -\frac{\rho}{\gamma}$$

Thus, for the thermal case we expect

$$\nabla \bullet \mathbf{T} = \frac{\rho_H}{k}$$

The heat source strength is

$$H = \frac{Q}{t} \text{ and the heat source density } \rho_H = \frac{H}{\text{Volume}}.$$

So, in 1-D the thermal divergence equation reduces to Fourier's equation.

For the magnetic case, we have $\nabla \bullet \mathbf{H} = 0$ (magnetic poles occur in pairs). So, in the thermal case, the analogue suggests that the sources of the $\mathbf{R}$ field will occur in pairs, too, so that $\nabla \bullet \mathbf{R} = 0$. This guess also suggests that the $\mathbf{R}$ field, whatever it is, is related to spin.

Suppose a source of $\mathbf{R}$ is labelled q. Then, since $\mathbf{R}$ is a force field, we expect

Force $\mathbf{F} = q\mathbf{R}$    The unit of q is K.s or $s^2$ in modified units.

We assume that the heat sources are fixed, so the $\mathbf{R}$ field does not arise. However, for thermal displacement effects (vibration or a varying heat signal), temporal values of $\mathbf{R}$ can occur.

Guided by Maxwell, we speculate that

$$\nabla \times \mathbf{T} = -p\frac{\partial \mathbf{R}}{\partial t} \quad \text{where the unit for p is } \frac{K.s^3}{cal.m} \text{ or } \frac{K.s^5}{kg.m^3}.$$

The term $p\frac{\partial \mathbf{R}}{\partial t}$ relates to thermal gradiometry $\frac{T}{m}$.

It is the analogue of the important subject of gravity gradiometry.

The parameter p (not to be confused with pressure) is the permeability of the solid to the phenomenon $\mathbf{R}$. In modified units, the temperature θ K has a dimension of time (unit s).

We assume that heat sources don't flow ($\mathbf{v} = 0$). Then from the two curl equations, $\nabla \times \mathbf{T}$ and $\nabla \times \mathbf{R}$, we can eliminate either $\mathbf{T}$ or $\mathbf{R}$ to derive 3-D equations just involving T or R. For 1-D, in the x-direction, the equation satisfied by the $\mathbf{T}$ field is

$$\frac{\partial^2 T}{\partial x^2} = \frac{\rho c}{k}\frac{\partial T}{\partial t} + pk\frac{\partial^2 T}{\partial t^2}$$

This looks like the heat diffusion equation with an added term. (Remember, $\rho$ is density, c is specific heat, k is thermal conductivity and p is the permeability of the solid to the field $\mathbf{R}$.) A check with electromagnetism shows that we have derived a thermal form of Maxwell's equation of telegraphy.

There will be a similar equation for R. Introducing the $\mathbf{R}$ field is a mathematical contrivance, but does the $\mathbf{R}$ field really exist? Given nature's penchant for analogues, it probably does. But we are unaware of the presence of the $\mathbf{R}$ field and certainly have no idea how to detect it.

The heat source conductivity $\sigma_H$ is given by

$$\sigma_H = \frac{\rho c}{pk} = \rho c c_s^2 \quad \text{where } c_s = \frac{1}{\sqrt{pk}} = \text{speed of disturbance in the material.}$$

For high frequency vibrations (hot atoms), $\sigma_H$ is the dominating factor and we get the equation of diffusion for a heat source. If we ignore the effect of diffusion (put $\sigma_H = 0$), we are left with a wave equation for T, where a vibration of T passes through the bar at the speed of sound $c_s$.

$$\frac{\partial^2 T}{\partial x^2} = \frac{1}{c_s^2}\frac{\partial^2 T}{\partial t^2}$$

We can compare this equation with the equation for the vibrations in a long, solid bar and it gives us a slightly different view of acoustics. It tells us that during the displacement x, the vibrating source, undergoes a temperature gradient T. With this view, acoustics may be treated as a pair of transverse $\mathbf{T}$-$\mathbf{R}$ waves travelling through the medium with virtually no loss of energy. Thus, the acoustic compression wave may be viewed as the result of a transverse $\mathbf{T}$ wave.

For solids, $c_s = \sqrt{E/\rho}$, so knowing k we can evaluate p, the permeability of the medium to the $\mathbf{R}$ field. For copper, p is of the order $1.72 \times 10^{-10}$.

Ignoring the acoustic waves, we are left with the heat diffusion equation. Heat is associated with short-wavelength atomic vibrations of very high frequency, but the thermal influence of a hot atom (heat source) is severely restricted to a length $\pm\delta$, called the diffusion length, either side of the atom. To solve the 1-D heat diffusion equation for a hot atom at some point within the bar, we assume that the temperature there is $\theta_0$ at time $t = 0$ and that the vibration of the atomic mass occurs with angular frequency $\omega = 2\pi f$. The vibration signal is rapidly attenuated as the distance from the atomic centre increases. The diffusion length $\pm\delta$ is where the amplitude of the vibration approaches zero. The speed at which the front of the heat influence spreads out from the hot atom is $v_D$.

$$\text{diffusion length } \delta = \sqrt{\frac{2k}{\omega\rho c}} \quad \text{and} \quad \text{diffusion front speed } v_D = \sqrt{\frac{2\omega k}{\rho c}} = \omega\delta.$$

Consider copper at 0°C with a thermal frequency of $10^6$ Hz.

| k | $\rho$ | c | $\delta$ | $v_D$ | E | $c_s$ |
|---|---|---|---|---|---|---|
| j/s.m.K | kg/m³ | j/kg.K | m | m/s | N/m² | m/s |
| 420 | 8954 | 386.4 | $6.23 \times 10^{-6}$ | 39.14 | $12.8 \times 10^{10}$ | $3.8 \times 10^3$ |

The electromagnetic analogue leads to the observation that for alternating current passing through a conductor, the movement of the vibrating charges is confined to a thin outer layer, known as the skin depth.

The usual model for acoustic (compression) waves is to assume longitudinal linear displacement of mass points (atoms in lattice). However, thanks to Maxwell's earlier insight, the acoustic waves in a solid medium can also be viewed as transverse **T-R** waves. So, acoustic waves may be polarisable, but the plane of polarisation is that containing the **T** wave, not the longitudinal direction. There's still the problem of how we might make an acoustic linear polariser. Following the light model, a thin screen with vertical lines of high conductivity might suffice. Most importantly, are the **T-R** waves strong enough to appear outside the confines of the medium containing the compression wave? According to the model (for a medium of infinite length), the acoustic energy is radiated away in a radial direction. The rate at which the acoustic energy crosses an area A is $(\mathbf{T} \times \mathbf{R}).A$. This is the analogue of the Poynting vector in electromagnetic radiation.

The purpose of the examination of the full thermal equation (diffusion plus acoustics) is to show that the analogue of the model for the electromagnetic field can be used to develop a field model for thermodynamics, suggesting the existence of a new dual force field **R**, currently unobserved, associated with the movement of heat sources.

Whether the model has any validity depends on the results of experiments to test predicted effects. For **T** this has been done, but an outstanding question remains about the existence of **R**.

Waves passing through a solid interact with the atomic lattice structure. Similarly, light waves passing through glass interact with the atomic lattice structure. Both are forms of conduction through a solid.

There is a duality between (continuous) waves and (discrete) particles. Thus, electromagnetic waves can be thought of in terms of streams of photons. These are quantum particles with discrete energy $E = hf$, where h is Planck's constant (the smallest discrete unit of angular momentum) and f is the electromagnetic wave frequency. Sometimes the symbol $\hbar$, called h-bar, is used, where $\hbar = h/2\pi$. The photon has a property called spin s with magnitude $s = \pm\frac{1}{2}\hbar$ (called spin $\pm 1$). So, a photon transports angular momentum as well as energy.

By analogy, we can extend the quantum approach to atomic and molecular vibrations in the lattice structure of a solid and replace the waves with phonons, each with energy $E = hf$. Phonons are also spin 1 particles. The vibrational energy of an atom can be represented by a swathe of phonons with a range of frequencies. Heat conduction is interpreted as the exchange of phonons between contiguous atoms, with the direction of the exchange dependent on the atomic energy levels.

In 1872, Ernst Mach (Mach 1 is the speed of sound in air) wrote *If then we are astonished at the discovery that heat is motion, we are astonished at something which has never been discovered*. What I think Mach was pointing out was that heat flow is not the flow of heat sources. But, now, we can model heat flow as the flow of energy particles called phonons. Thus, phonons are the modern-day equivalent of caloric particles and the analogue with fluids remains valid.

An acoustic wave in a solid bar can be thought of as a stream of phonons. Acoustic phonons are low frequency. As already noted, resonance arises when the length of the bar corresponds to an integer times $\lambda/2$. These are long wavelengths. By contrast, thermal phonons are associated with atomic vibrations, so their wavelengths associated with atomic dimensions are incredibly short. From the wave formula $c_s = f\lambda$, this means that thermal phonons have incredibly high frequencies.

Einstein's famous mass-energy formula $E = mc^2$ was derived from examining the energy in a photon beam (light ray). Einstein assumed that matter was frozen energy and extended his formula, without proof, to all matter. Subsequent experiments have proved him right. Theoretically, photons are massless, but since $E = hf$ and $E = mc^2$ they have an effective mass given by $m = hf/c^2$. The thermal analogue of the photon has led us to the phonon, which has energy $E = hf$, too. Following the Einstein approach, we might argue that examining the energy in a phonon beam leads to the mass-heat-energy formula $E = mc_s^2$, so that the effective mass of a phonon is $m = hf/c_s^2$. However, many scientists dispute this result and feel that we should still use $E = mc^2$ for the phonon, where c is the speed of light in a vacuum. However, it may be that $c_s = c$ within matter, although the value of c is not the speed of light in a vacuum.

The phonon model shows that low-frequency acoustic phonons have very little effective mass and, therefore, little inertia. Consequently, they suffer very little resistance to motion and, therefore, virtually no loss of energy. Thermal phonons, with high frequency, have greater effective mass and greater inertia, thus restricting their movement between the atoms and molecules of the lattice structure of the solid. This roughly explains why acoustic effects are relatively fast, while heat conduction (diffusion) is a relatively slow process. This suggests to me that acoustics is associated with gravity, while heat transfer is associated with electromagnetism. This picture is slightly modified with metal bars, where the outer orbital electrons of atoms can escape and convey heat energy, too.

The phonon model provides us with a different view of 1-D heat conduction. In Einstein's gravitational red shift, photons moving upward in a gravity field change frequency and potential with distance. The thermal analogue is that a phonon moving in a temperature gradient **T** changes its frequency and its temperature (potential) as it does so.

From Maxwell's analogue of electromagnetism we have developed an extended model of thermodynamics. Later, we will use the same analogue approach to extend Newton's model of gravity. The special case of a flow of heat sources is also a general case of a mass current, since all heat sources have inherent mass. We will devote a separate chapter to extending Newton's static model of gravity to look at the effect of mass movement.

# Acoustics, diffusion and extended thermodynamics

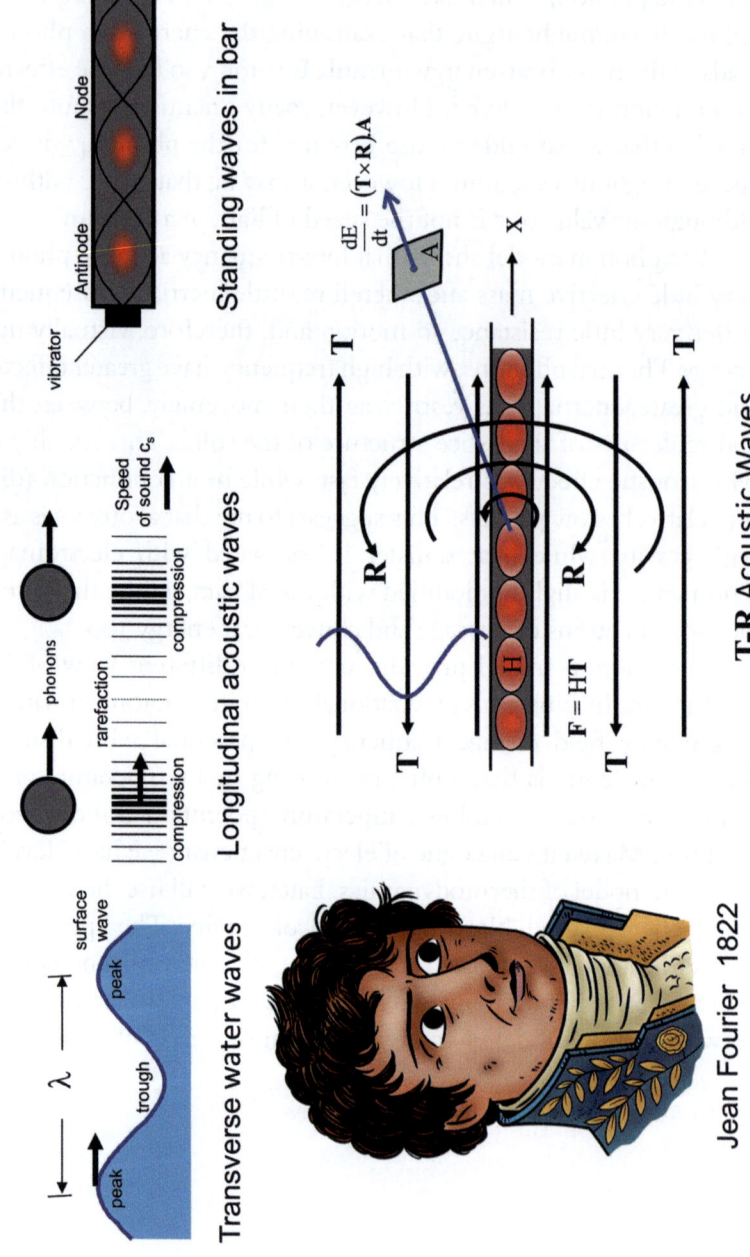

CHAPTER 31

# STARS, BLACK HOLES AND THERMODYNAMICS

The star's mass is a source of gravitational energy and its co-located heat source is a source of radiant energy. Newton queried whether gravitational matter and radiant heat might be compatible. Faraday also wondered whether a link existed between gravity and heat. Several centuries later, Einstein confirmed such a relationship existed via their energies with his $E = mc^2$. So, by virtue of its energy content, a heat source bends space-time, just like mass.

Stars form from vast clouds of hydrogen atoms. Gradually, atomic gravitational forces cause the hydrogen atoms to begin to coalesce, forming a compressed central region of atoms. During this inward movement, friction between atoms takes place and the central region begins to heat up until eventually the temperature is such that hydrogen atoms begin to fuse together to form helium atoms, radiating away heat and light in the process. A new star is born. The outward radiation pressure initially holds the inward gravitational force at bay.

Once all the hydrogen has been burnt, the star starts to burn its helium and temperatures further increase, leading to further fusion of atoms forming lithium, carbon, oxygen and other elements. Gravitational forces overcome radiation pressure and the star starts to shrink, with the core reaching a temperature of 4 billion degrees K, when iron and nickel start to form. From our earlier speculation, the inward gravity field may be enhanced by the outward temperature gradient. At this stage, further reactions taking place within the star depend on the strength of its gravitational field, which depends on its mass, M. Comparisons are made with the mass of the Sun, $M_s$. If $M < 1.4M_s$ then the star contracts to become a white dwarf. In each atom, the electric force of the electrons is just sufficient to oppose further gravitational contraction. If $1.4 M_s < M < 3M_s$, further gravitational contraction causes a further rise in core temperature resulting in a supernova explosion. The Universe is showered with elements from the star's outer mantle along with heavier elements formed during the explosion. And living beings are made of these elements! After the explosion, the gravitational force within the inner mantle is strong enough to overcome the electric force and push most of the electrons into the protons, leaving a core mostly composed of neutrons. The star has become a neutron star. The surface of the neutron star still contains some protons and electrons, so it can become charged.

A spinning charged sphere develops a dipole magnetic field. By analogy, we expect a spinning sphere of mass to develop a dipole gravitomagnetic field (Chapter 33). If the spinning mass is charged then we expect a combined magnetic and gravitomagnetic field to be created. Scaling things up to the size of the Earth, we know that it has a magnetic field, and a recent experiment by NASA (Chapter 35) has confirmed that it also has an extremely weak gravitomagnetic field.

Scaling things up considerably further, we expect a spinning neutron star to develop a strong dipole gravitomagnetic field. If the spinning neutron star is also charged then a strong magnetic field, combined with a gravitomagnetic field, is created. The combined field is strongest at the poles. Moving charged particles of matter in the polar regions would spiral away vertically from the neutron star. Astronomers have detected jets of matter emanating from the poles of neutron stars. If the neutron star wobbles about its axis (precesses), the combined gravitomagnetic and magnetic fields will rotate in space, creating an axial directed radio signal. An observer on Earth would detect a pulsating radio signal concomitant with the precession, rather like observing the flash from a lighthouse as the lamp rotates. Such rotating neutron stars are called pulsars.

If $M > 3M_s$, then gravity becomes even stronger and a black hole forms, made of supremely dense matter of unknown form. For example, suppose two neutron stars coalesce, then even the neutrons are squashed out of existence. Mathematically, a black hole occupies a single point in space.

Black holes are the largest sources of gravity in the Universe and are now assumed to reside at the centres of all galaxies. Professor John Michell (the inventor of the gravity torsion balance) of Cambridge University was the first scientist to predict the existence of black holes or, as he called them, dark stars. In a paper read at the Royal Society in 1783, Michell said that massive stars might exist which generated enough gravity to prevent corpuscles of light (photons) escaping from them so that they would be invisible. Completely independently, the French mathematician Pierre Laplace predicted the existence of invisible stars in 1796. We now call dark, or invisible, stars black holes, a term introduced by Professor John Wheeler of Princeton University in 1960.

On Thursday, 11 April 2019, many newspapers throughout the world carried a picture of the black hole at the centre of the galaxy M87, which is 55 million light years away from Earth. Eight telescopes around the world, working in unison with atomic clock accuracy, formed a large radio telescope array to obtain an image of the black hole or, rather, its periphery, called the event horizon. It is estimated that the mass of the black hole is $6.5 \times 10^9 \times M_S$. And the diameter of the black hole is $40 \times 10^9$ km. That's some point singularity! The mind just boggles at the enormous stupendous gigantic power of gravity of the black hole, a force over which we have no control.

During the editing phase of this book (October 2020) it was announced that Sir Roger Penrose, of Oxford University, was a joint winner of the Nobel Prize in Physics for his work proving that the existence of black holes formed by some collapsing stars stemmed from Einstein's theory of general relativity. The other winners were Professor Andrea Ghez, of UCLA in the US, and Professor Reinhard Genzel, of the Max Planck Institute in Germany.

For a mass to escape from the surface of a planet of mass M and radius r, its upward acceleration must be greater than its downward acceleration due to the pull of gravity. This leads to the escape velocity requirement

$$v > \sqrt{2gr} \quad \text{where } g = \frac{GM}{r^2}$$

On Earth's surface, $g = 9.81$ m/s² and $r = R_E = 6370 \times 10^3$ m, giving an escape velocity $v = 11180$ m/s.

For photons, moving with velocity c, to escape from the surface of the planet requires

$$c > \sqrt{2gr} \quad \text{so that for } r = \frac{c^2}{2g} \text{ 1}$$

the photons would just hover, not going up or down.

The spherical surface of radius r where the photons hover is called the event horizon. For all planets and stars, the event horizon is below the surface so light can escape from them, but not for black holes.

For modelling purposes, a black hole is assumed to be a point singularity surrounded by an external event horizon. The radius $r_s$ of the event horizon surrounding a black hole is called the Schwarzschild radius, named after the German scientist Karl Schwarzschild, who first drew attention to it in 1916. Photons leaving the black hole are trapped on the surface of the event horizon, so that the black hole is invisible from outside.

If the mass of the black Hole is M then another form for $r_s$ is

$$r_s = \frac{2GM}{c^2}.$$

In 1970, Jacob Bekenstein at Princeton suggested that there might be a link between thermodynamics and black hole physics, via entropy S. Bekenstein's suggestion was based on the idea that entropy describes the orderliness of a system. A low entropy system is more ordered than a high entropy system. Particles of mass m falling into a black hole make it more disordered. At first sight, the masses just disappear, but their arrival causes the surface of the event horizon to grow. So, there is a link between the total mass of the black hole and the surface of its event horizon. Bekenstein reasoned that the surface of the event horizon must be a depository of the information about the total mass below. In other words, the surface of the event horizon described the entropy of the region below, or the entropy of the black hole.

The surface area A of the event horizon is

$$A = 4\pi r_s^2$$

Bekenstein made use of the smallest quantum area $\Delta A = L^2$, where L is the Planck length, and argued that it contained one unit of information.

$$L = \sqrt{\frac{\hbar G}{c^3}} = 1.6 \times 10^{-35} \, m$$

Consequently, the number N of units of information on the surface A is

$$N = \frac{A}{\Delta A} = \frac{4\pi M c}{\hbar}$$

Suppose the mass M is dispersed into N unit masses m equal to M/N, each residing on a Planck area of the event horizon. The potential energy of each unit mass is $mgr_s$, where g is the gravity on the event horizon. We now make the assumption that if the unit mass fell towards the singularity, its energy would be $\frac{1}{2}k_B\theta$ (there may be a numerical factor missing here) so that the individual temperature of each unit mass m is

$$\theta = 2\frac{mgr_s}{k_B} = 2\frac{M}{N}\frac{gr_s}{k_B} = \frac{\hbar g}{4\pi c k_B}$$

This is the temperature of each unit mass m and, therefore, the temperature of the total mass M at the singularity. The temperature of a typical black hole is not much different from absolute zero K and can't be detected in the 3 K temperature of the background cosmic radiation left over from the Big Bang.

We use the modified expression for entropy, where heat Q is replaced with energy Q. Since the energy of the black hole is equal to $Mc^2$, the entropy of the system enclosed by the event horizon is

$$S = \frac{Mc^2}{\theta} = Mc^2 \frac{4\pi c k_B}{\hbar g} = \frac{Ac^3 k_B}{\hbar G} \quad \text{where } A = 4\pi r_s^2$$

A mass can pass through the surface of the event horizon of the black hole from the outside, but once inside the event horizon, it cannot escape from the gravitational clutch of the singularity into which it falls.

Originally, the entropy S of a black hole was thought to be constant, or to increase if further mass fell in. But then Professor Stephen Hawking considered the possibility that at the event horizon, some information about the conditions below might leak out. His solution linked classical gravitational field theory with the quantum vacuum.

Hawking introduced the idea of the quantum ether, where virtual particle pairs, one with positive energy and the other with negative energy, can spontaneously appear and disappear at any point in space, with the proviso that their time of existence satisfies Heisenberg's uncertainty principle.

Hawking showed that photons could escape from the black hole if one assumed that anti-particle pairs of virtual photons were created at the event horizon, with one of the photons being absorbed by the black hole and the other photon radiating away into space. In effect, it meant that the quantum nature of the vacuum allowed the fabric of space to be torn apart across the event horizon interface. Another view was that a positive energy virtual photon could quantum tunnel its way through the event horizon barrier and escape.

Each escaping virtual photon brought with it information about conditions below the event horizon. Furthermore, since it could not annihilate with its anti-particle pair, it became a real photon with a positive effective mass. This energy radiated away from a black hole is now called Hawking radiation.

Anti-photons swallowed by the black hole must have effective negative masses. So, in absorbing these anti-photons, the black hole must lose some mass and the negative energy stored in its gravitational field is also reduced. This phenomenon is termed black hole evaporation, but the rate of mass loss is virtually negligible in terms of the Universe's calculated age of 13.7 billion years.

Some scientists argue that Hawking radiation is the first step along the road to a theory of quantum gravity in the same way that Planck's quanta of radiation was the first step in the road leading to the theory of quantum mechanics. As yet, we don't know whether the $2^{nd}$ law of Thermodynamics applies to quantum mechanics under all circumstances. We are at the frontier of research for this subject.

In 1995, Professor Ted Jacobson, a theoretical physicist at the University of Maryland, showed that Einstein's field equations for gravity could be obtained from the theory of Thermodynamics. As far as I know, this theoretical study has not led to an understanding of how to control gravity.

While Project Greenglow was running, I met Dr Pharis Williams, a US Navy veteran, who had developed a 5-D theory combining gravitation with thermodynamics. But, again, as far as I know, this work has not led to a method of controlling gravity.

Many scientists believe that gravity and thermodynamics are closely linked. Investigating the linear link between thermodynamics and gravity via analogues is yet another, but more simplistic, approach in the continuous search for a means of controlling gravity.

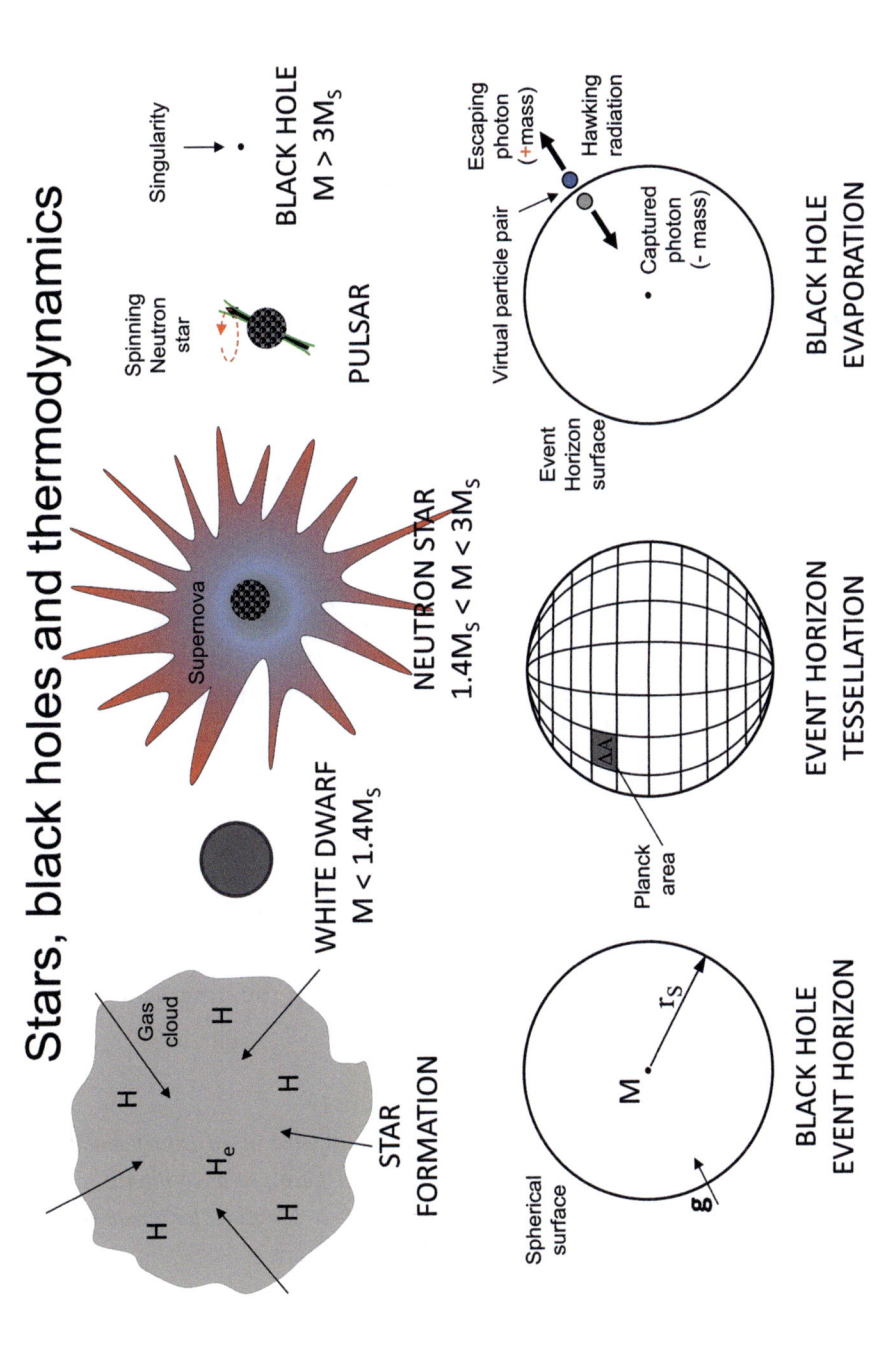

CHAPTER 32

# FARADAY'S GRAVITY EXPERIMENTS

Faraday began to formulate his gravity research programme in March 1849. Although it was a daunting task, since gravity then, as now, appears to be unchangeable, he was probably buoyed up by the success that he'd had in unravelling the phenomenon of electromagnetism in 1831.

In 1847, Helmholtz had published his conservation of force (linking work and energy) theory and, no doubt, Faraday was aware of it. Indeed, it fitted in with Faraday's view that one force could be turned into another. For this reason, Faraday felt that gravity was not only linked with the other natural forces known at that time, but that gravity could evolve into another force, and his guess was that electricity was such a force.

By today's standards, Faraday's gravity experiments seem rather crude and, ultimately, he was not successful in his endeavour. But his gravity experiments and his speculative thoughts behind them, noted as numbered paragraphs in his laboratory diary, are worth reading if we are to make any progress in understanding the enigmatic gravity phenomenon.

*Monday, 19th March 1849*
*(10018): Gravity. Surely this force must be capable of an experimental relation to Electricity, Magnetism and the other forces, so as to bind it up with them in reciprocal action and equivalent effect. Consider for a moment how to set about touching this matter by facts and trial.*

Faraday wondered whether gravity operated in unison with another force field, in the same way that electricity operates with magnetism.

*(10019): What in Gravity answers to the dual or antithetical nature of the forms of force in Electricity and Magnetism? Perhaps the to and fro, that is, the ceding to the force or approach of Gravitating bodies, and the effectual reversion of the force or separation of the bodies, quiescence being the neutral condition. Try the question experimentally on these grounds – then the following suppositions or suggestions arise.*

*(10020): Bodies approaching by gravitation, and bodies separated per force, whilst gravitating towards each other, may show in themselves or in surrounding matter or helices, opposite currents of electricity round the line of motion as an axis. But if not moving to or from each other, should produce no effect.*

High-speed aircraft become electrically charged, due to air friction, with a roughly radial electric field surrounding the body. Faraday's main idea was that an accelerating mass developed a circumferential electrical field around the line of motion of the body. Faraday's idea was focussed on a gravitational cause for the formation of the electrical field.

Faraday realised that any effect associated with change in gravity was likely to be small and that it might be necessary to use the Earth's mass in any gravity experiment.

*(10021): Though two ordinary masses of matter may not do for want of power, yet a mass of matter subject to the gravitating force of the earth as the other mass, may do. Still, the manifestation of induced lateral effect can only be very small, if any.*

*(10022): Then any motion up or down should do.*

*(10023): No motion other than that caused by gravitation should do.*

Faraday wondered whether a falling body would induce an electric current in a helix (coil). And that the helix might provide the mass of the falling body. Also, he thought about relative motions.

*(10028): A falling body may either fall through a helix or with it. Can that make a difference in the results? The falling body may either be the helix – or require to be independent of it. If the latter, it will show an inductive action at right angles to Gravity.*

Faraday's view of inertia was that it arose because of the conservation of force. However, he couldn't see how to explain it in terms of a force field theory. Strangely, he did recognise the possibility of altering the inertia of a body, although, for some reason, he didn't attribute it to a force field.

*(10029): The weight of a body having a current purposely passed round it ought to be affected, i.e. if falling, or rising, but not if still.*

Again, he considered a circumferential current.

*(10030): But that a body should have less weight whilst falling and more weight whilst rising than when still or moving only horizontal, is a strange conclusion and against the general notion. Still, it may be true, for I do not see as yet that natural conditions contradict it.*

Faraday glimpsed the idea that vertically accelerating bodies might change their weight. This idea was perceived long before lifts (elevators) became commonplace in tall buildings, allowing passengers to experience weight changes during periods of acceleration.

*(10037): The effect may be very small, even if it does exist.*

Faraday pointed out, again, that any effect was likely to be small.
Perhaps one of Faraday's most inspiring diary entries, written on the day that he started his gravity research, is the following.

*(10040): All this is a dream. Still examine it by experiments. Nothing is too wonderful to be true, if it be consistent with the laws of nature and in such things as these, experiment is the best test of such consistency.*

Faraday was convinced that gravity was not just a passive force but that it should, under changing conditions, interact with or evolve into other forces. Then, by reversing the process, it should be possible to alter gravity itself. Faraday, therefore, believed in the possibility of gravity control.
The first series of gravity experiments was carried out during the spring and summer of 1849, in the lecture room of the Royal Institution, in London.
Faraday placed a helix, or coil, around a mass rod, with the coil's centre

line vertical. A galvanometer was connected to the coil with trailing wires to form a complete circuit. The combined rod and coil was then allowed to fall freely in the Earth's gravitational field, from near the ceiling of the lecture room to the floor, a distance of 36 ft (11 m). Faraday's idea was that as the mass and the coil were released and accelerated downwards, a current might be induced in the coil, causing the galvanometer needle to move. Each drop test lasted about 1 second, so all he could hope for with his apparatus was a brief flicker of the galvanometer needle.

As Faraday recorded in the *Philosophical Transactions of the Royal Society* (page 1, 1851):

> *The thought on which the experiments were founded was, that, as two bodies moved towards each other by the force of gravity, currents of electricity might be developed either in them or in the surrounding matter in one direction; and that as they were by extra force moved from each other against the power of gravitation the opposite currents might be produced.*

Faraday was aware that the circular current for which he was searching was likely to be extremely small, if it existed at all. As we now know, gravitational forces are $10^{40}$ times weaker than electrical forces, so any induced current would, surely, be tiny.

Initially, Faraday got a reading, which must have been an exciting moment for him. However, being a brilliant experimentalist, he searched for other reasons first, rather than accepting that his idea was right. He traced the effect back to the pair of trailing wires connecting the coil to the galvanometer. As the wires fell through the Earth's magnetic field, they induced a current in the wire. Twisting the wires cancelled this effect. He repeated his experiment with various mass rods, including copper, bismuth and iron, but there was no effect.

In Faraday's free-fall experiments, his test period only lasted just over 1 second. For microgravity experiments, today, drop shafts have been developed which have extended the test period. Japan has a drop shaft at its Microgravity Centre on Hokkaido. The shaft is 710 m in depth, with a free-fall zone of 500 m, giving a 10-second period in which to carry out zero-g experiments.

Even while he had been carrying out his first series of gravity experiments, Faraday had been thinking about a different approach. As we will see, this led to a second series of gravity experiments with the Newman machine.

## Wednesday, 28th March 1849

*(10051): If there should happen to be any result of the kind imagined, then a body moving **up** would produce **one** current, and moving **down**, the reverse current. Now these may be converted by a commutator into one consistent current and that may be sent through a Galvanometer for the time of a **half vibration** of the needle, and they by a **second** commutator be sent for the second half vibration in the contrary direction – then back in the first direction, and so on continually. This would seem to be a good way of accumulating the induced force or current, if there be any.*

By the beginning of autumn 1849, Faraday had stopped working on his first series of gravity experiments and was getting ready to start a second series of gravity experiments. He remained optimistic that he might get a result.

## Saturday, 25th August 1849

*(10061): I have been arranging certain experiments in reference to the notion that Gravity itself may be practically and directly related by experiment to the other powers of matter, and this morning proceeded to make them. It was almost with a feeling of awe that I went to work, for if the hope should prove well founded, how great and mighty and sublime in its hitherto unchangeable character is the force I am trying to deal with, and how large may be the new domain of knowledge that may be opened up to the mind of man.*

Faraday's second series of gravity experiments involved relative motion between the mass and the coil detecting system, as he recorded in his laboratory diary.

## Thursday, 30th August 1849

*(10133): If, then, a falling helix or a falling body in a falling helix can produce any effect **by Gravity**, it must be very small in amount. But perhaps if a body were falling or rising in a helix at rest, the effect may be different and essentially so. Hence have this case to examine by the apparatus which Newman is making for me.*

*(10140): Also, when a helix carries a current, an iron core inside is affected, an iron tube outside not at all. Which shows the great difference **within** and **without**. Something of a like nature may occur in gravitation effects. So go on experimenting with the apparatus.*

According to his diary, Faraday received the Newman machine on Saturday, 1st September 1849 (10141). The machine enabled a mass to be rapidly vibrated within a coil. This was used in Faraday's second series of gravitational experiments. Periods of acceleration were still present during the oscillatory motion, but now there was relative motion between the rod test masses and the sensing coil, used to detect the presence of any surface currents.

By turning the great wheel, the V-shaped arm was made to vibrate through about 3 inches (7.5 cm). Fitted between the ends of the arm was a mass rod which oscillated to and fro through the fixed coil. Faraday used samples made from copper, bismuth, glass, sulphur and *gutta percha* (a form of rubber). The commutator, the device invented by William Sturgeon, summed the effect of any alternating current.

Faraday carried on testing right through to October 1849, but, as with his first series of experiments, no effect was detected. Faraday stopped working on gravity and turned his attention to other work.

Four years earlier, in 1845, Faraday had successfully completed a study on magneto-optics. The outcome of the work led to the discovery of Faraday rotation and the beginning of magneto-optics. When polarised light is shone through a glass block parallel to a magnetic field, the plane of polarisation of the light is rotated. We will say more about the Faraday Effect (FE) later, particularly with reference to a repeat of Faraday's first gravity experiment.

Ten years passed by before Faraday returned to the subject of gravity. He still held the view that gravity changes were accompanied by an induced electric field. But, since his first experiment had shown him that a closed electric current did not form in a coil surrounding a mass falling in a gravity field, he needed to reconsider his ideas.

*Thursday, 10th February 1859*
*(15785): Surely the force of gravitation and its probable relation to the other forms of force may be attacked by experiment. Let us try to think of some possibilities.*

*(15786): Suppose a relation to exist between gravitation and electricity, and that as gravitation diminishes or increases by variation of distance, electricity either positive or negative were to appear – is not likely, nevertheless try, for less likely things apparently have happened in nature.*

Faraday split his ideas for a new mass-gravity experiment into static and dynamic parts. Firstly, in the static part, he wondered whether the positioning of a mass in Earth's gravity field might result in it acquiring an electric charge and that in repositioning the mass, it might change charge. If so, he wondered, if a wire connected two masses at different heights, might an electric current (for a closed circuit) pass between them? Secondly, in the dynamic part, if a current did exist between two masses at different heights, then as the connected pair fell in Earth's gravity field, did the current increase? He also thought about the difference in discharge characteristics of conducting and non-conducting matter.

*(15788): Must not be deterred by the old experiments (10018, etc.) If there be any true effect of gravity, it may take much gravitating matter to make the effect sensible, and I had but very little. Moreover, the motion of a body with or against gravity ought not to form a current in a closed circuit, as tried in the former case, but perhaps give opposite states in lifted or depressed bodies, and though a current might be formed in a wire connecting two such, it would not be a current in a circuit. So may consider the imaginable effects under two views, **static** or **dynamic**. Take the former first and imagine as follows:*

*(15789): If an insulated body, being lifted from the earth, does evolve electricity in proportion to its loss of gravitating force – then it may become charged to a very minute degree either **positive** or negative. When thus charged it may be discharged, and then if allowed to descend insulated, it would become charged in the opposite manner, and so on...*

*(15790): Might not two globes (or masses, as pigs) of lead be attached to the end of a long rope passing over a large pulley at the top of the Clock tower, or in the whispering gallery of St. Paul's serve an experimental purpose. Starting with both balls insulated, discharged and balanced, then it would be easy to raise B and lower A, and examination by a very delicate static electrometer might shew A charged positive and B negative; then discharging both and reversing the motion, B would come down positive and A become negative and so on...*

*(15792): As to the character of the gravitating matter, it may be conducting or non-conducting – and probably an important difference might here arise. Non-conducting matter invested in conducting matter would be peculiar in its discharging action, etc.*

*(15795): The evolution of one electricity would be a new and very remarkable thing. The idea throws a doubt on the whole: but still try.*

*(15797): If the insulated masses assumed the state supposed, they would when connected give a current between them and thus dynamic effects; but it is possible that if connected whilst they were changing their position, more distinct dynamic effects might be produced...*

*(15799): Perhaps a jet of drops of water from a height would tell below – only water is a bad substance because of the discharging facility of moist air.*

*(15800): Probably a jet of lead would do better; the fall of shot in the shot tower. Might insulate the tub of water into which it falls below and so get traces of any evolution.*

*(15801): One of their pigs of lead, insulated and discharged below, then raised to the top and examined by the electroscope.*

During Faraday's time, it was not realised that the Earth's surface and the lower layers of the ionosphere, at an altitude of about 50 km, act as the two conducting plates of a gigantic spherical capacitor filled with air. (For electricity in the atmosphere, see Richard Feynman et al, *Lectures on Physics*, Volume II, Chapter 9.) At an altitude of 50 km, the electric potential is about 500 kv, while at the Earth's surface, the electric potential is zero. Thus, from the ground at zero potential, the electric potential increases vertically upwards by 100 v/m until at 50 km, it is 500 kv. This gives rise to a downward-directed electric E-field, through the atmosphere. In free space, an insulated discharged mass positioned at a height in the atmosphere would, given time, acquire a charge.

With reference to Faraday's idea that an insulated discharged mass may acquire an electric charge with height in Earth's gravity field, the following paragraph taken from the late Professor Feynman's book Volume II, Chapter 9, is apposite.

*How can we measure such a field if the field is changed by putting something there? There are several ways. One way is to place an insulated conductor at some distance above the ground and leave it there until it is at the same potential as the air. If we leave it long enough, the very small conductivity in the air will let the charges leak off (or onto) the conductor until it comes to the potential at its level. Then we can bring it back to the ground, and measure the shift of its potential as we do so. A faster way is to let the conductor be a bucket of water with a small leak. As the water drops out, it carries away any excess charges and the bucket will approach the same potential as the air.* So, Feynman supports Faraday's idea.

More from Faraday's diary:

*(15804): Let us encourage ourselves by a little more imagination prior to experiment. Atmospheric phenomena favour the idea of the convertibility of gravitating force into Electricity, and back again probably (or perhaps then into heat)…*

*(15806): So to say, even the changed force of Gravity as Electricity might travel above the earth's surface, changing its place then becoming the equivalent of Gravity.*

*(15807): Perhaps heat is the related condition of the force when change in Gravity occurs. Might associate a thermo electric pile or couple, to see if change of elevation from the earth causes any sensible change of temperature.*

*(15808): Perhaps almost all the varying phenomena of atmospheric heat, electricity, etc. may be referable to effects of gravitation – and in that respect the latter may prove to be one of the most changeable powers instead of one of the most unchanged.*

Then we have Faraday's exhortation for scientific research:

*(15809): Let the imagination go, guiding it by judgment and principle, but holding it in and directing it by* **experiment.**

*(15814): If anything results then we should have…*

*(15815): An entirely new mode of the excitement of either heat or electricity.*

*(15816): An entirely new relation of natural forces.*

*(15817): An analysis of Gravitation force.*

*(15818): A justification of the conservation of force.*

*(15819): If either heat or electricity evolved, would probably refer both them and gravitation to actions of the ether or medium in space.*

Faraday still thought in terms of the conservation of force (15818), as advocated by Helmholtz. But the new idea of the conservation of energy was gaining ground, championed by James Prescott Joule, a Manchester-based amateur scientist.

*(15834): In shot making – the descending shower of lead ought to charge the receiving tub below, it being insulated…*

*(15847): Have made two electroscopes…*

Paragraph 15834 and the earlier paragraph (15800) relate to transferring charge on falling lead shot to an insulated tub of water and detecting the change in charge of the water. If he had done the experiment and measured the temperature change of the water in the tub instead, he might just have detected an effect! A change of gravitational potential energy of the lead shot into kinetic energy and then into heat energy as the lead shot was brought to rest, might just have been detectable. But Faraday never did the experiment. This experiment can be viewed as the gravitational analogue of Ohm's law (see Chapter 29). Interestingly, Faraday was one of the referees of Joule's 1850 Royal Society paper on the mechanical equivalent of heat. But it would seem that Faraday hadn't fully grasped Joule's idea of the conservation of energy and still clung to the idea of conservation of force.

Faraday's third series of gravitational experiments began in March 1859 and involved raising and lowering masses in Earth's gravity field. They started in the stairwell of the Royal Institution but quickly relocated to the Shot Tower, on the south bank of the River Thames, next to Waterloo Bridge. The tower, built in 1826, survived the German bombs during the Second World War and featured in the 1951 Festival of Britain. It was demolished in 1963, and I doubt whether many people know of its links with Michael Faraday and his gravity research.

Faraday was trying to get his mind round a possible link between gravity and electricity and heat. He spent a considerable amount of time (from April through to mid-June 1859) carrying out several detailed temperature experiments.

In his experiment, Faraday raised and lowered a Casella mercury thermometer inside the Shot Tower, over a distance of 165 ft (50 m). The thermometer bulb contained about 1 lb (0.45 kg) of mercury and the vertical tube was marked with 7½ in (19 cm) of graduations to allow the height of the mercury column to be measured very accurately. The thermometer bulb was placed in a paper cup and placed in a deal box lined with cork. The rest of the space in the box was filled with mahogany sawdust to isolate the mercury from any external heating.

The thermometer was read at the bottom and top of the tower, to see whether any change of temperature occurred in the mercury mass with change in elevation, due to loss or gain of gravitational potential energy.

But Faraday detected no change in temperature of the mercury mass with the change of its height within the tower.

To improve the sensitivity of the temperature measurements, a differential air thermometer was used, containing just over 2 lb (0.9 kg) of mercury. Again, a box with sawdust packing was used to isolate the thermometer from any external heating. The previous procedure was repeated but, again, no change in temperature was noticed.

*Monday, 11th April 1859*
*(15913): If the temperature of bodies should be affected, then their capacity for heat should affect the amount of charge.*

*(15913): The following point is against electricity. There is no reason why, when two bodies recede or approach, that they should not change (if they change at all) in the same direction. But this would be against the idea of a dual power – though not against that of a single power (so to speak), as heat.*

*(15915): It would be strange if a body should heat as gravitation increases by nearness of distance. We conceive of heat as a positive force and of gravitation as a positive force, and then instead of being the inverse of each other, they would seem to grow up together. Or else heat must be negative to gravity or the converse of gravity and gravity must be in the same negative or converse relation to heat. This is against the expectation of any thing from the heat experiment. Nevertheless make it, for who knows. If gravitation depend upon forces external to the particles, such results might happen. Try.*

*(15916): Have constructed two apparatus for trials of any heat effects. One is a square box containing two differential chambers. Each one of these*

*contains an air thermometer bulb. One is to be surrounded by mercury, the other by air; and when all is arranged, they are to be connected externally by a scale and fluid tube, so as to form a differential thermometer, and this shew any change in temperature of the chambers...*

*(15917): Another instrument was constructed, first by Mr Ladd and then second by Mr Caselli. This was just a large delicate mercurial thermometer, containing about a pound of mercury...*

After completing his experiments, Faraday had to admit that he had detected no changes in temperature.

*(15985): So there is no evidence by either apparatus that any difference due to gravity varying by an elevation of 165 feet can shew a relation to heat by causing a change of temperature.*

Although Faraday was unaware of it, there is a drop in temperature with altitude in the troposphere. From the International Standard Atmosphere (first available during the 1920s) we can calculate that the change in temperature between ground level and the height (equal to that of the top of the tower) in free space is about 0.325°C. But inside the Shot Tower, conditions did not conform to those of free space.

Let us investigate Faraday's temperature experiment. The change in height $\Delta z$ of the mercury mass M in Earth's gravity field g (taken as constant) means that work $\Delta W$ is done in moving the mass. This is equal to the change in potential energy $\Delta E$ of the mercury mass M in the gravitational field.

$$-\Delta W = \Delta E = Mg\Delta z$$

Scientists assume that the change of static gravitational energy, the kinetic energy made zero, is directly related to an increase or a decrease of heat energy $\Delta Q$ and a change in temperature $\Delta\theta$. From Chapter 27, heat energy $\Delta Q = Mc\Delta\theta$, where c is the specific heat of the mass M.

With the benefit of Joules' work (Chapter 29) on the conservation of energy and his discovery of the mechanical equivalent of heat J, we know that the relationship between work done W and heat energy Q is given by $W = JQ$. Consequently, we might write $Mg\Delta z = JMc\Delta\theta$, where we have assumed that all the change in heat energy is confined within the mass.

This, in turn, means that we have assumed that the energy (kinetic and potential) resides within the mass. Thus, the temperature change is given by

$$\Delta\theta = \frac{Mg\Delta z}{JMc}$$

Here, M = 0.45 kg, g = 9.81 m/s², $\Delta z$ = 50 m, J = 4,2 Joules/calorie and the specific heat of mercury c = 33 calories/kg °C.

The formula predicts that in raising the mercury mass by 50 m, the mass temperature should fall by $\Delta\theta$ = 3.5 °C, whereas by lowering the mercury mass, the temperature should rise by $\Delta\theta$ = 3.5 °C.

There is no way that Faraday would have missed this temperature change. Therefore, there is something wrong with the assumptions made. The main flaw is that the mercury mass was raised and lowered by rope. So, the man working the rope expended the physical energy which led to the mercury mass gaining or losing gravitational potential energy. No energy was exchanged with the gravity field in the process, so there was no change of temperature. If the mercury mass had been allowed to fall (exchanging potential energy for kinetic energy) and be brought to a halt (somehow without smashing the bulb) with zero kinetic energy an inch from the ground, its temperature would have risen by 3.5°C. (See Faraday's idea for a lead shot experiment.)

Finally, Faraday turned his attention back to his idea that a mass might undergo a change in electrostatic charge with a change of its position in the Earth's gravitational field. Some preliminary tests were conducted in the stairwell of the Royal Institution, with a limited height variation of 39 feet. Within the Shot Tower, a mass could be raised, or lowered, a distance of 165 feet (50 m). Checks were made to see whether the mass under static conditions developed an electric charge as its gravitational potential energy changed. Initially, Faraday used a lead weight (a pig of lead) of about 170 lb (77 kg), later supplemented with a further lead weight, giving a total mass of 280 lb (126 kg).

Firstly, the insulated mass was charged at the bottom of the tower from a Zamboni pile (a version of Volta's pile). Steel spikes were driven into the lead masses and platinum wires attached. These were attached to an electrometer, to allow any charge to be measured. The mass was then raised to the top of the Shot Tower and the charge reading taken again. No change in charge was observed. The procedure was then reversed, with the mass being charged at the top. A null result, again. The experiment was then repeated, with the mass being discharged at the bottom of the tower, then raised to the top and checked to see whether a charge had developed.

*Saturday, 9th July 1859*
*(15986): At the shot tower, to try for Electricity...*

*(19587): A pig of lead of about 170 lbs had an iron pin and shackle put through one end...*

*(15988): The Electroscopes have platinum tops, and to make their contact with the lead, I had steel spikes prepared with each a piece of platinum wire brazed on to it. One of these spikes was driven into the lead, and the contact of the platinum cap of the electrometer made with the platinum attached to the spike.*

*(15990): The pig was charged by Zamboni – then sent to the top of the tower, 165 ft, then lowered and examined; it still gave a fair charge of electricity though the charge had in some degree diminished.*

*(15991): Now sent the pig to the top – there Chapman touched and uninsulated it – then it was lowered and examined for electricity; but there were no signs of any charge. Repeated the experiment and with exactly the same negative result.*

*(15992): Now charged the lead by Zamboni – sent it up – did **not** touch or uninsulate it – brought it down and again examined it. It was very fairly charged, shewing that there was little or no loss in the transit up and down.*

*(15997): The experiments were well made but the results are negative...*

Although there is a close analogue between the models for the gravity field and the electrical field of the Earth, there is a major difference. Air is very slightly electrically conducting, which allows tiny electric currents to flow where electric potential differences occur. This means that there is always a tiny current flowing down from the ionosphere to the Earth's surface, estimated to be about $10^{-12}$ Amp/m². Lightning is a very special case, with a massive transitory current of around $10^{11}$ Amp/s.

The electrical situation is more akin to the thermal analogue, where a body radiates, or absorbs, heat until it reaches thermal equilibrium with its surroundings. This is the 2nd law of Thermodynamics.

Air is not gravitationally conducting, so uncharged mass currents do not arise in a way analogous to the electrical case. However, temperature changes cause density changes in the atmosphere, so mass currents do occur, largely influenced by the Sun.

Thus, if an insulated mass is raised vertically, a distance d in free space in Earth's electric field E, it will gain or lose charge until it reaches the same potential as the Earth's field at that point. Eventually, the mass will attain a potential $\phi = Ed$.

Once static conditions prevail, we can estimate the size of the charge Q acquired by the insulated mass. Suppose the mass is spherical, with radius a, then we can treat it as a point charge Q. The potential at its surface is

$$\phi = \frac{Q}{4\pi\varepsilon_0 a}$$

Thus, for equal potentials

$$Q = 4\pi\varepsilon_0 a E d$$

An insulated sphere of lead weighing 77kg (170 lbs), if discharged at Earth's surface and then raised vertically in free space by 50 m (165 ft) and left to settle, will acquire a charge Q = 30 nC. The gold leaf electroscope can detect a charge of 0.01 pC = 0.00001 nC. So the electroscope would easily detect the charge on the lead globe.

So, why didn't Faraday detect the charge? Feynman probably has the answer. He points out that for a tall person at sea level, the difference in potential between the person's feet and head is 200 volts. But the body is a good conductor, which is earthed. So, within the body, the electric potential is zero. The Earth's electric field bends around people, similarly with buildings, especially those fitted with a lightning conductor. Within the Shot Tower, the electric potential was probably zero, so a mass earthed at the bottom and raised to the top would not develop a charge.

Faraday prepared a paper on his Shot Tower experiments, submitting it to the Royal Society in April 1860. It began,

> *Under the full conviction that the force of gravity is related to other forms of natural power, and is a fit subject for experiment, I endeavoured on a former occasion to discover its relation with electricity, but unsuccessfully. Under*

*the same deep conviction, I have recently striven to procure evidence of its connection with either electricity or heat.*

However, Lord Stokes, then the President of the Royal Society, persuaded Faraday to withdraw the paper.

Faraday never carried out an experiment to test the dynamic part of his new idea; that of dropping an insulated pair of vertically aligned masses connected by a wire to see whether an increasing current flowed between them. This would have been an update of his first gravity experiment of 1849.

As part of my reading about Faraday's gravity research, I asked Mrs McCabe, the librarian at the Royal Institution, whether I could see a copy of Faraday's paper on his gravity experiments carried out in the Shot Tower. It was agreed that I could, and I visited them in March 1988. I expected to be given some form of facsimile to read but, instead, I was given a large Victorian ledger containing Faraday's handwritten notes. It is hard to describe the thrill that it gave me (I, too, started my working life as a laboratory assistant) just to touch those pages on which Faraday had written down some of his thoughts about gravity, together with his little sketches in the margin. It gave me a real sense of being part of the continuing quest by scientists and philosophers to unravel the secret of gravity; that mysterious subject still awaiting a breakthrough in understanding to connect it with the other natural force fields.

# Faraday's gravity experiments

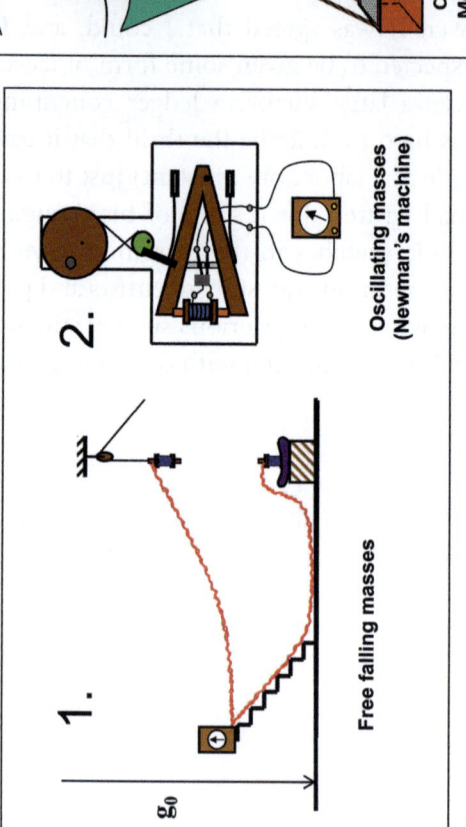

1849 At the Royal Institution

1859 At the Shot Tower

# CHAPTER 33

# EXTENDING NEWTON'S GRAVITY AND THE GRAVITY VORTEX

Newton introduced the idea that all masses have a static gravitational field. The simplest example is the radial gravitational field **g** of a point mass m.

$$\mathbf{g} = -\frac{m}{\gamma 4\pi r^2}\hat{\mathbf{r}} \quad \text{where Newton's gravitational constant } G = \frac{1}{\gamma 4\pi}.$$

For a solid spherical mass, calculating the internal gravity field $\mathbf{g}_{Int}$ is a little difficult. In Book 1 of the *Principia*, Newton began by determining the gravity field inside a mass shell (Appendix A.1). Because of the inverse square law, it turns out that for the mass shell, $\mathbf{g}_{Int} = 0$. This is an interesting result, since it is the analogue of the electro-static case of a perfectly conducting charged spherical shell where inside the shell the electric field is zero. That is, $\mathbf{E}_{Int} = 0$. This result was discovered by Faraday, in 1843, with his ice pail experiment, which is the basis for the Faraday cage.

Outside the mass shell, the gravity field is the same as that for a point mass concentrated at the centre of the shell.

Similarly, outside a mass sphere, the external gravity field is the same as that for a point mass. By treating a solid mass sphere as an outer mass shell and an inner mass sphere, we can determine the internal gravity field of the mass sphere (A.2).

Using the inverse square law, we can derive the internal and external gravity fields for an infinite line mass (A.3). Sometimes it is helpful, in modelling terms, to treat the surface of the Earth as being locally flat and the Earth's gravity field as constant (A.4).

We have seen that various analogies hold between gravity, fluid mechanics, thermodynamics and electromagnetism. Although the models are similar, none are exactly the same in all respects. For example, because of the absence of negative mass, we can't make a gravitational Faraday cage. An external gravity field will penetrate a mass shell.

Newton used his gravitational theory, in a semi-static way, to explain Kepler's laws of planetary motion. At any instant, he could assume that the planets were stationary. We would like to extend Newton's theory to include effects resulting from moving masses.

We started with gravity, where a point mass is a source of a radial gravitational field. Actually, in analogy with fluids, a point mass is a sink of acceleration. And a negative point mass, in theory, is the fluid analogue of a source of acceleration. In fluid mechanics, we introduced streamlines to visualise the fluid flowing out of sources and into sinks. Gravity field lines in the ether are analogous to streamlines.

Then we looked at the analogy of positive and negative electric charges with the fluid source and sink. Electric field lines in the ether emanate from positive electric charges and end at negative charges. They are, also, the analogue of fluid streamlines. Positive and negative electric charges, fluid sources and sinks and positive (ordinary) and negative masses all obey the inverse square law. The analogy also applies to positive and negative magnetic poles (although in reality they only occur in pairs), which also obey the inverse square law.

In thermodynamics, we noted that a stationary thermal field was analogous to a gravity field. We learnt that Ohm's law, concerning the flow of electric charges (an electric current) in an electrical conductor, was based on an analogy (mismatched, because heat sources are stationary) with the flow of caloric in a thermal conductor.

In fluid mechanics, we introduced the idea that a fluid could have some stickiness, or viscosity. This led to the idea of the line vortex surrounded by circular streamlines. Reading across to the electric case, one analogy suggested the idea of a line of moving electric charges (an electric current) surrounded by circular magnetic field lines.

We are now ready to go full circle; to consider gravity as an analogue of electromagnetism and look for the vortex associated with moving mass.

We start by taking mass as the analogue of charge and the gravity field **g** as the analogue of the electric field **E**. Consequently, gravitational permittivity $\gamma$ is the analogue of electrical permittivity $\varepsilon$. Since moving electric charge creates a magnetic field **H**, the analogue suggests that moving mass creates a gravitomagnetic field **h**. Thus, the gravitomagnetic field **h** is the dual of the gravity field **g**. The idea that gravity had a dual force field was first proposed by Faraday.

Analogous to the magnetic permeability $\mu$, we let $\eta$ be the gravitomagnetic permeability. This leads us to a set of 'Maxwell-type' equations for gravitation. According to the tensor calculus experts, these equations are called the GEM equations (see Chapter 14) and arise from the linearisation of Einstein's field equation for gravity.

The equations are given here for comparison with the equations for the other analogues. The mathematical hieroglyphs may make the equations meaningless to some of you. In which case, treat the equations as a pattern and note the absence of terms and the symmetry (or duality) of terms. Clearly, if you are reading this, you must be interested in the subject of gravity, so please read on.

$$\nabla \bullet \mathbf{g} = -\frac{\rho}{\gamma}$$

$$\nabla \bullet \mathbf{h} = 0$$

$$\nabla \times \mathbf{h} = -\rho \mathbf{v} - \gamma \frac{\partial \mathbf{g}}{\partial t}$$

$$\nabla \times \mathbf{g} = \eta \frac{\partial \mathbf{h}}{\partial t}$$

For static Newtonian gravity, we already have $\nabla \bullet \mathbf{g} = -\rho/\gamma$. Based on the analogue with magnetism, we would expect gravitomagnetic mono-poles to be paired, so they cancel out, leaving us with $\nabla \bullet \mathbf{h} = 0$. This also explains the absence of the gravitomagnetic mono-pole velocity in the $\nabla \times \mathbf{g}$ equation. Note the new factor $\eta$, which is the permeability of the medium to the gravitomagnetic field.

The analogue of flow of electric charge is flow of mass. For current flow in a wire, the electric field is neutralised, leaving us with just the magnetic field. In the gravitational case of mass flow along a pipe, with density $\rho$ and speed **v**, the analogue suggests the existence of the gravitomagnetic field **h**, but the gravitational field of the mass is not neutralised. For steady mass flow we get $\nabla \times \mathbf{h} = -\rho \mathbf{v}$, where the negative sign arises because masses attract. This is the gravitational analogue of Oersted's discovery of a magnetic field around a wire carrying a steady current of electricity. So the gravitomagnetic field **h** surrounding the mass current forms the vortex associated with gravity. The mass current equation is mathematically valid, but a gravitomagnetic field **h** has never been detected around a steady current of mass. So, is **h** a real force field or is it merely a theoretical construction?

For unsteady conditions we get the added term $\gamma \partial \mathbf{g}/\partial t$. Since **g** is equivalent to acceleration **a**, then $\partial \mathbf{g}/\partial t$ is equivalent to the rate of change of acceleration, known as jerk **J**. From Newton's 2$^{nd}$ law, a body of mass m under a constant acceleration **a** experiences a constant force $\mathbf{F} = m\mathbf{a}$. If the acceleration changes with time, the force changes and the mass m experiences jerk **J**, given by $\mathbf{J} = (1/m)(\partial \mathbf{F}/\partial t)$. Since $\gamma$ is large, the amplified jerk term $\gamma \partial \mathbf{a}/\partial t$ may be large and cause the creation of a strong but short-lived gravitomagnetic field **h**. People subjected to a spike-like jerk find it a most unpleasant experience, while for machinery it can be very damaging.

Because of the absence of gravitomagnetic mono-poles (we assume that they always occur in ± pairs), there is no steady equation for $\nabla \times \mathbf{g}$. But, the analogue suggests that while the unsteady term $\eta \partial \mathbf{h}/\partial t$ prevails, it creates a gravity field, the magnitude of which is dependent on $\partial \mathbf{h}/\partial t$ and the value of the gravitomagnetic permeability $\eta$. So, jerk can create a strong gravity field, if $\eta$ is large.

The two curl terms (those containing the × symbol) are missing in Newton's static gravitational theory.

The first scientist to consider extending Newton's theory of gravity with a Maxwell-type theory of gravity was Maxwell himself. When a mass is introduced into space then its gravity field **g** spreads throughout space, creating a tension in the ether. This tension holds gravitational energy. In the same way, hitting a nail into a block of rubber creates a tension in the block. The compressed rubber, pushed out of the way by the nail, contains elastic energy. Stored energy contains the potential to do work. That is, to force an object to move. Suppose a mass m is placed in a gravity field of an existing mass with field $\mathbf{g} = -\nabla \phi$, where $\phi$ is its gravitational potential. The change in energy is given by $E = m\phi$ (A.5). We speculate that the energy stored in the ether by a single point mass is $-mc^2$, although this value is open to question (A. 6).

The Universe contains myriads of masses which fill space with an unending cobweb of interacting gravitational **g**-fields. As well as creating its own gravitational field, a mass is subject to the gravitational fields of other masses, and they are all trying to expend gravitational energy and move each other. Due to the gravitational attractive property of masses, the potential energies are negative. Since space is so huge, the negative gravitational energy it contains will be infinite. So we resort to gravitational energy densities determined over specified volumes V of space. The energy density over a volume V of space containing one mass in the field of another is $E/V = -\gamma g^2$ (J/m$^3$), where g is the total gravitational intensity at any point in the volume (A.7). For a distribution of mass the energy density is $-\frac{1}{2}\gamma g^2$ (J/m$^3$), where the ½ arises because each mass in the distribution sits in its own gravitational field and we must remove that contribution from the energy sum (A.8). As examples, we can evaluate the gravitational energy of a section of an infinite line mass (A.9) and the gravitational energy of a solid spherical mass (A.10).

We are used to the idea that energy is a positive quantity. So, where is all the infinite positive energy to counteract the enormous amount of negative gravitational energy? Maxwell was perplexed by this problem and didn't develop his model any further.

The energy problem was partly resolved by Einstein, who stated that a mass m has intrinsic positive energy E given by $E = mc^2$.

Oliver Heaviside was not so inhibited as Maxwell and in a short comment in Appendix B in Volume 1 of his book *Electromagnetic Theory*, published in 1893, he wrote:

> *Now what is there analogous to magnetic force in the gravitational case? And if it has an analogue, what is there to correspond with electric current? At first glance it might seem that the whole of the magnetic side of electromagnetism was absent in the gravitational analogy. But this is not true.*

In a paper with the title *A Gravitational and Electromagnetic Analogy*, published in 1893 in Volume 31 of the journal *The Electrician*, Heaviside expanded on his idea. He proposed that the dual of gravity was a hitherto unknown force field associated with moving mass which he didn't name, but to which he gave the symbol **h**. Heaviside didn't dwell on the subject, as he realised that its effect across space would be tiny.

Faraday had, earlier, carried out his series of moving mass experiments in his search for the dual field of gravity. At the time, apart from gravity, scientists were only aware of two other force fields, namely electricity and magnetism. Nature gave no hint of the existence of any other force fields. Thus, Faraday guessed that the dual of gravity might be electricity, but this wasn't substantiated by his experiments. The discovery of more force fields, the strong and weak force fields hidden within the atom, came half a century after Faraday's time.

Maxwell's extension of the Newtonian model of gravity supported Faraday's idea that gravity had a dual force field, but it didn't support Faraday's guess of electricity. Based on the electromagnetic analogue, both Maxwell and Heaviside predicted the dual of gravity to be a hitherto unknown force field, now called gravitomagnetism, with the symbol **h**. Maxwell's brief foray into extending Newton's model of gravity came too late to help Faraday with his ideas for experiments. In any case, there were no known effects in space which could be attributed to gravitomagnetism, so whether it was a real force was questionable. Furthermore, without any idea of the effect that gravitomagnetism might induce in matter, how could a detection system be built? And as Faraday, and later Heaviside, observed, any effect was likely to be tiny.

What form might the gravitomagnetic field **h** take? At this stage, all we can do is guess. In dimensional terms of mass M, length L and time T, the dimensions of **h** are M/LT. So, the field strength, or intensity, of **h** at any point in space is measured in kg/m.s. This is the same as the dimensions of fluid viscosity, suggesting that the ether has some stickiness. This should not come as a surprise if we view the ether as being analogous to a fluid. The equation for vorticity $\zeta$ is $\nabla \times \mathbf{v} = \zeta$. Therefore, the curl of acceleration is $\nabla \times \partial \mathbf{v}/\partial t = \partial \zeta/\partial t$, which may be compared with $\nabla \times \mathbf{g} = \partial(\eta \mathbf{h})/\partial t$, suggesting that $\eta \mathbf{h}$ is analogous to $\zeta$, or angular velocity.

In his examination of the motion of the planets around the Sun, Kepler discovered that a planet's angular momentum is constant. In vector form, angular momentum is written as **H** (not to be confused with magnetic intensity) where (A.11)

$$\mathbf{H} = M(\mathbf{R} \times \mathbf{v}) \quad \text{J.s}$$

In planetary terms, M is the planet's mass, R is its distance from the Sun, **v** is its orbital velocity and its angular momentum **H** is constant.

Angular momentum has dimension $ML^2/T$, so that angular momentum density **H**/V has dimensions M/LT. Thus, **H**/V has the same dimensions as those of **h**, where the volume V is a volume of space containing the mass M. With this as a clue, we guess, further, that at any point in space a distance **R** away from a mass M moving with speed **v** the intensity of the gravitomagnetic field **h** there is equivalent to the density of angular momentum (**H**/V).

Gravity fields **g** associated with the huge mass of the Sun and the smaller masses of the planets fill the ether of the Solar System. Moving and rotating masses give rise to gravitomagnetic fields **h** filling the Solar System, too. Perhaps this view is more in line with Descartes' idea of ethereal vortices.

Our clue suggests that we consider the gravitomagnetic intensity **h** at any point, as $\mathbf{h} = \mathbf{H}/V = \rho(\mathbf{R} \times \mathbf{v})$, where $\rho$ is the density of mass in the volume V. But from our extended gravitational equations for steady conditions (velocity **v** constant), we require $\nabla \times \mathbf{h} = -\rho\mathbf{v}$. For this result to be satisfied, we must take $\mathbf{h} = -\tfrac{1}{2}\mathbf{H}/V = -\tfrac{1}{2}\rho(\mathbf{R} \times \mathbf{v})$ (A.12).

First let us consider the case of a point mass M moving with constant linear velocity v in the z-direction. The moving mass M will generate a gravitational disturbance in 3-D space, moving outwards from it at the speed of light c. The difficulty of modelling a point mass in 3-D is that the gravity and gravitomagnetic fields are moving with time, so that the problem is unsteady.

It is better to start with a 2-D case, where the gravity and gravitomagnetic fields are steady. Using cylindrical polars (R, θ, z), we consider an infinite line mass moving with constant speed v along the z-axis.

Suppose an element of the line mass has a mass M. The cross-sectional area of the element is $\pi a^2$ and its length is $\ell$. The mass current $I = Mv/\ell$.

External to the line mass element, where R > a, we have

$$\mathbf{h}_{Ext} = -\frac{1}{2}\frac{M(R\hat{\mathbf{R}} \times v\hat{\mathbf{z}})}{V} = -\frac{1}{2}\frac{MRv}{\pi R^2 \ell}\hat{\boldsymbol{\theta}} = -\frac{1}{2\pi}\left(\frac{Mv}{\ell}\right)\frac{1}{R}\hat{\boldsymbol{\theta}} = -\frac{I}{2\pi R}\hat{\boldsymbol{\theta}}$$

Since the line mass has a circular cross-sectional area $= \pi a^2$ then the internal mass current is reduced by the ratio $R^2/a^2$, so that

$$\mathbf{h}_{Int} = -\frac{IR}{2\pi a^2}\hat{\boldsymbol{\theta}}$$

The formulae for $\mathbf{h}_{Ext}$ and $\mathbf{h}_{Int}$ for a line mass moving with speed v are the gravitational analogues of the Biot-Savart formulae for the magnetic field around a straight wire carrying a steady electric current.

Let us concentrate on a particular small length $\ell$ of the line mass, with mass M. At time t = 0, suppose that this mass is at O on the z-axis. Then, using cylindrical polars (R, θ, z) at time t, the radius of the outward gravitational disturbance from the mass at O will be R = ct in the radial R-direction (perpendicular to the z-axis). The perimeter of the circle of disturbance represents a gravitomagnetic field line in the θ-direction.

Suppose at time t we choose a point P on the circle of disturbance. The centre of the circle is O, where the mass was at time t = 0. But, in time t, the mass M has moved forward a distance vt to position O'. Viewed from P, the angular direction of the mass M at O' appears to be rotating, or twisting, in space as time progresses by an amount

$$\frac{d\alpha}{dt} = \Omega_R = \frac{v}{R} = \left(\frac{v}{c}\right)\frac{1}{t} \quad \text{radians/s}$$

If the disturbance takes place instantaneously then $c = \infty$ and $\Omega_R = 0$, so that there is no twisting in space.

Thus, our model suggests that the **h**-field has something to do with angular velocity $\Omega_R$, or rotation. So, introducing the field **h** would account for space having a twist, or torsion, property.

For R > a

$$\mathbf{h}_{Ext} = -\frac{I}{2\pi R}\hat{\theta} = \frac{Mv}{\ell}\frac{1}{2\pi R}\hat{\theta} = -\frac{\rho_{Ext}\left(\pi R^2 \ell\right)v}{2\pi R \ell}\hat{\theta} = -\left(\frac{\rho_{Ext} R^2}{2}\right)\Omega_R \hat{\theta}$$

For R < a

$$\mathbf{h}_{Int} = -\frac{IR}{2\pi a^2}\hat{\theta} = \frac{Mv}{\ell}\frac{R}{2\pi a^2}\hat{\theta} = -\frac{\rho_{Int}\left(\pi a^2 \ell\right)R v}{2\pi a^2 \ell}\hat{\theta} = -\left(\frac{\rho_{Int} R^2}{2}\right)\Omega_R \hat{\theta}$$

An induced magnetic field is given by $\mathbf{B} = \mu\mathbf{H}$, so we expect the gravitomagnetic analogue to take the form $\mathbf{b} = \eta\mathbf{h}$. The term $\eta$ is the permeability of the medium to allowing it to twist due to a moving gravity field. The dimension of **b** is $1/T$, the same as angular velocity, so **b** is related to spin.

In the magnetic case, the values for $\mu_{Ext}$ and for $\mu_{Int}$ are determined experimentally. Unfortunately, we are not able to do this in the gravitomagnetic case.

We let

$$\eta_{Ext} = -\frac{2}{\rho_{Ext} R^2} \quad \text{so that} \quad \mathbf{b}_{Ext} = \eta_{Ext}\mathbf{h}_{Ext} = -\Omega_R$$

And we let

$$\eta_{Int} = -\frac{2}{\rho_{Int} R^2} \quad \text{so that} \quad \mathbf{b}_{Int} = \eta_{Int}\mathbf{h}_{Int} = -\Omega_R$$

## EXTENDING NEWTON'S GRAVITY AND THE GRAVITY VORTEX

Looking at $\eta_{Ext}$. As R tends to infinity, then $\rho_{Ext}$ tends to zero and the limit is of the form $1/(0 \times \infty^2)$, which is zero. We can confirm this value if we accept that $\gamma\eta = 1/c^2$. In free space $\gamma = 1.193 \times 10^9$ kg.s²/m³ and c = $3 \times 10^8$ m/s so that the gravitomagnetic permeability $\eta_{Ext} = 4\pi \times 0.74 \times 10^{-27}$ m/kg = $0.9314 \times 10^{-26}$ m/kg, which is essentially zero.

To detect the **h**-field we must induce a **b**-field in a device. Since $\eta_{Ext} \approx 0$ this means that it is highly unlikely that we can detect an **h**-field in free space.

We must now consider $\eta_{Int}$ within the moving mass element, where R < a.

$$\eta_{Int} = -\frac{2}{\rho_{Int} R^2} = -\frac{2\pi\ell}{M}\frac{a^2}{R^2} \text{ m/kg.}$$

It looks odd but at this exploratory stage we have to work with it.

Internally, R < a, the induced gravitomagnetic field is given by

$$\mathbf{b}_R = \eta_{Int}\mathbf{h}_R = \left(-\frac{2\pi\ell}{M}\frac{a^2}{R^2}\right)\left(\frac{Mv}{\ell}\right)\frac{R}{2\pi a^2}\hat{\boldsymbol{\theta}} = -\frac{v}{R}\hat{\boldsymbol{\theta}} = -\Omega_R\hat{\boldsymbol{\theta}}$$

If the line mass accelerates in the z-direction then a gravity field $g_i$ is induced in the opposite direction. This is Newton's 3rd law in action.

$$\frac{\partial \mathbf{b}_R}{\partial t} = -\frac{1}{R}\frac{\partial v}{\partial t}\hat{\boldsymbol{\theta}} = -\frac{\partial \Omega_R}{\partial t}\hat{\boldsymbol{\theta}} = \nabla \times \mathbf{g}_i$$

The induced gravity field is in the opposite direction to the acceleration.

$$\mathbf{g}_i = -\frac{\partial v}{\partial t}\hat{\mathbf{z}} = -R\frac{\partial \Omega_R}{\partial t}\hat{\mathbf{z}} = R\hat{\mathbf{R}} \times \left(-\frac{\partial \Omega_R}{\partial t}\right)\hat{\boldsymbol{\theta}}$$

This is the same gravity field pushing back on you that you experience when you accelerate forwards in a car. If the acceleration of the line mass M is the result of a static gravity field **g** acting in the z-direction, then the two gravity fields cancel and the line mass M is in free-fall.

During mass movement, we get another form of energy called kinetic energy. Since gravitational energy is stored in a gravity field, it is likely that kinetic energy is stored in a field, too. The candidate is the gravitomagnetic field.

From the duality with gravity, the energy density at any point of a gravitomagnetic field (energy associated with mass movement) is given by $E/V = -\eta h^2$ (J/m³), or by $E/V = -\frac{1}{2}\eta h^2$ (J/m³) for a distributed mass, so the energy density of the gravitomagnetic field is negative, too.

The gravitomagnetic energy generated by a mass m moving with speed v in an external gravity field is equal to ½mv² (A.13).

The gravitomagnetic energy of a mass element M of a line-mass moving with speed v is ½Mv² (A.14).

So, although we might locate kinetic energy with the moving mass, in fact, it is distributed throughout space in the gravitomagnetic field of the moving mass. We have Gottfried Leibnitz to thank for introducing his vis-viva (mv²), which Lord Kelvin tweaked to give us kinetic energy (½mv²). The use of kinetic energy with Newtonian statics allowed Newtonian dynamics to be developed without the need for gravitomagnetism (A.15).

Another feature of the magnetic field that can be 'read-across' to the gravitomagnetic field is the gravitomagnetic vector potential **A**, where **b** = ∇×**A** (A.16). For a mass falling freely in a static gravity field, it turns out that a gravity field ∂**A**/∂t is created, which is equal in magnitude but opposite in direction to the static gravity field. The result is that the gravity fields cancel and the mass is in a force-free environment (A.17). This is the root of equivalence between gravity and acceleration. The induced gravitomagnetic field **b** = ∇×**A** = –**Ω** (A.18). If a mass is accelerated by non-gravitational means then a gravity field ∂**A**/∂t is created but is not cancelled. This is the cause of inertia.

Equivalence is an important phenomenon of gravitation. It is impossible to distinguish between the effects that gravity and imposed acceleration have on a mass. From duality there must be an equivalence effect for gravitomagnetism. We expect that it is impossible to distinguish between the effects that induced gravitomagnetism and an imposed rotation have on a mass. Note that gravity and acceleration act in opposite directions, so we must expect that induced gravitomagnetism and angular velocity act in opposite directions, too.

Although acceleration is equivalent to gravity and induced gravitomagnetism is equivalent to angular velocity, neither of the equivalences possess field properties. They have to be imposed to produce an effect.

In terms of relativity, when a mass moves with speed v relative to an observer, we expect a gravitomagnetic field to arise in the observer's frame of reference (along with an altered gravitational field). However, if the observer travels with the mass, moving at uniform speed, then the observer is at rest with respect to the mass and no gravitomagnetic field arises (and the gravitational field of the mass is unaltered). If the mass accelerates that is a different matter; the gravitomagnetic field changes, inducing a gravitational force which opposes the force causing the acceleration.

If the Lorentz transforms of special relativity are used to model the force **F** exerted by one mass m on another mass m, both moving with uniform speed **v** along parallel paths, the result in a stationary reference (0)-frame can be written in the form $\mathbf{F} = m\mathbf{g} + m\eta(\mathbf{v} \times \mathbf{h})$, showing that apart from the gravity field g, there exists a gravitomagnetic field **h** around a moving mass.

When a mass M moves through an external gravitomagnetic **h**-field we must expect the reverse effect to occur and the mass will be induced to rotate about the gravitomagnetic field line. This idea stems directly from the magnetic analogue.

In the magnetic case, a charge Q moving with speed **v** across a magnetic field **H** will circle around the magnetic field line. This is due to the magnetic part of the Lorentz force, $\mathbf{F} = Q\mu(\mathbf{v} \times \mathbf{H})$. Since charge has mass, there will an accompanying twinning effect. This is the gravitomagnetic analogue of the Lorentz force $\mathbf{F} = m\eta(\mathbf{v} \times \mathbf{h})$. So, in the case of an uncharged mass entering a strong gravitomagnetic field, we expect the mass to rotate, or circle, round a gravitomagnetic field line. In fact, we are very familiar with the equivalence form, of a mass rotating about an axis; it is the Coriolis force. If we imagine the equivalent angular momentum density field, we would expect to see the same pattern as that exhibited by a vortex in a free stream, or a current-carrying wire in a magnetic field.

Newtonian gravito-statics plus kinetic energy must give the same result as Newtonian gravity plus gravitomagnetism. As an example, consider a point mass m at the end of a light horizontal rigid rod of length R. We use cylindrical polar coordinates (R, θ, z) to describe the model. Suppose the mass m rotates about the vertical z-axis with angular velocity $\Omega$.

First, let us look at the Newtonian approach. From Newton's 2$^{nd}$ law the force **F** on the rotating mass m is given (A.19) by

$\mathbf{F} = -mR\Omega^2 \hat{\mathbf{R}}$  where $\hat{\mathbf{R}}$ is the unit vector in the radial direction.

This is the centripetal force. Note that the inward acceleration $-R\Omega^2$ is equivalent to gravitation in the outward direction, giving rise to the centrifugal force exerted on the mass m.

Now let us view the model in terms of gravitomagnetism. As the mass m moves with tangential velocity v, it generates a gravitomagnetic field **h**. But in rotating, the mass is subject to an angular velocity $\Omega$ which is equivalent to an induced gravitomagnetic field **b** in the opposite direction. The two gravitomagnetic fields interact. The gravitomagnetic field of the moving mass is exaggerated in the illustration to show the effect.

We can rewrite the Newtonian result as

$\mathbf{F} = -m(R\Omega\hat{\boldsymbol{\theta}} \times \Omega\hat{\mathbf{z}})$ where $\mathbf{v} = R\Omega\hat{\boldsymbol{\theta}}$ is the tangential velocity and $\Omega\hat{\mathbf{z}}$ is the angular velocity about the vertical axis.

Ignoring the weight term m**g**, this is in the form of the gravitational analogue of the Lorentz force $\mathbf{F} = m(\mathbf{v} \times \mathbf{b})$ where $\mathbf{b} = -\boldsymbol{\Omega}$ is the equivalent induced gravitomagnetic field. It is the interaction of the gravitomagnetic fields that gives rise to the centripetal acceleration $\mathbf{v} \times \mathbf{b}$.

A linearly accelerating body is accompanied by a changing gravitomagnetic field. Maxwell's linearised gravity theory predicts that a gravity field is induced in the body, in opposition to the acceleration. We can speculate that the induced gravity field forms a mass dipole within the body. The negative mass component of the dipole is created by converting some positive mass into positive energy which is radiated away. In the case of a body rotating with uniform angular velocity, the body experiences radial acceleration but the gravitomagnetic field remains constant. Consequently, no gravity field is induced in the body and no positive energy is radiated away.

In electromagnetism, changes in the **E**-field and the **H**-field take place at the speed of light c in a medium, where $c^2 = 1/(\varepsilon\mu)$. In Newtonian gravity, changes in the **g**-field take place instantly at every point in space, which is unrealistic. With our extended Newtonian model of gravity, we assume that changes in the **g**-field and the **h**-field also take place at the speed of light c, only in this case where $c^2 = 1/\gamma\eta$. So, the assumption is that electromagnetic and gravitational disturbances travel across space together. Furthermore, the analogue suggests that gravitational disturbances create gravitational waves in the ether (see Chapter 36). The linearised gravitational wave equations are

$$\frac{\partial^2 \mathbf{g}}{\partial x^2} = \frac{1}{c^2}\frac{\partial^2 \mathbf{g}}{\partial t^2} \text{ and } \frac{\partial^2 \mathbf{h}}{\partial x^2} = \frac{1}{c^2}\frac{\partial^2 \mathbf{h}}{\partial t^2}.$$

Recent detection of gravitational waves from a massive gravitational disturbance, combined with astronomical observation of the same event, has confirmed that electromagnetic waves and gravitational waves both travel in free space at the same speed c across the universe. That is

$$c^2 = \frac{1}{\varepsilon\mu} = \frac{1}{\gamma\eta} = (3\times10^8 \text{ m/s})^2 \text{ in the vacuum.}$$

Since $\gamma = 1.193 \times 10^9$ kg s²/m³ (taken to be a constant, at least for the Solar System), then $\eta = 0.9314 \times 10^{-26}$ m/kg. So, $\eta$ is very nearly zero in free space, but not quite. The induced effect that an **h**-field has on a body is given by $\mathbf{b} = \eta \mathbf{h}$, which means that only when the **h**-field is huge, say due to a rotating astronomical body, will there be any noticeable induced effect **b**. In general, we will be unaware of external gravitomagnetic fields acting on bodies across space.

In Newtonian gravitation, speed c is infinite. Since $\gamma$ is constant, this means $\eta = 0$ and there is no induced gravitomagnetic field. Classically, as found in text books on Newtonian mechanics, a point mass moving at a velocity v with respect to a reference (0)-frame has kinetic energy ½mv², where the mass m is constant. If we allow for a change in mass, as predicted by special relativity where c is finite, we get $m_1 = m_0/\sqrt{1 - v^2/c^2}$, where the (1)-frame is moving with speed v relative to the stationary (0)-frame. Expanding the expression under the square root sign as an infinite series and neglecting insignificant terms of order $v^4/c^4$ and higher, we get $m_1 = m_0(1 + \frac{1}{2}v^2/c^2)$. Multiplying both sides by $c^2$, we find that the rest energy $m_1 c^2$ in the (1)-frame is equal to the rest energy $m_0 c^2$ in the (0)-frame plus the Newtonian kinetic energy $\frac{1}{2} m_0 v^2$ as observed in the (0)-frame. So, apart from introducing rest energy, special relativity doesn't alter Newtonian mechanics.

Experiments with electrons of charge -e and mass $m_e$ being deflected by a magnetic field have confirmed the special relativity mass-velocity relationship. But, we might argue that the mass m remains constant and that the energy ½mv² is the gravitomagnetic energy of the moving mass m stored in space. With this view, the magnetic field **H** of the moving electric charge is combined, or twinned, with the gravitomagnetic field **h** of the moving mass $m_e$. The claim that the deflection experiments support the increase in mass of the electrons might be better interpreted, as the experiments support the presence of a gravitomagnetic field. The apparent increase in mass of an electron shows up as an increase in its inertia. During the deflection of the mass, its gravitomagnetic field changes and, according to the gravitational analogue of Faraday's law of induction, a gravity field is created. It is this gravity field acting on the electron mass which is the source of the increased inertia.

Since Newtonian dynamics has been so successful, what is the point in developing a theory of gravitomagnetism? The point is that with our current understanding of gravity, we have no idea how to control this force. There's something missing from the theory of gravity that holds the key to controlling it, and that key is gravitomagnetism. Once we can manipulate gravitomagnetism, we will have the key. So, first of all, we must learn all we can about gravitomagnetism. That's the easy part. From our knowledge, we must then devise some Earth-based laboratory experiments to investigate the phenomenon of gravitomagnetism. The crucial part of the experiments will be the method of detecting gravitomagnetism. That's the hard part given that for gravitomagnetism to induce an effect, we are dependent on the gravitomagnetic permeability η, which is nearly zero in free space. If the experiments are successful then we must probe gravitomagnetism and discover how we can manipulate it. Only then will we be able to control gravity.

The simple pendulum provides us with an example to examine the presence of gravity and gravitomagnetism. Suppose that the pendulum is of length ℓ and that the mass of the bob at the end is m. Due to Earth's gravity field **g**, when the bob is released it will swing down and then oscillate to and fro.

For small angles of oscillation θ, the differential equation of the motion is (A.20)

$$\frac{d^2\theta}{dt^2} + \sqrt{\frac{g}{\ell}}\theta = 0$$

This is an equation for simple harmonic motion with resonant frequency:

$$f_{Res} = \frac{1}{2\pi}\sqrt{\frac{g}{\ell}} \text{ Hz}$$

We now look at the simple pendulum in terms of an electromagnetic analogue (A.21). The moving bob creates a mass current I around which is a gravitomagnetic field. Due to the angular velocity dθ/dt of swing, the bob is also moving in an equivalent induced gravitomagnetic field. Based on the electromagnetic analogue, the self-inductance of the simple pendulum is

$$L = \frac{(\ell\theta_0)^2}{m} \text{ m}^2/\text{kg}$$ where $\theta_0$ is the starting angle where m is at rest.

And the capacitance C of the simple pendulum is given by

$$C = \frac{m}{g\ell\theta_0^2} \text{ kg}^2/\text{J}$$

From the analogue of an electromagnetic tuning circuit, the resonant frequency of the pendulum oscillation is

$$f_{Res} = \frac{1}{2\pi}\frac{1}{\sqrt{LC}} = \frac{1}{2\pi}\sqrt{\frac{g}{\ell}} \quad Hz$$

Thus, our gravitomagnetic model of the simple pendulum coincides exactly with the Newtonian mechanical model.

In obtaining the above results, we have treated the pendulum bob as though it was a point mass. In reality, the bob has dimensions, and the gravitomagnetic permeability $\eta$ within the bob might depend on its substance and be different to the free space value $\eta$. In which case, the induced gravitomagnetic field $\mathbf{b} = \eta\mathbf{h}$ inside the bob would wax and wane in a manner different to that in free space. So, the gravitomagnetic model suggests that tests using bobs of different substances might lead to experimental results different from those predicted using the Newtonian point-mass model. But this idea flies in the face of Newton's experimental results, when he examined the properties of inertial and gravitational mass (see Chapter 5) and found that using bobs of different substances made no difference. So we must proceed with the utmost caution. Nevertheless, Faraday suspected something of this sort might occur with his comment on *the great difference **within** and **without*** (See Chapter 32, diary entry: 10140, 30 August 1849). In fact, during the 1920s, Dr Charles Brush reported that just such an effect did occur (see Chapter 37), but his results were never taken seriously, no doubt because of Newton's results. But, if we could alter the induced gravitomagnetic field within a body, it would give us the first sign that gravity can be manipulated, via the gravitational analogue of Faraday's law of induction. Careful experiments need to be undertaken.

The tuning circuit, connected to an antenna, enables a radio receiver to lock into a particular external electromagnetic wave. In the gravity case, natural gravity waves are extremely weak, so we can't use our simple pendulum to tune into (resonate with) them. But, resonance is a very common occurrence within structures, and if we could look inside them, we might 'see' the changing gravitomagnetic field if we knew how to detect it. Due to the twinning between magnetism and gravitomagnetism, one possibility is to use magneto-optics.

Not long after Faraday's discovery of magneto-optics in 1845, the French physicist Marcel Émile Verdet carried out a series of experiments with different transparent materials. He derived an approximate formula linking the rotation $\theta$ of the plane of polarised light with the intensity H of the magnetic field and the path length $\ell$ of the polarised light ray through a particular material. The simple formula is

$$\theta = VH\ell$$

V is the Verdet constant for the medium, measured in minutes of arc (1/60 of a degree) per Ampere, H is the magnitude of the magnetic field measured in Amperes per metre (A/m) and $\ell$ is the length of the material measured in metres (m). The rotation $\theta$ is in minutes of arc.

These days, the rotation of a linearly polarised light beam encountered in the Faraday Effect (FE) may be simply demonstrated with a length of fibre optic. The fibre optic length is placed along the axis of an active solenoid. A laser light beam is then shone through a linear polariser onto one end of the fibre. At the output end, the laser light beam emerging from the fibre is shone through another linear polariser, which is rotated until a maximum light signal is received. Ideas for experiments to detect the gravitomagnetic field using the Faraday Effect are discussed in more detail in Chapter 40.

There is also an inverse Faraday Effect (IFE); a phenomenon whereby circularly polarised light passing through a gyrotropic medium (in which incident light and its mirror reflection do not match), such as a plasma, induces an axial magnetic field. During the 1970s, using low-powered lasers, relatively small magnetic fields were generated in this way. Since then, very large magnetic fields **B** have been induced in plasmas using the IFE method with the Vulcan laser at the Rutherford Appleton Laboratory, near Oxford.

When Faraday developed his vision of electromagnetism experimentally he introduced the idea of interacting electric and magnetic fields. Maxwell and Heaviside turned Faraday's field idea into a mathematical model for electromagnetism. A quantum theory for electromagnetism came later after J. J. Thomson's discovery of the electron in 1897. In extending Newton's static theory of gravity we have based the idea on an analogue of electromagnetic field theory. Since an electron has charge $-e$ and mass $m_e$ we can be certain that a quantum theory of gravity will follow an analogue with quantum theory for electromagnetism. An electron also has spin so it has a magnetic and a gravitomagnetic dipole. We expect that a theory of quantum gravity will explain how a gravity field is created within a mass made of quantum particles (atoms and molecules) when the mass accelerates.

The fundamental particles of matter within the atom all have tiny masses. The neutrons are uncharged but the protons have charge $+e$, while the quarks, which make up the protons and neutrons, have a fractional charge of $\pm e$. The nuclear strong force holds the charged and uncharged particles in the atom together. They all have mass, so it is likely that the strong force is some form of extremely strong gravitational force. The electrons in the shells orbiting the nucleus have charge $-e$. Overall, the charges of the sub-particles cancel out, leaving the atom with zero charge, but non-zero mass.

With our guess, we might expect that the continuous gravitomagnetic field **h** (J.s/m$^3$) is linked with discrete quantum theory and Planck's constant h. After all, h = $6.626 \times 10^{-34}$ J.s is the smallest unit of angular momentum that can exist. When Bohr extended Rutherford's Solar System model of the atom, he suggested that the angular momentum of the orbiting electrons must be an integer multiple of Planck's constant h. This meant that the electrons orbiting the nucleus could only occupy discrete orbits. Another way of viewing Bohr's suggestion is that the orbiting electrons are subject to discrete values of orbital angular velocity. But angular velocity is equivalent to induced gravitomagnetism. For Bohr's model of the hydrogen atom, the induced gravitomagnetic field **b** within the atom is of order $10^{14}$ rads/s. So, inside the atom, the equivalent induced gravitomagnetic field is huge. Furthermore, the Newtonian kinetic energy of the orbiting electron, used to calculate the atomic radiation frequencies, is the gravitomagnetic energy of the electron. As far as I am aware, particle physicists pay no heed to gravitomagnetism in their atom-smashing experiments. Perhaps, taking gravitomagnetism into account within the atom might explain the weak and strong forces within the atom.

Let us now consider the discrete quantum mechanical property of spin. All the sub-particles of mass (fermions) within the atom possess spin and, therefore, have intrinsic angular momentum, so we can think of each of them as gravitomagnetic dipoles. But it's a strange form of rotation. Spin is linked to Planck's constant h. If the particle rotates clockwise (as seen in your imagination), its spin is $+\frac{1}{2}(h/2\pi)$, known as spin s = $+\frac{1}{2}$. If it rotates anti-clockwise, its spin s = $-\frac{1}{2}$. Not only is spin s incredibly small, it is discrete and only has two values.

The electron, with charge $-e$ and mass $m_e$, has spin s = $\pm\frac{1}{2}$, so the electron rotates and develops a microscopic magnetic dipole as well as a microscopic gravitomagnetic dipole. The nucleus of an atom is surrounded by a number of electron shells. When an electron shell is full, it has an even number of electrons and their spins cancel out. If the outer electron shell of an atom contains an odd number of electrons, then the atom will have spin. But, since the electric field is $10^{40}$ greater than the gravity field, the magnetic dipole of the electron is the dominant characteristic of the atom. Within the atomic nucleus, other magnetic and gravitomagnetic dipoles exist, but what part they play is unknown.

At the macroscopic scale of uncharged matter, the random values of particle spin making up a mass leave it with no apparent external spin. However, rotating the mass about an axis with a range of angular velocities will give it a continuous range of angular momenta, or spin. Thus, the rotating Earth will form a gravitomagnetic dipole with an external gravitomagnetic field, but its strength is very, very small.

We can apply the electromagnetic analogue to a spinning mass ring. For ordinary-sized spinning rings, the gravitomagnetic field will exist inside the ring but, again, outside the ring, the induced gravitomagnetic field is so small that it can be approximated to zero for most cases.

In the electrical current ring case, the virtual dipole acts like a magnetic compass needle and a suspended ring will experience a turning force causing it to line up with an external magnetic field, which may be created by another ring current. The analogue with the current ring suggests that the rotating mass ring has virtual positive and negative gravitomagnetic poles. However, it seems most unlikely that virtual gravitomagnetic poles from different rotating mass rings will interact across space. But it depends on the sizes and angular velocities of the rotating mass rings. In general, gravitational attraction between the rings (whether rotating or non-rotating) will be the dominating factor, unless gravity is neutralised by having one ring in free fall in the gravity field of the other. In the NASA Gravity Probe-B experiment, a spinning sphere (think of it as a layer of mass discs) is used to detect the Earth's gravitomagnetic field (Chapter 35).

Based on the electromagnetic analogue that a rotating ring of electric charge creates a magnetic dipole, we expect that a rotating ring of mass will create a gravitomagnetic dipole with virtual positive and negative gravitomagnetic poles. We can be fairly sure that such things do exist, but even for extremely large rotating masses, the strengths of such dipoles are likely to be extremely small.

Let us model a gravitomagnetic dipole as a ring of mass M rotating with angular velocity $\Omega$. We let the face area of the ring be $A = \pi a^2$. The moment of inertia of the ring about its centre is $Ma^2$ and the angular momentum **H** of the ring is $Ma^2\Omega$. The mass current of the ring is given by $I = (1/2\pi)M\Omega = H/(2A)$. From the magnetic analogue we expect that the gravitomagnetic dipole will have a moment $\kappa$ with strength given by $\kappa = -\eta IA = -\eta \mathbf{H}/2$.

On placing a magnetic dipole in an external magnetic field it will experience a twist, or a couple C, as the dipole axis tries to align with the magnetic field. This phenomenon is explored in more detail in Larmor precession. The analogue indicates that when a gravitomagnetic dipole is placed in an external gravitomagnetic field $\mathbf{h}_{Ext}$ it, too. will experience a couple $C = \kappa \times \mathbf{h}_{Ext}$, equal to $d\mathbf{H}/dt$, as the dipole axis tries to align with the gravitomagnetic field. This gives rise to the gravitomagnetic analogue of Larmor precession. Any point on the axis of the rotating mass ring will perform a circle about the $\mathbf{h}_{Ext}$ field line with constant angular velocity $\omega = \kappa \mathbf{h}_{Ext}/H = \frac{1}{2}\eta \mathbf{h}_{Ext} = \frac{1}{2}\mathbf{b}_{Ext}$. If $\mathbf{h}_{Ext}$ increases with time then $\omega$ increases, too. This means closer alignment of the axis of the mass ring with the direction of the $\mathbf{h}_{Ext}$ field. During this time a point on the axis will follow an accelerating upward spiral. Since $\partial \mathbf{b}_{Ext}/\partial t = \nabla \times \mathbf{g}_i$, we can view the acceleration as being due to the creation of the induced gravity field $\mathbf{g}_i$.

We are now ready to investigate a simple quantum model of what happens within a mass when it is made to accelerate, From field theory, when the gravitomagnetic field changes with time, $\partial \mathbf{b}_{Int}/\partial t$, within a mass then an induced gravity field $\mathbf{g}_i$ is created which opposes the acceleration. We can think of the mass as a framework of atoms, or molecules, held in a lattice which gives them freedom to vibrate or spin. Since the atoms, or molecules, have mass and spin we can represent them as gravitomagnetic dipoles. From Larmor precession, a gravitomagnetic dipole in a changing gravitomagnetic field creates a gravity field which rotates about the $\partial \mathbf{b}_{Int}/\partial t$ field line such that $\partial \mathbf{b}_{Int}/\partial t = \nabla \times \mathbf{g}_i$. Thus, for mass acceleration field theory indicates a changing gravitomagnetic field within the mass, while a quantum theory indicates the creation of a gravity field in response. The two go hand in hand.

Since there is an equivalence between induced gravitomagnetism $\mathbf{b}$ and angular velocity $\mathbf{\Omega}$ there will also be a Larmor precession for a rim flywheel. Suppose we attach the moment arm of a flywheel, via a pivot, to the centre of a horizontal turntable. As the table is turned (ignoring gravity) the moment arm will try to align with the angular velocity of the turntable.

The photon is a virtual particle (boson) which carries the electromagnetic force. It is an uncharged quantum unit with effective mass $hf/c^2$ and spin $s = \pm 1$. Consequently, we may think of the photon as a gravitomagnetic dipole. According to Pauli's exclusion principle, particles of matter (fermions) cannot overlap, but virtual particles conveying the effect of a force field (bosons) can. This means that photons can be crowded together. So we can think of a beam of laser light moving through the vacuum (non-gyrotropic medium) as made up from millions of photonic gravitomagnetic dipoles all moving along at the speed of light c. In the beam of laser light, the axis of spin of all the photons aligns with the axis of the beam. The axis of spin constitutes an angular velocity vector which is equivalent to an induced axial gravitomagnetic field **b**. This indicates a link between gravitomagnetism and the IFE.

Suppose we have an annulus, or ring, containing either a vacuum or uncharged gas. If a polarised laser beam is formed in the ring then an induced gravitomagnetic field **b** will form along the centre line within the ring. Suppose a pulse of circularly polarised laser light is added to the laser beam. Does it result in an impulsive change in the **b**-field? If so, an induced impulse gravity field **g** would be generated through the centre of the ring, perpendicular to the plane containing the ring of laser light (since $\nabla \times \mathbf{g} = \partial \mathbf{b}/\partial t$). Has any experimenter thought to try this?

For more than a century after Maxwell proposed his model to extend Newton's theory of gravity, which included a dual force field of gravity, his work remained largely unknown, or ignored, by scientists. Following the publication of Einstein's theory of gravity, Maxwell's version lay hidden in its shadow. Today, experts in general relativity refer to the Maxwell form of gravity as a linearisation of Einstein's equation for gravity (or GEM model), where gravity is weak and velocities are small compared with the speed of light. But the electromagnetic analogue of gravity and the linearisation of Einstein's gravity equation are not quite the same.

The conclusion that gravity is due to the curvature of space-time has not led to a breakthrough in understanding how to manipulate gravity in the way that we can manipulate electromagnetism. Over the last 50 years, a number of scientists have re-examined the Maxwell electromagnetic analogue of gravity in the hope that, being much less complex than Einstein's theory of gravity, it might shed some light on the gravity phenomenon. Note that in the analogue we assumed that induced gravitomagnetism **b** was torsion, but torsion is missing from Einstein's model. Those who follow Heaviside's original observation refer to the gravitomagnetic field as the Heavisidean field. Dr Robert Forward, a research scientist at Hughes Aircraft Company, referred to it as the protational field. The version of gravitomagnetism derived from special relativity by Professor Paul Lorrain, of the University of Montreal, has been labelled the gyron field. Professor Laithwaite called it the kinemassic field. Dr Oleg Jefimenko of the University of West Virginia called it the co-gravitational field. My own choice would be to call it the Rondor field, which is shorter to write and gives the dual field its own personality.

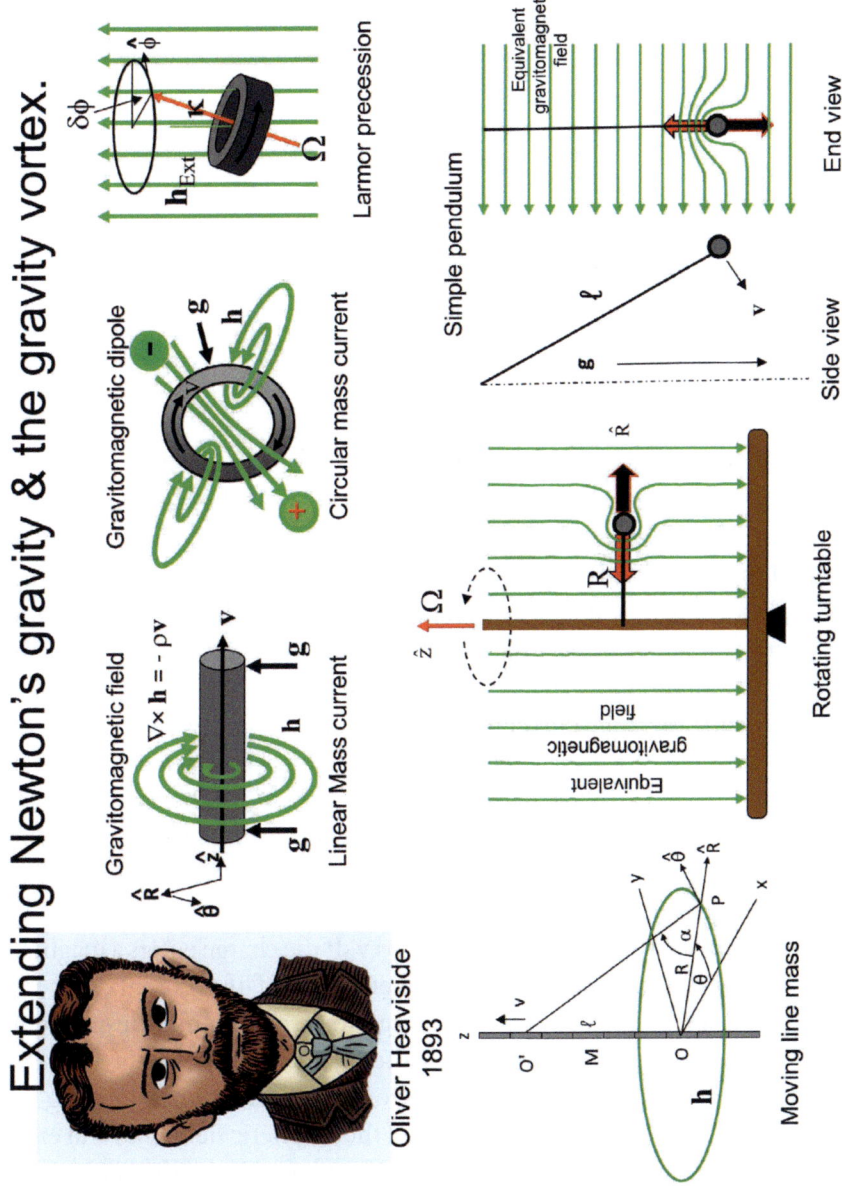

CHAPTER 34

# DETECTING MAGNETISM AND GRAVITOMAGNETISM

To understand how we might detect gravitomagnetism, we will start by looking at the analogue with magnetism and see how that field is detected.

The most obvious way to detect the presence of an external magnetic field **H** in space is with a compass, where the magnetic needle experiences a couple and aligns with the magnetic field. The Earth acts like a gigantic magnetic dipole, creating a magnetic field around the globe allowing the compass to be used for navigational purposes.

The coil, or ring, current creates an electromagnet with a virtual magnetic dipole. If we suspend a current-carrying coil from a point on its perimeter in the Earth's magnetic field then the axis of the coil will try to align with the Earth's field, indicating its presence. We can use a suitably mounted coil, as a search-coil, to plot out the magnetic field lines of an external magnetic field.

Suppose a charge Q moves with speed **v**. If the charge enters a magnetic field **H**, then it experiences the magnetic part of the Lorentz force, **F** = Q(**v** × $\mu$**H**), which causes the charge to circle around the magnetic field lines. The spiralling motion of the charge detects the presence of the magnetic field. Charged particles from the Sun on encountering the Earth's magnetic field corkscrew around the magnetic field lines and enter the atmosphere near the poles, causing the green glow of the auroras.

We can detect a magnetic field using the Faraday Effect (FE). When a linearly polarised light beam, of length $\ell$, runs parallel to a magnetic field **H**, the plane of polarisation is rotated by an angle $\theta$ (see Chapter 33). Measuring $\theta$ allows us to determine the magnitude of the magnetic field **H**.

The magnitude of a magnetic field **H** can be measured very accurately using a quantum mechanical device called a SQUID. The device has an extremely small aperture which allows quantum amounts of magnetic flux h/2e to pass through the hole, where h is Planck's constant and e is the charge of an electron.

We now consider the gravitomagnetic field **h** and the equivalence of the induced field **b** with spin $\Omega$. The Earth can be thought of as layers of concentric mass rings forming a sphere. Consequently, the rotating Earth creates a gravitomagnetic dipole giving rise to a gravitomagnetic field. The analogue with magnetism suggests that (if our theoretical model is true) we might be able to use a rotating mass ring to detect the Earth's gravitomagnetic field.

The spinning top, with a whip to keep it spinning, was a favourite Victorian toy, with its origins in ancient Greece. It's merely a question of scale, but the spinning top and the spinning Earth are both spinning masses. The simplest spinning mass is the flywheel, where the mass is mostly concentrated in the rim of the wheel. The flywheel has an axial direction associated with it called the angular velocity of rotation. The electromagnet is a magnetic dipole which has an axial direction, from a virtual south-seeking pole to a virtual north-seeking pole. The analogue indicates that the flywheel is a gravitomagnetic dipole with a virtual positive (+) gravitomagnetic pole and a virtual negative (-) gravitomagnetic pole. The direction of the gravitomagnetic field is from the negative pole to the positive pole. Thus, the angular velocity vector of the flywheel is in the opposite direction to the perceived gravitomagnetic field.

The first person to consider turning the flywheel into a scientific instrument was the famous French scientist Pierre Laplace. He mounted the axle of a flywheel in a gimbal. The gimbal is a pivoted support that allows the flywheel to rotate about a single axis. The French scientist Léon Foucault took the gimbal idea a step further and introduced a nested set of three gimbals, one mounted on the other orthogonally (at right angles), allowing a flywheel, once it had been set spinning, to continue rotating about the same axial direction even when the support was rotated. Foucault called his device a gyroscope.

In 1852, Foucault was the first scientist to demonstrate the rotation of the Earth using a gyroscope. He suspended the gyroscope from a string, initially with the flywheel's axle horizontal. Once he had started it spinning, it would only keep running for about 10 minutes (due to bearing friction), so he had to make his observations during this time.

Foucault fixed a long, very light, horizontal pointer to the outer gyroscope frame (to amplify the effect of movement), the end of which moved in azimuth over a scale marked in degrees.

The magnitude of the Earth's angular velocity $\Omega_E$ in degrees is 360° per day, or 15° per hour, or 15/60 degrees per minute. Since angular velocity is a vector, we can resolve $\Omega_E$ into locally horizontal and vertical components for any latitude $\lambda$. Foucault conducted his experiment in Paris, where $\lambda$ = 48.9°North. The flywheel pointer registered the rotation of the flywheel's axle about the vertical axis as seen by an observer standing next to it.

From his mathematical model, Foucault knew that in 5 minutes the rotation of the flywheel axle direction would be 1 degree. It was a small change and might have been missed if Foucault hadn't been looking for it. That's the importance of having a mathematical model to work with!

A year earlier, Foucault had used a simple pendulum with a very long string and shown that the plane of swing gradually rotated around the vertical, demonstrating that the world was rotating. Since rotation $\mathbf{\Omega}$ is equivalent to induced gravitomagnetism $\mathbf{b}$, this can be seen as the gravitational analogue of the magnetic part of the Lorentz force, namely $\mathbf{F} = m(\mathbf{v} \times \mathbf{b})$.

The property of the gyroscope axle always to point in the same direction in space (regardless of any twisting and turning of the support), coupled with the electric motor to keep the flywheel spinning, led to its use as an inertial navigation system, replacing the magnetic compass. Note that if the gyro moves over the surface of the Earth, that is tantamount to a rotation, for which correction must be made.

Suppose we go to the North Pole and suspend a 3-gimballed frame there, containing a spinning flywheel, or gyro, with its axle horizontal. An observer standing by the suspended gyro would see the flywheel axle rotate by 360° during one sidereal day. That is, one complete rotation of the Earth with respect to the fixed stars. In fact, the flywheel axle stays pointing in the same fixed direction in space that it started with, or so it seems. It is the observer who rotates.

If the gimballed frame is fixed to the ground, then the flywheel axle will try to align with the axis of Earth's rotation. This is an example of the equivalence between induced gravitomagnetism $\mathbf{b}$ and enforced angular velocity $\mathbf{\Omega}$.

Suppose we suspend a flywheel at the equator with its axle initially pointing in the west-east direction. After 6 hours, the axle would be vertical; then 6 hours later, it would lie in an east-west direction; then after a further 6 hours, the axle would have turned upside down until finally, after a total of 24 hours, the axle would return to its original west-east setting. In fact, the axle direction remains fixed in space and only appears to change to an observer on the rotating Earth.

But this seems to show that the spinning flywheel is not able to detect Earth's gravitomagnetic field. Is this because the flywheel doesn't move in relation to the Earth? No, if we keep the flywheel fixed on Earth, then we have to imagine all the mass of the heavenly bodies in space rotating around the Earth, so relative mass motion still occurs and the Earth should have a gravitomagnetic field about it, but the model becomes more complicated. The idea of the relative motion of a spinning mass was considered by Newton, with his rotating bucket of water experiment. What Newton wanted to know was why did the rotating water rise up the inside of the bucket wall? We can guess that in Newton's mind he saw water as mass and mass responded to gravity. So where did the gravity field arise in the rotating water? He wondered whether, if the bucket of water was rotated in truly empty space, the water would still rise up the inside of the bucket. During Newton's time, Huygens had shown that a rotating mass developed an inward radial acceleration. Therefore, from equivalence, a gravity field would arise in the outward radial direction. This would seem to answer Newton's question; even in empty space, mass rotation involves mass acceleration. With the introduction of gravitomagnetism, we can see that the gravitational force experienced by the rotating water is part of the gravitational analogue of the Lorentz force $F = m(\mathbf{v} \times \mathbf{b})$.

The answer to the question of whether a spinning flywheel can detect a gravitomagnetic field depends on the sensitivity in measuring any change in axle spin direction. Theoretically, it should be possible, but due to the extreme smallness of the gravitomagnetic permeability of space, $\eta$, any change in axle direction due to gravitomagnetism is exceedingly small and is usually undetectable. From equivalence, the flywheel always responds to induced gravitomagnetism arising from imposed angular velocity.

Galileo was the first scientist to realise that a pendulum, with its constant period of oscillation, could be used to measure time. This idea led to pendulum- regulated clocks. If Galileo had realised that he could use a long pendulum to demonstrate the Earth's rotation to the Church authorities, one wonders what his fate might have been.

No establishment likes having its authority challenged. When Giordano Bruno, at one time an Oxford University lecturer, returned to Rome, he was accused by the Inquisition of teaching the Copernican view of astronomy, rather than the fixed-Earth-at-the-centre-of-the-Universe dogma of the Church. He refused to recant and was burnt at the stake in 1600. Galileo held the same view as Bruno, but knowing of the latter's fate, sensibly recanted when he appeared before the Inquisition in 1632. Even so, he was placed under house arrest for life.

We now know that the fixed Sun model proposed by Copernicus leads to a simpler model of the Solar System, but the fixed Earth Ptolemaic model is equally legitimate, although more complicated, since acceleration is involved.

Over time, the severity of dealing with scientists with controversial views has changed. Today, if a scientist is deemed to have stepped out of line with the accepted orthodoxy then he, or she, is likely to be ostracised, or ridiculed, by his, or her, peers. Unfortunately, most scientists (including me) are guilty of these initial reactions. We feel uncomfortable if the firm foundation of our understanding is disturbed. This doesn't just apply to science but is a natural reaction to all major changes. But, it's a curious fact that some breakthroughs in scientific understanding start out based on ideas which, at first sight, are unbelievable, illogical or just plain daft. Who would have thought that dancing frogs' legs would lead to the electric battery? As Albert Einstein observed, "If at first the idea is not absurd, then there is no hope for it."

Never mind the idea of Galileo using a long pendulum to show the Church officials that the Earth rotated; suppose he had suggested that a saint's halo, or ring of light, could be used to measure the Earth's rotation. "That's enough. Where's the tinder box?"

We have assumed that induced gravitomagnetism is equivalent to enforced angular velocity, or rate of rotation (inverse time). And that in the dual case, gravity is equivalent to enforced acceleration. Note that gravity and gravitomagnetism both have fields and can influence matter across space without direct contact, while the equivalent properties of acceleration and angular velocity must have direct contact with matter to cause an effect.

How do we detect enforced rotation? As already mentioned, the simplest devices are the swinging pendulum and the spinning flywheel. As we noted above, gravitomagnetism is so weak that we can, generally, ignore its effect across space influencing another mass. The theoretical model for the motion of the flywheel is based on Newtonian mechanics and begins with the statement that the rate of change of angular momentum of the flywheel is equal to the moment (force exerted on the flywheel times the distance from the flywheel pivot) of the applied force. This leads to Euler's equations of motion for the flywheel. The mathematical model explains why the flywheel rises during precession. The weight of the flywheel is not a factor.

In much simpler terms, rotation is a vector and to add vectors one must use the parallelogram rule. So, if an attempt is made to rotate a flywheel about some other axis, the flywheel will move so that its axis of spin aligns with the resultant sum of the two rotation vectors. The axial movement indicates the actual rotation.

In 1913, the French physicist Georges Sagnac wanted to detect the rotation of the Earth, like his compatriot Foucault had done 60 years earlier. He had the strange idea that a ring of light could be used to detect rotation. His idea was based on interfering light waves, or interferometry.

Sagnac's experiment involved a horizontal turntable with four 45° plane mirrors placed at the four corners of a square and a coherent, monochromatic (single frequency) source of light. The first mirror was half silvered so that the light beam split and two light beams bounced around the square in opposite directions. The perpendicular to Sagnac's turntable was coincident with the vertical component of Earth's rotation, which is $\Omega_E.\sin \lambda$ for latitude $\lambda$. Sagnac based his idea on light moving in the ether. He assumed that the rotational axis of one of the beams of light going around the square would point in the same direction as the vertical component of the Earth's rotation, while the other would point in the opposite direction. So he assumed, mistakenly, that one beam of light would be slightly speeded up, while the other would be slightly slowed down. On viewing the combined waves, Sagnac expected to see a moving interference pattern due to the vertical component of Earth's rotation. However, he couldn't detect any movement of the interference pattern.

To check that enforced rotation really did cause a change, he rotated the turntable at just two revolutions per second (720°/s = 12.57 rads/s) and, sure enough, the interference pattern moved.

We now know that Sagnac's apparatus was not sensitive enough to detect the Earth's vertical component of rotation. In Paris, $\Omega_E.\sin \lambda \approx$ 0.0032°/s ($5.5 \times 10^{-5}$ rads/s), which was far too small for Sagnac to detect. But, although his premise that the speed of light would change was not true, he had shown a way of detecting enforced rotation in general.

It wasn't until the advent of the laser in the 1960s that scientists could really make use of Sagnac's idea, and that's when the development of the active Ring Laser Gyro (RLG) began.

Imagine a skipping rope tied at one end and the handle at the other end vibrated. Waves travelling along the rope are reflected at the fixed end and travel back towards the handle. For the length of skipping rope, there are particular frequencies of vibration which can cause patterns of waves to appear in the rope which remain stationary. Where the two waves travelling in opposite directions exactly overlap, they combine to form an antinode that has twice the amplitude of the separate waves. When the two waves are exactly out-of-phase, they cancel each other out and the combined wave has zero amplitude, forming a node. For light waves, the antinode gives maximum intensity, resulting in a bright spot of light, while the node gives minimum intensity, resulting in darkness. Heinrich Hertz used standing waves in his experiment to detect wireless waves (electromagnetic waves with frequency lower than light), placing his detecting ring at an antinode and looking for a spark.

An RLG is formed from a square-shaped sealed cavity, with 45° mirrors at the four corners, filled with helium-neon gas. A laser beam, input at one corner, splits into two identical beams which pass in opposite directions around the square cavity. Minor adjustments are made to get a resonant cavity length and then standing waves form. From special relativity we assume that the light waves, travelling in opposite directions, both have the same speed c. So, when the waves combine, they form standing waves which remain stationary in space. Consequently, any rotation of the cavity about its central axis makes the nodes and antinodes appear to move. A peephole in the side of the cavity allows a photodiode to detect bright antinodes as they pass by.

The square encloses a planar area $A = a^2$ and has a perimeter length $P = 4a$. The total number of nodes, or antinodes, in the cavity is $P/\lambda$, where $f = c/\lambda$ is the laser frequency. The distance between two nodes or two antinodes is $\lambda/2$.

Suppose the cavity turns around its vertical axis with an angular velocity $\Omega$. Then the photodiode at the peephole will detect the antinodes of the standing waves passing by. In time t, the distance moved by the peephole is equal to peephole speed $(a/2)\Omega$ times the time t. This distance is equal to the length of the standing waves which have moved past the peephole photodiode, which is $n(\lambda/2)$, where n is the number of antinodes. Rearranging this equation, we get

$$\frac{4A}{P\lambda}\Omega = \frac{n}{t} = f_B$$

The Sagnac Beat frequency $f_B$ is the frequency at which the antinodes pass the photodiode.

An Inertial Measurement Unit (IMU) is a combination of three orthogonal accelerometers and RLGs which provides attitude, velocity and position data. All BAE Systems aircraft have an IMU fitted near to their centres, linked with a GPS receiver, to keep a constant check on their flight parameters.

The Earth's rotation can be detected with an RLG, if it is sensitive enough and, generally, this means increasing the enclosed area A.

The Fibre Optic Gyroscope (FOG) is a smaller and lighter development of the RLG, where the cavity of the RLG is replaced with a fibre optic ring. The FOG detects rotation $\Omega$ about the axis of its fibre optic ring and is a lot cheaper than an RLG but not as accurate.

Nowadays, the wave-like nature of matter has been established for atomic particles. Following the success of RLGs and FOGs for navigation purposes, some researchers have turned their attention to developing atom interferometers. That is, replacing beams of light particles (photons) with beams of matter particles. Atom interferometers are expected to vastly improve the sensitivity of detecting rotation by an order of $10^9$ over RLGs and FOGs.

We are now ready to look at trying to detect the effect that gravitomagnetism has on matter. From the analogue with electromagnetism we expect that the gravitational form of the Lorentz force will take the form

$$\mathbf{F} = m\mathbf{g} + m(\mathbf{v} \times \eta\mathbf{h})$$

Therefore, analogous to a charge moving in a magnetic field, we expect a mass m crossing a gravitomagnetic field $\mathbf{h}$ to spiral around. Because the gravitomagnetic permeability $\eta$ of space is so small, we are only likely to detect gravitomagnetic effects in space in the vicinity of large rotating astronomical bodies, such as stars and planets. To try to detect the gravitomagnetic field in space around the rotating Earth, we have two options. We can try with an Earth-based experiment or with a space-based experiment.

We start with an Earth-based experiment. Ultra G-1 (UG-1) was the world's biggest active RLG, when it was built in 1999. It is sited at the University of Canterbury, in Christchurch, New Zealand. UG-1 is rigidly fixed to the Earth's surface with the perpendicular to its area A pointing vertically upwards. The detection sensitivity of angular velocity of UG-1 is estimated to be $10^{-13}$ rads/s or about $6 \times 10^{-12}$ degrees per second. Christchurch is at the latitude of 44° south, so the vertical component of Earth's rotation there is 10°/hr, which is about $4.8 \times 10^{-5}$ rads/s. So detecting the Earth's rotation with UG-1 is not seen as a problem.

The Beat frequency for UG-1 is $f_B = 1.513$ kHz. This would be constant if the Earth's rotation was smooth and if UG-1 didn't suffer any disturbances. But the Earth's rotation is not quite smooth and there are some disturbances. The most important disturbance is due to the tides around Christchurch tilting the land mass near the shore ever so slightly. The slight tilt in the land results in the perpendicular to the face of UG-1 moving slightly out of alignment with the local vertical. Corrections have to be made for this and any similar effect caused by any local seismic activity.

Once these corrections are made, scientists start to see a more regular pattern for the Beat frequency, but it is not uniform. The Earth's axis of spin wobbles very slightly, like a child's spinning top. This is called the Chandler wobble, which has a period of about 432 days. In this time, the tip of the angular velocity vector $\Omega_E$ at the Poles traces out a circle with a diameter of between 4 and 6 metres. Small indeed, but it is detectable by UG-1. Also, due to the frictional effects of the Earth's tides, the magnitude of $\Omega_E$ is reducing and the length of a day is increasing by about 20 microseconds each year. This has also been detected in UG-1 data.

The vertical component of the Earth's induced gravitomagnetic field in Christchurch is predicted to be $0.35 \times 10^{-14}$ rads/s. Given the sensitivity of UG-1 is $10^{-13}$ rads/s, the detection of Earth's gravitomagnetic field using UG-1 was known to be borderline. During 2004, our Greenglow adviser, Professor Robin Tucker, with a small team from Lancaster University, spent 3 months at the University of Canterbury, NZ, analysing the UG-1 Beat frequency data. They attempted to separate out the Earth's gravitomagnetic field component from the Earth's rotation, but without success.

We now consider a space-based experiment to try to detect Earth's gravitomagnetic field. Briefly, we need a gravitomagnetic dipole, made from a rotating mass, with its axis of rotation initially set in a known direction. Detection of the Earth's gravitomagnetic field is then made if the direction of the dipole's axis moves due to interaction with the Earth's gravitomagnetic field. The magnetic analogue is the axis of an electromagnet moving in an external magnetic field. This was the basis of the US NASA Gravity Probe-B satellite experiment started in the early 1960s. The experiment took 50 years to complete. We will look at the experiment in more detail in Chapter 35. To date, the NASA space-based experiment is the only successful experiment to detect gravitomagnetism.

The NASA GP-B experiment collected dipole alignment data over a year before it was able to detect any axial movement. Technology has already moved on with the advent of atomic interferometers which are expected to be able to measure the Earth's gravitomagnetic field instantaneously.

In quantum mechanical terms, if the gravitational analogue of a SQUID is ever realised, then it might be possible to detect the gravitomagnetic flux in discrete units of $h/2m_e$, where h is Planck's constant and $m_e$ is the mass of an electron. This would be linked with a quantum theory of gravity which has yet to be formalised.

The major difficulty with detecting an induced gravitomagnetic field **b** in matter, due to an external gravitomagnetic field, is that the gravitomagnetic permeability η of space is so incredibly small. NASA made use of the rotating mass of the Earth to do it. However, all is not lost. Rather than trying to detect the presence of an external gravitomagnetic field via its interaction across space with a gravitomagnetic dipole, we might look within moving matter itself for the presence of a gravitomagnetic field.

Let us look at the magnetic case. At the Earth's surface, the induced magnetic field **B** = μ**H** is about $5 \times 10^{-3}$ Tesla, where in space $\mu = 1.257 \times 10^{-6}$ H/m. But materials exist with large magnetic permeabilities $\mu$ which enable induced magnetic fields **B** of 20 Tesla to be created, a thousand times stronger than the natural value associated with the Earth's induced magnetic field. At the moment, we don't know what the gravitomagnetic permeabilities η within matter are. There are reasons to suppose that they might be large within some materials.

For gravity control, we need to be able to create a changing induced gravitomagnetic field $\partial \mathbf{b}/\partial t$ within matter. It seems more likely that this will only be achieved by operating within matter, not across space from outside. But, first we must be able to detect gravitomagnetic fields, both steady and changing, in Earth-based experiments. In the long-term future, once we can generate a gravity field within matter, it may be possible to project gravity beams across space for attracting and repelling matter.

At the moment, the most likely way of detecting an internal gravitomagnetic field associated with moving matter is to use Faraday rotation. This assumes that gravitomagnetic Faraday rotation exists. That is, linearly polarised light in a fibre optic aligned with a gravitomagnetic field undergoes a rotation of its plane of rotation.

Other ideas exist for creating and detecting gravitomagnetic fields in Earth-based experiments. For example, it may be possible to separate a gravitomagnetic field harboured within a magnetic field, allowing other detection methods to be tried. A disputed Russian idea to detect a gravitomagnetic field is that a current in a wire aligned with a gravitomagnetic field experiences an increase in its conductivity or a reduction in its resistance. A number of experimental ideas to detect gravitomagnetism are considered in more detail in Chapter 40.

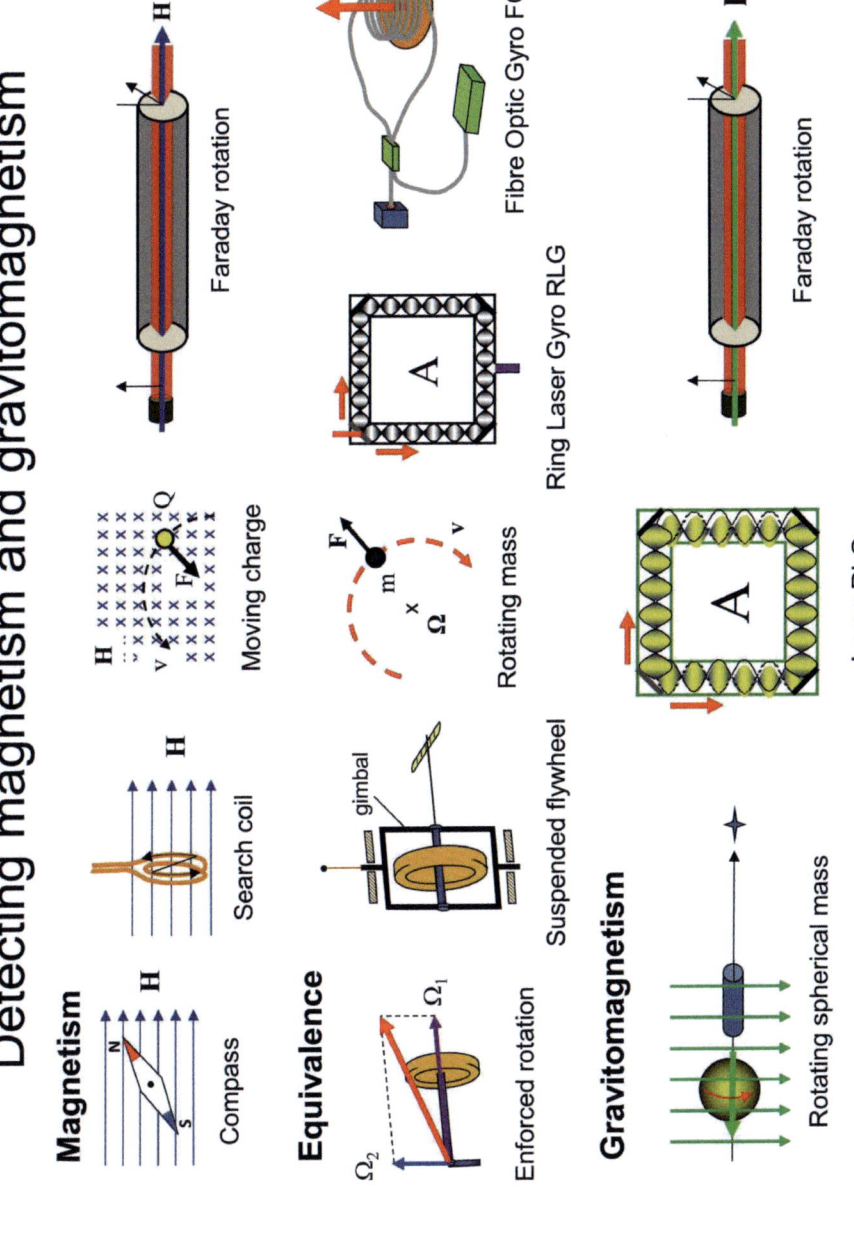

CHAPTER 35

# NASA GRAVITY PROBE-B

From our gravitational analogue of electromagnetism, we expect the rotating mass of the Earth to have a gravitomagnetic field. To try to detect the field, we need to place a spinning flywheel, or gyro, above the Earth and check whether its axis of spin moves. To keep the flywheel above the Earth, it must be placed in orbit. So, although the gyro axial direction is initially fixed, the gyro circles the Earth in an orbit, the plane of which doesn't turn but remains fixed with reference to the fixed stars. So, any spin axis interaction will be due to an averaging of the exposure of the gyro to the Earth's gravitomagnetic field.

According to the experts, Einstein's theory of general relativity predicts that the spin-axis of a spinning gyro will experience two orthogonal effects, namely:

1. Interaction of the spin-axis of the gyro with the Earth's gravitomagnetic field. This follows from the gravitational analogue of magnetism. The effect was also predicted by two Austrian scientists, Josef Lense and Hans Thirring in 1918, who derived it from a linearisation of Einstein's equation for gravity. They called the effect frame-dragging. It was as though the vacuum of space is slightly viscous, so that when a large astronomical mass was rotated it seemed to drag nearby space around with it.
2. Interaction of the spin-axis of the gyro due to the Earth's curvature of space-time. This is a non-linear effect predicted by Einstein's theory of gravity, but not by the analogue with electromagnetism.

Professor Leonard Schiff, of Stanford University, first began to think about an experiment to test Einstein's theory of general relativity in 1960. Schiff calculated that with an extremely sensitive gyroscope, orbiting close to the Earth in a fixed plane, it would be possible to measure the two effects predicted by Einstein's theory of gravity.

Schiff discussed his idea with other physicists and engineers at Stanford University and gradually the ambitious gravity experiment was conceived, with funding initially being obtained from NASA in 1964. Professor Frances Everett was appointed the Principal Scientific Investigator in 1966. The experiment took more than 40 years to prepare, with nearly 80 PhDs being awarded on the way. In simple terms, the NASA Gravity Probe-B (GP-B) experiment involved placing a satellite containing a weightless spinning ball-shaped flywheel, or gyro, in a polar orbit at an altitude of 650 km, above the rotating Earth. The centre of the orbit corresponded with the Earth's centre but, as the Earth orbited the Sun, the orbital plane of the satellite remained fixed with respect to the fixed stars. The gravity experiment was very expensive and several times NASA considered cancelling it. But, finally, the GP-B experiment module was launched from the Vandenberg Airbase in California on 20$^{th}$ April 2004.

General relativity experts predicted that the Earth's gravitomagnetic field would cause the spin axis of the GP-B gyro to move (precess) by 0.042 arcsec per year, and that space-time curvature would cause the GP-B gyro to precess by 6.6 arcsec per year. Both effects were exceedingly small (which is why we generally don't notice them). The effect due to space-time warping was several orders of magnitude greater than the effect due to gravitomagnetism, in line with Einstein's non-linear theory of gravity.

At the heart of the GP-B module is a 2.3 m long cylindrical vacuum chamber of diameter 250 mm which contains four gyros. Although only one gyro is needed for detection purposes, four are used to improve the measurement accuracy. Each gyro, about the size of a table tennis ball, is a quartz sphere of diameter 38 mm (1.5 in). They are coated with a thin layer of niobium and kept at a temperature of 1.8 K, where they become superconducting. A 356 mm (14 in) long tracking telescope occupies the end of the chamber containing the gyros. A reference star was chosen in the constellation of Pegasus and helium thrusters on the module are used to lock the telescopic sightline on to the star and, with it, the axis of the cylindrical chamber. The weightless gyros are spun up to 170 Hz (~

10,000 rpm). Initially, all their axes are aligned with the telescope sightline, two rotating one way and two the other.

When a superconductor rotates, it creates a current in its surface which generates a magnetic dipole field, the strength of which is called the London moment (after the German scientist Fritz London who first discovered it). The dipole axis always coincides exactly with the spin axis. The superconducting GP-B spherical gyros have a loop of wire around their circumference. Any change in the axial direction of spin is accompanied by a change in the direction of the magnetic dipole, resulting in a change of magnetic flux through the loop, which is detected by a SQUID magnetometer. The movement of the gyro rotation axes relative to the telescopic sightline can be measured to an accuracy of 0.0001 arcsec.

A magnetic field generally harbours a gravitomagnetic field, but the assumption is that superconductivity does not affect the gyro's gravitomagnetic field. The gyros are shielded from the Earth's magnetic field by a lead bag which, itself, is surrounded by a mu-metal shield. Masses cannot be shielded from gravitational fields, so it is assumed that rotating masses cannot be shielded from gravitomagnetic fields, either.

In April 2007, NASA revealed that analysis of the GP-B experimental data had confirmed that the presence of the Earth did warp space-time. The results showed that the predicted geodetic effect agreed with the experimental measurement to within 1%. This is a most important result as it confirms that gravitation is a non-linear phenomenon.

Trying to extract the effect of frame-dragging due to the Earth's rotation from the GP-B experimental data proved to be much more difficult than extracting the effect of space-time warping. It wasn't until April 2011, after much data processing, that scientists at Stanford University were finally able to claim that the GP-B experiment had detected the Earth's gravitomagnetic field and that its magnitude agreed with the theoretical prediction given by NASA.

When NASA began funding the Gravity Probe-B programme in 1964, the use of a small spinning, ball-shaped gyro was the state-of-the-art means for detecting the Earth's gravitomagnetic field. Since then, the Ring Laser Gyro has been developed and it will, most likely (see Chapter 34), provide a more sensitive means of detecting gravitomagnetism in the future.

The result of the NASA GP-B experiment has confirmed Einstein's non-linear theory of gravity. However, we should bear in mind that for weak gravity fields and low velocities compared with the speed of light, Einstein's model agrees with Newton's linear model of gravity. Furthermore, it was Newton's model that laid bare the underlying mechanism behind the motion of the individual planets in the Solar System. So, in the same way, the gravitational analogue of electromagnetism, being a linearised and much simpler version of Einstein's gravitational theory, may uncover hidden connections with gravity and other phenomena that the non-linear theory is too complex to reveal.

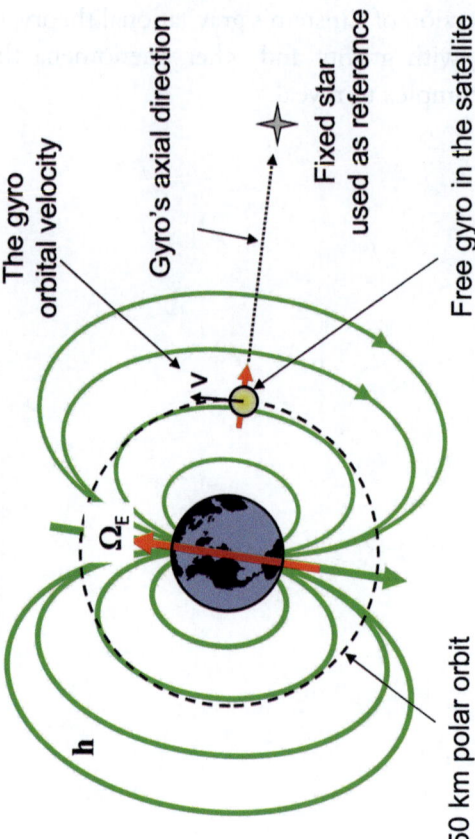

CHAPTER 36

# GRAVITY WAVES AND THEIR DETECTION

In 1916, only a year after he had published his theory of general relativity, Einstein showed that his model predicted the existence of gravitational waves in the space-time medium. A few scientists, conversant with Einstein's theory, were sceptical and suggested that such waves were merely a theoretical artefact and did not represent a real phenomenon. The idea of gravitational waves remained a physical curiosity for many years. In 1936, even Einstein expressed some doubt about the possible existence of space-time gravitational waves. But, in February 2016, US scientists at LIGO, the Laser Interferometer Gravitational-Wave Observatory, announced that they had detected gravitational waves.

Suppose we have a positive and a negative source of gravity linked together, somehow, so that they can oscillate vertically in opposition, changing ends in the process. As they pass in the middle, their changing gravitomagnetic fields combine. At a high enough frequency of oscillation, gravity waves would be broadcast into space. Viewed in 3-D, the outer edge of the broadcast waves (defined by power level) takes the shape of a doughnut with the vertically oscillating positive and negative masses at the centre. In 2-D, the cross-section of the broadcast gravity waves forms a two-lobed dipole field. The analogy fits perfectly with the dipole generation of electromagnetic waves. For the dipole generation of gravity waves, we need astronomically sized oscillating bodies made of positive and negative masses. But a search of the Universe has not revealed any astronomical bodies made of negative mass. As we will see later, when we consider negative mass in more detail, it does not form conglomerates, unlike positive mass. This seems to rule out the possibility of natural dipole-type gravity wave generation.

However, quadrupole-type gravity wave generation is possible, with two positive masses oscillating in opposition. In this case, as the masses pass in the middle, their gravitomagnetic fields cancel. In 2-D, the cross-section of the broadcast gravity waves forms a four-lobed pattern. The analogy holds for generating electromagnetic waves, too.

In the generation of electromagnetic waves, the background electric field of the oscillating charges is usually neutralised. In the gravity case, the background gravity field of the oscillating masses is always present and the gravity wave disturbance is imposed on top. Now, from Newton's inverse square law, we know that the strength of the gravity field of a mass is attenuated with the inverse square of distance from the mass centre. However, the strength of the gravity field disturbance, or gravity radiation, is only dependent on the inverse of the distance from the mass centre. So, in the far field, gravity radiation is the dominant effect.

Another point to be made is that since charge has mass, all electromagnetic waves contain extremely weak gravitational waves. So, electromagnetic disturbances might interfere with gravitational wave detection. Seismic disturbances are a more obvious problem. This means that at least two gravitational wave detection systems, vastly separated in distance, are needed to weed out local disturbances.

Studies predicted that quadrupolar gravitational waves, with a frequency range 10–300 Hz, would be broadcast continuously from a pair of closely orbiting stars as their masses changed position in space. In the case of a cataclysmic change in astronomical mass, such as stellar mass disintegration during a supernova, gravitational waves are expected to be more like blast waves, having a massive leading edge, containing gravitational waves of many frequencies predicted to be in the range 1–10 kHz, which then trail away to nothing.

In 1974, radio telescope measurements of a particular binary star system, a star orbiting an unseen neutron star, showed an increase in the visible star's orbital speed, indicating a decaying orbit which implied that energy was being radiated away from the binary pair in the form of gravitational waves. This stimulated the search for gravitational waves. The US physicists Russell Hulse and Joseph Taylor received the Nobel Prize in Physics in 1993 for this observation.

At any point in the gravity field surrounding a mass, the gravitational potential $\phi$ is negative and inversely related to the distance from the mass. The gravitational energy density in any region surrounding the mass is negative, being $-\frac{1}{2}\gamma g^2$. In the region closer to the mass, where g is stronger, the energy density is more negative. Now suppose that we have two masses. As they move closer together, the gravitational energy density of the region between them becomes more negative. To achieve this, positive energy must be displaced away from the inner region, making the gravitational energy density of the outer region less negative. The process of displacing positive energy is what forms the gravitational wave.

From the idea of wave-particle duality, the information about a disturbance in force-field geometry broadcast by waves can equally be thought of as being distributed by virtual particles, called bosons. Force-field particles are characterised by having integer amounts of elementary angular momentum $h/2\pi$, where h is Planck's constant.

The carrier of electromagnetic information is the photon, which has intrinsic angular momentum of $\pm 1(h/2\pi)$, labelled as spin $s = \pm 1$. The photon has energy $E = hf = mc^2$, where f is the frequency of radiation. We assume that the photon has effective mass $m = hf/c^2$. To picture the photon, imagine it as a sphere of effective mass moving in the direction of propagation with speed c and spinning about an axis aligned with the direction of propagation as it does so. It has kinetic energy $\frac{1}{2}mc^2$ and rotational energy $\frac{1}{2}hf$, where $\omega = 2\pi f$. As the photon moves forward one wavelength $\lambda$, the sphere rotates once.

Now the moment of inertia of a sphere of mass m about a diameter is $\frac{2}{5}mr^2$, where r is the radius of the sphere. And the rotational energy of the sphere is a half times the moment of inertia times the square of the angular velocity $\omega$. Thus, we see that

$$\frac{1}{2}\left[\frac{2}{5}\left(\frac{hf}{c^2}\right)r^2\right]\omega^2 = \frac{1}{2}hf$$

Since $c = f\lambda$, we get $r = \sqrt{\frac{5}{2}\frac{\lambda}{2\pi}}$

So, for light waves ($f \approx 10^{15}$ Hz), the photon radius is of order $10^{-7}$m, making it a compact virtual particle. But for BBC Radio 4 long wave transmission ($f = 198 \times 10^3$ Hz), the photon radius is about 380 m, making it a very large virtual particle with a very soft interior. Our model of a photon as a spherical package of energy is just to help us imagine what is going on and must not be taken too seriously.

For gravitational waves, by analogy with the photon, the carrier of information is taken to be the graviton. If the waves are created by an oscillating mass dipole, then the graviton has spin s = ±1. So, as the graviton sphere moves forward a distance λ, it rotates once. But, if the waves are created by a pair of masses oscillating in opposition, creating quadrupolar radiation, then the graviton has spin s = ±2 and in moving forward a distance λ, it rotates twice. However, without knowing the full radiation pattern, I can't see how one receiving station can distinguish between the two types of gravitons. Quadrupolar radiation of frequency f might be interpreted as dipolar radiation of frequency 2f.

Gravity is exceedingly weak so it seems reasonable to consider a linearised form of Einstein's gravitational theory. In the far field, we assume that the gravitational wave front is planar. This leads to the idea that a gravitational wave has two orthogonal components: an oscillating gravity field **g** and an oscillating gravitomagnetic field **h**. Analagous to the equations for the electromagnetic wave, in the linearised form of gravity, the 1-D wave equations are

$$\frac{\partial^2 \mathbf{g}}{\partial x^2} = \frac{1}{c^2}\frac{\partial^2 \mathbf{g}}{\partial t^2} \quad \text{and} \quad \frac{\partial^2 \mathbf{h}}{\partial x^2} = \frac{1}{c^2}\frac{\partial^2 \mathbf{h}}{\partial t^2} \quad \text{where wave speed } c = \frac{1}{\sqrt{\gamma\eta}}.$$

Suppose a test mass m is placed in the path of a gravitational wave. As the wave passes by, the mass will be subjected to two orthogonal force components given by $\mathbf{F} = m\mathbf{g} + m(\mathbf{v} \times \eta\mathbf{h})$. This is the gravitational analogue of the Lorentz force for electromagnetism. The oscillation due to the **g** component causes the mass m to move, giving it a changing velocity **v**. However, all masses accelerate together, so there is no way of detecting a relative motion of the test mass in the laboratory. This leaves the other component of force $m(\mathbf{v} \times \eta\mathbf{h})$, which squeezes and stretches the test mass.

During the 1960s, Professor Joseph Weber, of the University of Maryland in the USA, was the first scientist to try to detect gravitational waves using a bar antenna, rather like trying to detect the electromagnetic waves generated by lightning using a copper rod as an aerial.

Weber's gravitational wave receiver antenna was a cylindrical bar of aluminium, 2 m in length and 0.5 m in diameter. Based on the experiment of an electric bell in an evacuated glass jar, which we can see ringing, but can't hear, we conclude that sound needs a medium through which to travel. So, Weber suspended his bar antenna, by wires, inside a vacuum chamber to isolate it from external sound vibrations and from seismic effects. He kept the bar at a uniform temperature to avoid any thermal vibrations.

# GRAVITY WAVES AND THEIR DETECTION

We are aware that one resonating system can cause another similar system to resonate in sympathy, via energy transmitted in waves through the medium. Bonded around the circumference of Weber's bar, near to its centre, was a band of piezo-electric transducers which could detect any circumferential strain, due to squeezing or stretching. Based on the radial resonance of his bar, Weber hoped to detect, or tune into, gravitational waves having a frequency around 1 kHz.

Weber had two bar antennae for comparison purposes: one at Maryland University and the other at the Argonne National Accelerator Laboratory, near Chicago, nearly 1,000 km away. Filtering out known spurious effects, the two bars initially seemed to confirm that gravitational waves were arriving from space.

Unfortunately, attempts by other groups of scientists to repeat Weber's gravity wave experiment all failed and it was eventually concluded that bar antennae are not sensitive enough to detect gravitational waves. A more sensitive detection method was needed.

I wonder whether any one has tried a variation of Weber's experiment, using two bar antennae in vacuum chambers and causing one bar to resonate to see whether the other bar resonates in sympathy. The medium is the ether and the gravitational disturbance passing through the ether is akin to a sound wave moving at the speed of light.

Professor Weber's postgraduate researcher, operating the gravitational wave antenna, was Bob Forward, from the Hughes Aircraft Company. I met Dr Forward, in August 1989, at his holiday home, near Dounreay in the far north of Scotland. I consider the late Dr Forward to be one of the pioneers of modern gravity research and it was a pleasure to meet him.

Professor Weber and Dr Forward proposed a new gravitational wave detection system. Since a gravitational wave distorts a test mass, the distance between the surfaces of two test masses, several kilometers apart, will alter as a gravitational wave passes by them. This could be measured with a coherent laser beam. The crux of the new detection system was the use of a Michelson interferometer to measure any changes in the distance between the two test masses, where such changes could be measured in terms of a fraction of the wavelength of light. This idea led to the development of the laser interferometer gravity wave detector.

The apparatus consisted of a pair of suspended test masses fitted with mirrors. A laser beam was bounced back and forth between the mirrors with the resulting interference pattern being detected by a photodiode. Any subsequent change in the interference pattern signalled a difference in the distance between the test masses, possibly due to the distortion caused by the passage of a gravitational wave through the apparatus. Dr Forward built the first such device, in 1973, at the Hughes Aircraft Company site at Malibu in California.

In January 1988, I visited Professor Jim Hough and Dr Norna Robertson in the Department of Physics and Astronomy at Glasgow University, to see their prototype gravity wave laser interferometer. I was interested to learn that the optical quality mirrors used in the prototype interferometer were made by British Aerospace.

To detect gravitational waves from a supernova explosion, with wavelengths of several kilometres, the distance between the mirrors has to be of similar length. To reduce interference to the laser beam, the mirrors are placed at each end of an evacuated tube. This forms an arm of the laser gravity wave detector. Since gravitational waves can be elliptically polarised, just like electromagnetic waves, two arms at right angles are needed to receive a complete gravitational wave. Combining the two arms, only two test masses are needed, with one at the end of each tube.

Over the last 20 years, about half a dozen large gravitational wave interferometers have become operational around the world. Due to the very high costs involved, British and German scientists have now collaborated and jointly run the GEO 600 gravity wave interferometer at a site near Hannover, in Germany.

The biggest laser interferometer gravitational wave detectors are the two in the USA, with arms 4 km long. The two detectors, based 3,000 km apart, form the Gravity-Wave Observatory (LIGO). It began operating in 2002, but no gravity waves were detected. This is not so surprising, given that the predicted change in beam length between the two 4 km arms they hoped to detect was of order $10^{-18}$ m!

As an aside, I have wondered whether the Faraday Effect applies to the gravitomagnetic h-field component of a gravitational wave, as it does for a magnetic H-field. If so, this would cause the angle of the plane of polarisation of the 4 km light beam to rotate. If applicable, the effect of the h-field oscillation on the rotation of the plane of polarisation would be additive. This means that the light beam would rotate, so that the effect would contribute to the apparent change in beam length between the reflecting mirrors. In fact, detecting a change in the polarisation angle may be tantamount to gravity wave detection. The effect is an increasing one, whereas the squeezing and stretching by the $\mathbf{v} \times \mathbf{h}$ term is oscillatory. Clearly, this effect has not been seen.

# GRAVITY WAVES AND THEIR DETECTION

Work on an advanced version of LIGO, with greater detection sensitivity, was begun in 2010 and completed in 2015. In September 2015, scientists at LIGO detected gravitational waves from two black holes colliding about 1.3 billion light years away from Earth. But the official announcement was not made until February 2016. In 2017, the Nobel Prize in Physics, for the discovery of gravitational waves, was awarded to the LIGO members Professor Rainer Weiss, Professor Kip Thorne and Professor Barry Barish.

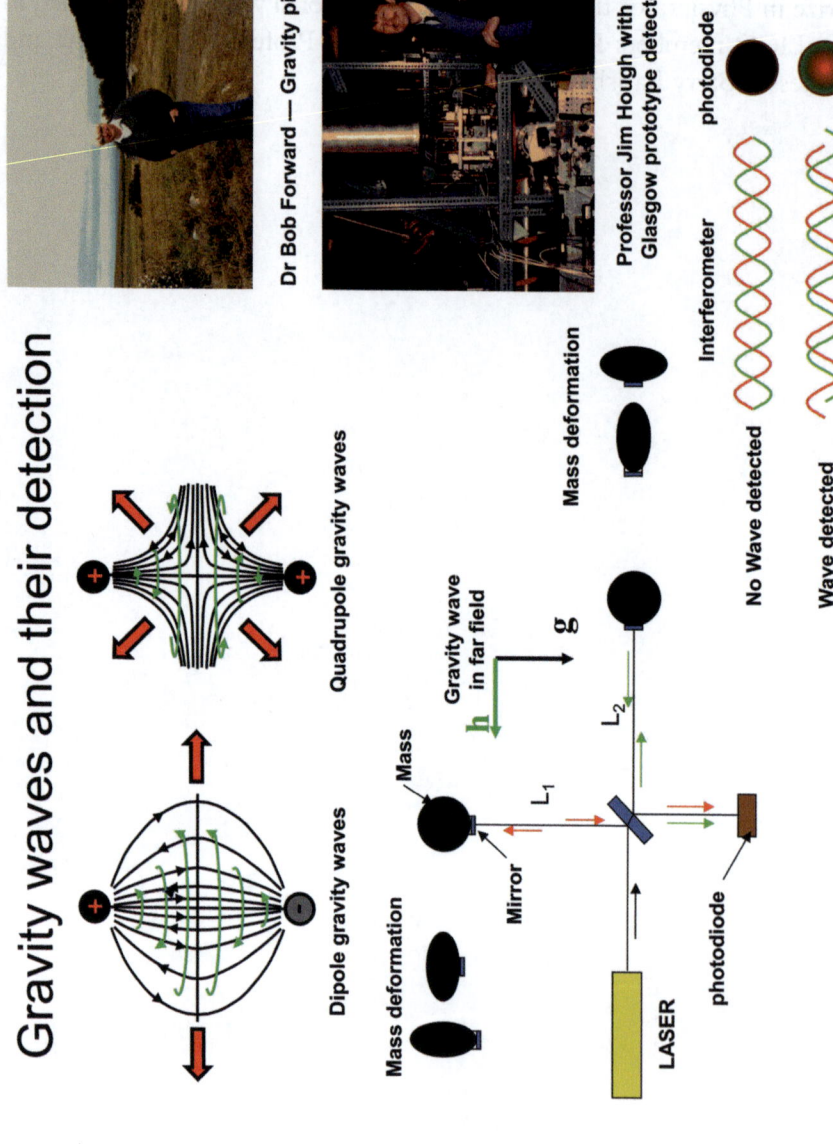

CHAPTER 37

# SOME SPECULATIONS CONCERNING GRAVITY

When Einstein produced his theory of general relativity, in 1915, he assumed that the size of his space-time Universe, containing many galaxies, although infinite, was bounded in extent. Within the boundary, matter could move and space-time could be deformed, one acting on the other. To ensure that the boundary of the Universe remained fixed (or static), Einstein introduced his cosmological constant $\Lambda$ (Chapter 14). However, in 1929, Edwin Hubble at the Mount Wilson Observatory in the USA discovered that the galaxies are mostly moving away from each other at, what appeared to be, a steady rate. So, the Universe was increasing in size. Tracing this fact backwards in time led to the idea that the Universe began as a tiny point mass of unimaginable density (the primeval atom) which exploded some 13.8 billion years ago. This is the Big Bang theory.

Before the Big Bang, the Universe did not exist. The primeval atom contained all the matter there was. For the primeval atom to remain intact, the internal gravitational forces between all particles of matter must have been infinite. In other words, before the Big Bang, the value of the gravitational factor G within the primeval atom was infinite. So, before the Big Bang, the gravitational permittivity $\gamma = 1/4\pi G$ was zero. A tiny increase in g would have led to a slight expansion of the primeval atom. Once started, the expansion was unstoppable and led to the primeval atom exploding and creating the Universe, at a temperature above $10^9$K, filled with sub-atomic particles and hydrogen atoms.

In some regions of the Universe, gravitational attraction between hydrogen atoms led to the formation of hydrogen gas clouds. As the gas clouds attracted more hydrogen atoms and grew in size, they experienced gravitational compression and compaction. During this process, extreme frictional heating occurred leading to the formation of more elements. Eventually, gravitational attraction led to huge astronomical bodies of matter forming, many of which burst into light to form stars. Over eons of time, various forms of matter collected together to form separate galaxies, seen through telescopes as swirling collections of stars. After billions of years of expansion, the temperature of the Universe has cooled to about 3K.

Stars may be viewed as sub-primeval atoms. In turn, many will undergo their own Big Bang, or supernova explosion, showering the Universe with more complicated, although less stable, elements. In exploding, we can speculate that they create new spatial regions of ether, too, adding to the size of the Universe.

We have two ways of viewing the medium of the Universe. We can think of it in terms of 4-D space-time or in terms of 3-D space with separate time. In the latter case, the medium is the ether, which supports the gravitational fields of masses extending their influence throughout the Universe. The same ether supports electrical fields but is not included in 4-D space-time. The present view of the ether is that at every point of the Universe (in the vacuum of space, and within matter) virtual particles of energy $\Delta E$ constantly emerge from nowhere but last only for a short time $\Delta t$. These quantum fluctuations of discrete energy satisfy Einstein's form of Heisenberg's uncertainty principle, which states that $\Delta E.\Delta t \geq \hbar$, where $\hbar = h/2\pi$. Generally it is assumed that $\Delta E.\Delta t \approx \hbar$, where $\hbar$ is of order $10^{-34}$ J.s. So, if $\Delta E$ is large then $\Delta t$ is incredibly small.

The name for this new version of the ether is the Quantum Vacuum. The British physicist Paul Dirac, who played a large part in developing Quantum Mechanics in the 1930s, referred to the fleeting virtual particles of the vacuum of space as the quantum foam. When associated with electromagnetism the virtual particles are called photons. Photons have spin 1; that is to say they have angular momentum $\hbar$. When associated with gravitation the virtual particles are called gravitons. Whether gravitons have spin 1 or spin 2 is open to question. When just referring to the ether we mean the universal medium of space which supports both gravitational and electrical fields.

In the absence of masses and charges the ether of space, at near 0K, has an ambient energy called the Zero Point Energy (ZPE). A change of reference frame makes no difference to the ambient energy level of the empty ether. We know that the quantum fluctuations are there since the phenomenon of the Lamb Shift and the Casimir Force have demonstrated their existence. Laser light suffers from shot noise, which is due to the quantum fluctuations of the ether. Scientists have devised a scheme to suppress these fluctuations in a technology called squeezed light. So, the ether is open to manipulation.

If the ether is stretched it will contain more quantum vibrations and more (positive) energy above ZPE. If the ether is compressed, or squeezed, it will contain less quantum vibrations and less (negative) energy below ZPE. The values of $\gamma$ and $\varepsilon$ will change in regions of space where the energy density is different to ZPE.

In case you think that the ZPE of the vacuum is insubstantial, scientists have estimated its energy density to be of order $10^{114}$ J/m$^3$, which is much greater than the energy density of an atom. Whether we can tap the ZPE of the Quantum Vacuum is a question that scientists have been investigating for several decades.

Accepting that the Universe is expanding suggests that the ether is being stretched. Outside the Solar System, the ether might be more, or less, stretched, suggesting that the free-space electromagnetic and gravitational variables ($\varepsilon$, $\mu$, $\gamma$, $\eta$) in these regions might be different to those in the Solar System.

If we had the means to stretch, or contract, the ether, we could propel a spacecraft (made of matter or charge) across space. The ether in front of the craft would be contracted and the ether behind the craft expanded, so that the spaceship would be carried along by an ether wave, like a surfboard rider. The force on the spaceship would arise from the energy gradient created in the ether across the spaceship.

Perhaps changing the free-space values of $\varepsilon$ and $\gamma$ is equivalent to stretching or contracting the ether. A conducting box forms a Faraday cage so that if the box is charged then inside the box the electric E-field is zero. Aerospace engineers make sure that their airframes form a Faraday cage to avoid any problems from lightning strikes (see illustration, Chapter 22). Suppose one face of the box is coated with dielectric material with a very high permittivity $\varepsilon_1$. Elsewhere around the box, the electrical permittivity is that of free space, $\varepsilon_0$. Suppose that the box is charged, creating an E-field around the box. The energy density of a volume of electric field is $\frac{1}{2} \varepsilon E^2$. In the direction perpendicular to the coated face there is an energy gradient created across the box, since $\varepsilon_1 >> \varepsilon_0$. Since force is equal to minus an energy gradient, theoretically the coated box will experience a thrust in the direction of the coated face. In effect, the ether ahead of the box has been compressed, creating an electrical ether wave which carries the box along. This effect is probably related to the Biefeld-Brown effect (see later in chapter). There is a rumour that the wing leading edge of the B2 bomber is coated with a high dielectric material so that when the wing is charged it creates an extra thrust. By sectioning the wing leading edge to allow separate conducting panels to be charged, some steering without using aerodynamic control surfaces might be possible. But it is just a rumour.

When it comes to creating a gravitational ether wave, we have a problem since it appears that $\gamma$ is constant. And yet we know that gravity waves exist. In terms of space-time, in 1994, Miguel Alcubierre proposed his space-time warp to carry a spaceship (of mass) along.

Big G has been determined from experiments on Earth, based on Newton's inverse square law of gravitation. Although G has only been determined for gravitating matter in free space over short distances, when the value is used in Newton's law to predict the movements of the planets around the Sun, the results are highly successful, so, naturally, G is taken to be a constant within the Solar System. Thus, the gravitational permittivity $\gamma$ is taken to be a constant, too. Similarly, free-space measurements of electrical permittivity $\varepsilon_0$ and magnetic permeability $\mu_0$ have been determined. The speed of light c across the Solar System, where $c^2 = 1/\varepsilon_0\mu_0$, agrees with measurement made on Earth, so $\varepsilon_0$ and $\mu_0$ are also taken as being constant within the Solar System. In the past, the Solar System and the Universe were treated as the same. So, by extension, $\gamma$, $\varepsilon_0$, and $\mu_0$ were taken to be universal constants throughout the ether of free space. But we now know that the Solar System is an incredibly tiny part of the Universe, which means that $\gamma$, $\varepsilon$ and $\mu$ may vary in other parts of the Universe.

The fact that Big G (and $\gamma$) seems to be constant has not stopped scientists from conducting small-scale gravity experiments on Earth searching for changes in G. These include looking for variations in G depending on temperature, magnetisation, radioactivity and electric charge of the gravitating matter, but to no avail. Negative results are seldom published, unless they are quite contrary to what is expected, so details are sparse. Nevertheless, we cannot say that no effects on G occur, only that no effects have been detected and confirmed, so far. Scientists have also looked for anisotropy effects on G. That is, whether the axial direction of the gravitating masses with respect to the fixed stars affects G, but found nothing. And yet, collecting many Earth-made experimental measurements of G together does indicate that G varies very slightly with time, over a cycle of about 5.9 years, perhaps linked to the sunspot cycle of 11 years.

Electromagnetic waves and gravitational waves appear to travel at the same speed c in free space. This suggests that there might be a link between the permittivities and permeabilities of the two phenomena, which is not really surprising. We might guess that $\varepsilon$ and $\gamma$ are linked and that $\mu$ and $\eta$ are linked. In 1920, Sir Oliver Lodge suggested that $\varepsilon$ and G might be linked. In 1990, at a British Aerospace-sponsored gravity meeting (pre-Greenglow) with UK university academics, Professor Roger Jennison, of Kent University, outlined an experiment to search for such a link between $\varepsilon_0$ and G. It involved a pair of vacuum parallel-plate capacitors at the ends of a bar rotating about a horizontal axis. The idea was that it might be possible to detect a cyclic change in capacitance (dependent on $\varepsilon_0$) with change in gravitational potential (dependent on height and G). It was not carried out and, as far as I am aware, no one has ever discovered any such links.

Following the electrical analogue, we can imagine a gravitational capacitor. The Universe is still expanding and as the gravitational ether of free space is stretched, perhaps $\gamma$ changes. In a similar way, since the ether of space is also a dielectric, perhaps $\varepsilon$ changes with spatial distortion. Although not noticeable on a small Solar System scale, over really immense galactic distances of space, $\gamma$, $\varepsilon$ and $\mu$ may change slightly.

In 1936, the Swiss-US astronomer Fritz Zwicky compiled a list of galaxies and galaxy clusters. He investigated the motion of stars in some of the large rotating galaxies and surmised that the compaction of stars at the centre moved like a solid body held together by strong gravitational attraction. However, in the outer sparser regions, with much lower stellar gravitational interactions, he couldn't understand why the stars there, with their high rotational speeds and high centrifugal forces, didn't hurtle off into space. He concluded that there must be some other gravitational attraction existing in the outer regions which held these stars to their galaxies. Since the source of this unknown gravitational attraction was not visible, he called it dark matter.

During the 1970s, the US astronomer Vera Rubin examined the motion of the stars in the Andromeda galaxy in close detail. The stars near to the galactic centre followed a solid body rotation, where increasing distance from the centre resulted in increasing rotational speed. But the stars in the outer region all moved at roughly the same speed, not depending on their distance from the centre. It seemed odd. She examined other galaxies and found the same result.

This seemed to confirm Zwicky's surmise that the motions of stars in the outer regions of rotating galaxies are gravitationally held in check by a pervading disc of dark matter. But no one knows what dark matter is.

A gravitational field contains energy and from $E = mc^2$ we see that a gravitational field contains some form of effective negative mass, or matter. We can think of it as natural dark matter. In fact, the term dark is a misnomer. This natural form of dark matter is invisible and transparent, but we know it's there because it bends stellar light rays. If we allow the gravitational permittivity of space $\gamma$ to reduce in the outer (stretched?) reaches of the galaxies then the gravitational attraction involving the outer stars must increase. We might say that the increased strength of the gravitational fields in these regions harbours enhanced dark matter.

One might have expected that in the future, due to gravitational attraction, the expansion of the Universe might eventually slow down, stop and then reverse. In time, all the galaxies would then come together, forming a single point mass of unimaginable density again, sometimes referred to as the Big Crunch. Then the Big Bang process could start all over again. However, in 1998, two teams of astronomers reported that, at the moment, according to their separate findings, the galaxies are mostly moving away from each other with increasing speed. Something in the vacuum of the Universe is forcing the galaxies apart.

Due to the presence of mass (stars, planets, dust, gases, etc.), the Universe is filled with negative gravitational energy. Suppose that the ether of space is endowed with some form of positive energy. If within a galaxy the overall energy balance is negative, then the stars in the galaxy will stick together. However, if between galaxies the overall energy balance is positive, then the galaxies will push apart.

Scientists have made use of Einstein's cosmological constant $\Lambda$ to introduce added energy into the vacuum. For empty space, where for flat space-time $g_{pq} = 1$, the expression for the added energy density at every point is

$$\frac{1}{2}\gamma c^4 \Lambda \quad \text{J/m}^3 \quad \text{where } \gamma = \frac{1}{4\pi G}$$

Astronomical observations show that the added energy density of the ether is about $10^{-9}$ J/m$^3$. It is very small, but significant. Although this energy has a gravitational influence, we can't see it. It is called dark energy and is separate from dark matter.

From the gravitational form of the Bernoulli equation (Chapter 16), the pressure at any point in a gravitational field in space created by bodies of mass is

$$p = -\frac{1}{2}\gamma g^2 + \frac{1}{2}\gamma \Lambda c^4$$

If space is compressible, so that $\gamma$ can change, the first term is associated with dark matter. We have speculated that $\gamma$ might decrease with the stretching of the Universe, but it might also decrease with the age of the Universe. After all, we know that gravity affects time, so why shouldn't time affect $\gamma$? The second term is associated with dark energy.

Based on their current ideas, astrophysicists have estimated that about 70% of the mass of the Universe is due to dark energy; about 25% of the mass of the Universe is due to dark matter, with ordinary mass making up only 5% of the mass of the Universe. For the moment, in our search for a way to control gravity, we must work with the 5% of ordinary matter.

Following the same idea as that suggested for explaining dark matter, but at the other end of the scale, a large reduction in the magnitude of $\gamma$ (tending towards zero) within the nucleus of a conventional atom, might be the source of the strong nuclear force, holding the negatively charged, repelling protons and the neutrally charged neutrons together by means of gravitational attraction.

If we could find out how to alter $\gamma$, it would allow us some control of gravity.

The gravitational factor G has also been measured within the Earth, at the bottom of a mine shaft, by George Airy, the Astronomer Royal, in 1856. But, to within experimental measurement error, he found no difference in G from the free-space value of Big G. Perhaps extreme matter compression may affect the internal value of G (as it must have done in the primeval atom). Compression plays a part in electromagnetism and results in some elements becoming superconducting at very high pressures. High compression of matter can cause changes, as exemplified by the laboratory production of diamonds from carbon.

Many of the experiments to measure G have been carried out under quasi-static conditions. Some have been carried out under dynamic conditions.

Dr Charles Brush was an eminent US scientist who won many

prestigious awards for his inventions of electrical devices during the end of the 19th century. He was also a very successful businessman at the birth of the electrical industry, competing with the likes of Edison. He founded the Brush Electrical Company, later absorbed by GEC, and the Linde Air Products Company.

In 1911, Brush published a paper in *NATURE* describing his 'Kinetic theory of gravity'. In Brush's view, the kinetic energy of mass resided in the ether. This view corresponds with the idea that the kinetic energy of mass lies in the gravitomagnetic field. During Brush's retirement, from 1914 till 1926, he investigated his theory of gravity by carrying out a number of experiments in his private laboratory at home. He began with an experiment to test Newton's inverse square law using Boy's apparatus, which is a miniaturisation of Cavendish's original experiment. From his experiments with different substances, Brush concluded that the universal gravitational constant G was not constant but was dependent on the atomic weight of the substances used. This was completely at odds with Newtonian theory. Brush next carried out a simple pendulum experiment with hollow bobs containing different substances. Again, contrary to accepted physics, Brush found that the period of oscillation was very slightly greater for those bobs with higher atomic weights.

In a final series of free-fall experiments (over 4 ft. /1.22 m), Brush dropped an identical pair of hollow aluminium containers filled with the same weight of different substances in the Earth's gravity field. According to his measurements, most bodies accelerated at $g_0$ (Earth's surface gravity), but a few with low atomic masses did not. Moreover, Brush noted that these same bodies spontaneously generated heat during free fall, which was not of a radioactive nature. In Brush's view, his experiments called into doubt the accepted principle of weak equivalence between gravitational mass and inertial mass.

In free fall in a surface gravity field $g_0$, theory shows mass acceleration is resistance free. The gravity field $g_0$ is cancelled by an induced gravity field $g_i$ so that the movement is free of inertia. Now heat can disturb the magnetic field in some materials, so 'reading-across' suggests that heat might disturb the gravitomagnetic field in some materials, too. So if heat is generated within a falling body, the magnitude of the changing internal gravitomagnetic field might be degraded, resulting in the induced gravity field $g_i$ being less than $g_0$. In this case, the body will experience some inertia and its acceleration will be less than $g_0$. Resisted motions are associated with heat radiation. So, a free-falling mass at high temperature (lagged, so no change of surface temperature) might have a slight change in its acceleration from $g_0$. If such a heated mass is a pendulum bob, then the period of swing in a gravity field $g_0$ might change.

Although Brush's results hinted at a new revelation with regard to gravity, as far as I am aware, his experiments have not been repeated by other scientists. His work has largely been ignored (and forgotten), probably because it does not fit in with the standard theory of Newtonian gravity, and his strange results dismissed as probably being due to experimental measurement errors.

Within matter, the change in ε is associated with displacement currents which arise when matter is subjected to an external electric field. This is because both positive and negative electric charges exist within matter and move in opposite directions in response to an external electric field. Different materials respond in a different way to an external electric field, giving rise to a range of ε values. In the gravitational case, assuming we only have positive mass, mass displacement currents within matter are not created by an external gravity field, so that γ does not change. Even if traces of negative mass do exist within matter, we will still not get mass displacement currents, because positive and negative masses move in the same direction in response to an external gravitational field (see later in chapter). However, we can simulate the effect of mass displacement currents by compressing and expanding matter, and this is what happens in a dielectric and, perhaps, in space, too, as the ether is a dielectric.

From our analogue approach, we have paired equations for electromagnetism (**E** & **H**), for extended thermodynamics (**T** & **R**) and for extended gravity (**g** & **h**). The three sets of equations are separate from each other. However, scientists are convinced that the force fields are all linked in some way. Mass, as the source of gravity, appears as density in one of the gravity equations. Although charges and heat sources have mass, it doesn't appear in their equations. Might this be a route to explore to see whether it throws up any ideas for experiments to find a link between the force fields? Only an experiment can confirm such a link.

In dimensional terms (using mass M, length L and time T), the above variables can be grouped to form a pattern for the electron. Each of the groups has the same dimensions ($1/L^2$).

$$\frac{\gamma g}{m_e} \quad \frac{\varepsilon E}{-e} \quad \frac{kT}{h_e}$$

The mass of the electron is $m_e$, the electron charge is -e and $h_e$ is assumed to be a fundamental electron heat source, undefined as yet. Having the same dimension doesn't imply that any link exists between the groups, but it raises the possibility that such links might exist within materials. For example, perhaps

$$\mathbf{g} = -\left(\frac{m_e}{e}\right)\left(\frac{\varepsilon}{\gamma}\right)\mathbf{E}$$

It is up to experimenters to determine whether such links exist macroscopically.

In 1928, while experimenting with an X-ray tube, US laboratory technician Thomas Townsend Brown made a curious discovery. Although a directed beam of X-rays imparted no noticeable force to objects, when the tube was switched on, it experienced an impulse, or thrust, of its own. Brown traced the impulse to the very high voltage developed across the plate electrodes. Experimenting further, he built a multi-parallel-plate capacitor, interspersed with glass dielectric plates, that developed a thrust towards its positive electrode when it was highly charged. The thrust only lasted during the charging period, while the dielectric plates became highly compressed. Brown thought that he had discovered a link between gravity **g** and electricity **E** and he discussed the effect with Paul Biefeld, a professor of physics. Over the ensuing years, the curious phenomenon has become known as the Biefeld-Brown effect, or the study of electro-gravitics.

The effect is real and, since its chance discovery, the phenomenon has been of continuing fringe interest to scientists and engineers, who are not quite sure what to make of it.

In July 2001, Trans-Dimensional Technologies (TDT), a research company based in Alabama, USA, amazed many people with their demonstration of a levitating device which they called a Lifter. This device was the brainchild of Jeff Cameron, the Chief Scientist at TDT. The basic, triangular-shaped lifting cell consisted of three vertical thin balsa wood poles supporting a wire strung from their tips, with a strip of aluminium foil wrapped around their lower part. When high positive and negative direct current (dc) voltages were applied to the wire and the foil, respectively, the Lifter levitated. To obtain a greater lifting effect, several basic lifting cells were ganged together. Typically, a Lifter weighing 250g, including a 50g payload, would levitate when a voltage difference of 30kV was applied. The Lifter is another form of the Biefeld-Brown device. Clearly, the Lifter develops thrust which is directly linked to acceleration,

# SOME SPECULATIONS CONCERNING GRAVITY

which through equivalence is linked with gravity, but no external gravity field is apparent. Unlike the transient Biefeld-Brown effect, the Lifter levitated continuously while the dc voltage was maintained.

Within some bodies, the magnetic permeability $\mu$ can be quite large, say 1,000 times greater than the free-space value (the same is true for $\varepsilon$). Perhaps something similar is true for the gravitomagnetic permeability $\eta$ and possibly its value inside some bodies is greater than its free-space value.

When considering $\eta$, the situation is different because we assume that gravitomagnetic dipoles exist, so that within matter the value of $\eta$ may change.

This raises an important issue. If we accept that $\eta$ might change within matter then we must consider the possibility that $\gamma$ might change within matter, too, meaning a change in Big G. After all, in the electromagnetic phenomenon $\varepsilon$ and $\mu$ change within matter, so why shouldn't $\gamma$ (the gravitational analogue of $\varepsilon$) and $\mu$ (the gravitomagnetic analogue of $\eta$) change in the gravito-gravitomagnetic phenomenon?

Even if nature does not allow for natural changes in $\gamma$ and $\eta$, we still have the possibility of developing gravitational metamaterials with unnatural properties. If we look at the electromagnetic analogue, electromagnetic metamaterials have been designed and built with negative refractive indexes, which do not occur in nature. These are of particular interest to engineers developing invisibility cloaks for military vehicles. Experiments have already confirmed that these stealth materials work. We need to get chemists and metamaterial engineers to investigate ways of altering $\gamma$ and $\eta$. The advent of 3-D printing will be of considerable help. Being able to alter $\gamma$ and $\eta$ will result in the blossoming of the gravitational phenomenon which, until now, has appeared to be unchangeable.

During the 1940s, the Russian astrophysicist Nikolai Kozyrev studied the rotation of stars and thought of them as stellar gyroscopes. This led him to carry out a series of laboratory tests on gyroscopes. During the 1950s, Kozyrev claimed that his results showed that the weight of a gyroscope measured in Earth's gravity field depended on its angular velocity and its axial direction of rotation. Although the weight changes measured were extremely small, the claim was, nevertheless, extremely controversial. Other Soviet scientists concluded that, while spinning, a gyroscope generated a chronal (linked to time) dipole field which interacted with the Earth's chronal dipole fields. We might interpret the chronal field as a torsion field or an induced gravitomagnetic field (linked to inverse time).

The analogue of the current ring with a rotating mass ring suggests that a spinning gyro has virtual positive and negative gravitomagnetic poles which may interact with the gravitomagnetic poles generated by the rotating mass of the Earth. The analogue (assuming gravitational attraction is neutralised) indicates that like gravitomagnetic poles attract, while unlike gravitomagnetic poles repel.

In 1989, a team of scientists at Tohoku University in Sendai, Japan, led by Professor Hideo Hayasaka and Professor Sakae Takeuchi, used a chemical balance to measure the weight of a spinning gyro. The mass of the non-spinning gyro was 175 g. In their published results, they claimed that when the gyro was spinning, with its angular velocity vector pointing vertically downwards, it experienced a very tiny reduction in weight proportional to the frequency of its rotation. Looking downwards on the gyro, the rotation was to the right, or clockwise. For an angular velocity of 13,000 rpm, the weight loss amounted to 13 mg. The effect was attributed to a tiny upward acceleration generated by the spinning gyro. For left, or anticlockwise, rotation, there was no change in weight.

Later on, the same research team carried out experiments with a gyro in free fall in Earth's gravity field $g_0$. The time taken for a non-spinning gyro to fall over a vertical distance L was measured for comparison purposes. For a spinning gyro with an angular velocity pointing downwards of 18,000 rpm, they measured a tiny increase in the time taken for the spinning gyro to fall the vertical distance L compared with the time taken by a non-spinning gyro. This indicated a tiny decrease in the downward acceleration of the gyro, giving rise to a weight loss of 25 mg. Again, for a gyro with an upward-pointing angular velocity vector, no difference in the time of fall was noticed compared with the time of fall of the non-spinning gyro. Is there any link here with Dr Brush's observations?

Research groups in the USA (1990), France (1990), China (2001) and Germany (2015) have replicated the Japanese experiments. All groups concluded that, within experimental error, spinning gyros in a gravity field did not change weight. The opinion of mainstream scientists is that claims of weight changes of spinning gyroscopes, especially those of a non-symmetrical nature, are due to experimental measurement errors.

During the Cold War, intelligence services in the West heard rumours about research by several groups of Soviet physicists working on torsion fields, including the generation and detection of such fields. There was concern that the work might underlie technology for a new method of communication, including telepathy and ESP, or a new weapon system against which the West had no understanding. In case you think the reference to telepathy and ESP is a joke, the Russians have spent millions of roubles and the US has spent millions of dollars on speculative research into these subjects.

The Anatoly Akimov torsion field generator (details on the web) relies on an electromagnet with a ferrite core contained within a flat plate capacitor, which is contained in a screened metal box. One pole of the magnet is centred directly under a small cone-shaped metal cavity. The claim is made that by applying an oscillating charge to the capacitor plates, the resulting resonant interaction (pumping) with the ferrite core causes their electron spins to oscillate and radiate torsional energy. The shape of the conical cavity is supposed to direct the radiant energy in particular directions.

Now, it is possible to screen against a magnetic field, but not (in theory) to screen against a gravitomagnetic field, so it might be possible to separate the two fields. Thus, if torsion is synonymous with gravitomagnetism, there appears to be an element of support for the Akimov device. However, the estimate of the induced gravitomagnetic (torsion) field **b** generated by an induced magnetic field **B** is incredibly tiny, which seems to rule out the feasibility of such a device, unless the Russians have discovered how to greatly increase $\eta$.

With the demise of the Soviet Union and the opening up of Russian science to the world, most scientists (including many Russian scientists) have concluded that the claims made about torsion field generation and its detection were based on pseudoscientific ideas, not supported by mainstream theories.

During the early 1990s, Dr Evgeny Podkletnov, a Russian materials scientist, began working at the Tampere University of Technology, in Finland. He was an expert on the manufacture of large high-temperature superconducting YBCO (Yttrium-Barium-Copper-Oxide) discs. The main purpose of his research study was to investigate the use of spinning YBCO discs, contained in a cryostat, to store very large current flows which could then be used to supply electrical power in an emergency. Meissner levitation was used to float the YBCO disc and three 3.6 MHz letterbox-shaped solenoids were used to rotate the disc.

In 1992, by chance, Dr Podkletnov noticed that the smoke from a colleague's pipe, as it drifted over the cryostat, suddenly moved upwards to the ceiling. He was convinced that the movement was not due to a pre-existing air current. Out of curiosity, he suspended a small mass from the end of a balance arm above the cryostat and was surprised to discover that the mass lost weight by a tiny amount. So began a study to see whether the YBCO discs created a shield against the Earth's gravity. Because we only know of positive matter, most scientists assume that gravity screening is impossible.

One thing a lifetime in research has taught me is that breakthroughs in science often happen following a chance observation. It may start with an anomalous result, but it needs an observant and tenacious scientist to unravel its meaning. For a while, it is exciting and, if the scientist is lucky and it really is a breakthrough, then euphoria. More generally, it's just a glitch which can't be repeated, or explained, and the scientist is left feeling emotionally drained and very frustrated.

The *Sunday Telegraph* published an article (1$^{st}$ September 1996) about Dr Podkletnov's anti-gravity discovery that created a furore among many scientists who didn't believe the result. The furore was so intense that Dr Podkletnov withdrew his latest paper on the subject just as it was due to be published. In a subsequent article, written by Robert Matthews, in the *New Scientist* magazine (21$^{st}$ September 1996), it was mentioned that Dr Ning Li and Dr Douglas Torr, at the University of Alabama, had already published details of a theory which might explain the phenomenon observed by Dr Podkletnov. The article stated that *According to the General Theory of Relativity, rotating matter can generate a new force of nature, known as the gravitomagnetic interaction, whose intensity is proportional to the rate of spin.*

NASA in the USA, in 1999; Hathaway Consulting Services in Canada, in 2003; and ESA in Europe, in 2006 all carried out comprehensive studies to try to repeat Dr Podkletnov's experiment, but all failed to detect any sign of gravity screening.

Thinking about the result recently, I wondered whether the gravitational effect was due to a red shift above one of the radio-frequency solenoids, with a non-rotating YBCO core, operating at 3.6 MHz. Since, effectively, photons have mass then in undergoing red shift they nullify gravity. Although photons are subject to electromagnetic screening, their gravitational change is not. Perhaps, in this way, a powerful vertical radar beam (operating in the GHz microwave region) might provide a gravity-free environment for astronauts, wearing their radiation protected spacesuits, allowing them to experience weightlessness, as though in orbit. Indeed, there are rumours that NASA has a microgravity chamber (not to be confused with a drop shaft) near to its Houston site, but nothing is known of its operating principle, nor whether it really does exist.

# SOME SPECULATIONS CONCERNING GRAVITY

The late Eric Laithwaite, Professor of Heavy Electrical Engineering at Imperial College London, was an advocate of analogues and an excellent presenter of science topics. He is well remembered for his work on the Maglev rail system (see *Magnetic rivers* on YouTube), based on his development of the linear induction motor. Naturally, Laithwaite was well versed in Maxwell's theory of electromagnetism. What intrigued him, though, was the possibility that gravity could be treated as an analogue of electromagnetism. He was well aware of Maxwell's attempt to model gravity in this way (Chapter 33). In December 1986, I visited Professor Laithwaite to discuss the Maxwell form of gravity with him. I was particularly interested in the two curl terms that arise in Maxwell's dynamic theory of gravity but which are missing in Newton's static theory of gravity. Laithwaite called the gravitomagnetic field the kinemassic field, in deference to experiments by Henry Wallace (US aeronautical engineer) who investigated the effect of macroscopic change in nuclear spin of a body during the 1960s and 1970s. Laithwaite referred to the gravitomagnetic permeability as the inertial permeability. He was interested in the possibility that a spinning flywheel (rotating mass ring), or gyroscope, might be a source of the gravitomagnetic field. In fact, at that time, NASA was working on an experiment based on this assumption which eventually led to the Gravity Probe-B experiment. The size of the rotating mass is very important.

In 1974, at a lecture at the Royal Institution in London, Professor Laithwaite had demonstrated the ease with which a heavy spinning flywheel could be lifted above the horizontal during precession and suggested that it might have lost weight during the process (see *Laithwaite and gyro* on YouTube). To support this idea, Professor Laithwaite suggested that the weight loss of the spinning flywheel might be explained if the motion of the gyroscope was treated as an analogue of electromagnetism. This provoked criticism from some academics who maintained that gyro theory, which is based on Newtonian mechanics, explains why the flywheel rises or falls and it doesn't mention weight loss. The theory is based on the principle that the rate of change of angular momentum **H** of a rotating wheel about an axle pivoted at its end is equal to the length **R** of the axle (called the moment arm) times the force **F** exerted on the wheel. That is, $d\mathbf{H}/dt = \mathbf{R} \times \mathbf{F}$. Providing the angular velocity of flywheel precession is small compared with the angular velocity of the flywheel, a simple model can be derived which describes the counterintuitive motion of the flywheel. A more complicated model leads to a set of equations called Euler's equations.

Dr H. Ron Harrison, formerly Senior Lecturer in the Department of Mechanical Engineering & Aeronautics at City University, London, has published his textbook, *Advanced Engineering Dynamics*, co-authored with T. Nettleton, which contains the theory of gyroscopes. Dr Harrison criticised Professor Laithwaite's speculations about gyroscopic motion, particularly his suggestion of flywheel weight loss. Dr Harrison was also critical of Professor Laithwaite's idea that Newton's $2^{nd}$ law might be modified by adding a jerk (rate of acceleration) term to it. Since Laithwaite was an electromagnetics expert, it's surprising that he didn't propose that Newton's $2^{nd}$ law should be replaced with the gravitational analogue of the Lorentz force. But at the time, gravitomagnetism was hardly known about, although Laithwaite knew of it. Although critical of Professor Laithwaite's speculations about gyroscopic motion linked with gravity, Dr Harrison is interested in gravity dynamics and has published his own theory on the subject in his book, *Gravity – Galileo to Einstein and Back*.

Laithwaite was not the first person to demonstrate a gyroscope in action in the lecture theatre of the Royal Institution. The Reverend Baden-Powell, the Savilian Professor of Geometry at Oxford University (whose son, Robert Baden-Powell, founded the Boy Scout movement), was the first to demonstrate the weird properties of the gyroscope (invented by Foucault in 1852) there, in March 1854. Trying to turn a gyro about one axis, results in it turning about an axis at right angles. Michael Faraday, who was in attendance, afterwards made his own gyroscope, carrying out a series of experiments to familiarise himself with its properties.

In December 1974, during the $145^{th}$ Christmas Lecture at the Royal Institution, Laithwaite followed up his earlier heavy flywheel demonstration with a twin flywheel device mounted on a balance. When it was activated, the balance reading indicated that it had lost weight. It was clear for all to see that the device generated vertical pulsations, and Laithwaite claimed that the up-and-down forces were unequal, resulting in the device experiencing an upward inertial thrust. The device was called an Inertial Thrust Machine, or ITM. A number of patented force-precessed gyroscopic devices already existed, which their inventors claimed produced inertial thrust. Laithwaite suggested that the thrust effect might be explained if gyroscopic theory was viewed in electromagnetic terms. Laithwaite's suggestion was met with a barrage of criticism. The mechanical engineering experts maintained that the apparent anti-gravity effect arose because the weighing system was too crude and couldn't react quickly enough to the large vertical oscillations of the device. Laithwaite's demonstration resulted in a scientific uproar of disapproval.

Laithwaite told me that as a result of making his speculative ideas about gyroscopes public, he had been shunned by some UK academics, who had found them just too controversial to consider. The popular press had decided that his twin gyroscope device was a magical 'anti-gravity' machine, while the scientific establishment branded him a heretic for bringing science into disrepute. He said that he did not wish to exacerbate the situation any further, so he had decided to continue his research work in private.

In the laboratory, Professor Laithwaite showed me parts of an experiment he was preparing, whereby he hoped to show that a moving flywheel could induce an effect across space on another rotating flywheel a very small distance away. He called the effect 'inertial radiation'. Looking back, I wonder if the arcing flywheel, with its radial acceleration linked to gravity through equivalence, might subject the other flywheel to a large $\gamma \partial \mathbf{g}/\partial t$ pulse as it passed across its face. As far as I know, the experiment was never completed.

During 1986 the UK media paid a lot of attention to an ITM built by a Scottish instrument maker, named Sandy Kidd. The essential feature of his device was a pair of flywheels at the end of rotor arms connected to a vertical shaft. In forcing the flywheels around the shaft, the flywheels flew up and down creating large up-and-down thrusts. Kidd demonstrated his ITM to Professor Laithwaite, who suggested that the novelty of using loose linkages between the rotor arms holding the flywheels might explain why the up-and-down thrusts weren't balanced, giving an overall thrust in one direction. The Kidd ITM also appeared on BBC's *Tomorrow's World*. At British Aerospace, several of us were intrigued by the Kidd ITM and I asked Sandy if we could test his machine. He readily agreed to this. A series of trials were carried out by engineers from the Wind Tunnel Department at Warton Airfield, near Blackpool. The Kidd ITM was suspended from a load cell by an elastic cord, used to dampen the vertical thrusts. Our interest was in detecting any weight change of the device, not in measuring the magnitude of the vertical thrusts (accelerations). During one sequence of test runs, the engineers detected a sudden, very brief, change of weight. We were all amazed. However, during many repeated test runs, we never saw any change of weight again and the engineers assumed that there had been a glitch in the electronic instrumentation. So, it was concluded that the ITM did not change weight and that the inertial thrust was zero. Sandy Kidd has described his visit to Warton in his book, *Beyond 2001*. That was our only foray into investigating a mechanical means of overcoming the force of gravity.

Nearly 50 years after Professor Laithwaite's lecture, where he pondered on the idea that gyroscope behaviour might be modelled as an analogue of electromagnetism, perhaps we can begin to see what was in his mind. Clearly, Laithwaite was right to insist that there is an analogue between Newtonian mechanics and electromagnetism. This analogue was explored in Chapter 33. A mass m at the end of a moment arm of length R precesses (moves cross-radially in the θ-direction) with angular velocity ω about a vertical z-axis. The mass speed v = R ω. We saw that the centripetal force $mv^2/R$ could be attributed to the interaction between the gravitomagnetic field of the moving mass and the equivalent induced gravitomagnetic field of the angular velocity of precession (see illustration in Chapter 33).

Let us extend the above example and replace the mass at the end of a pivoted moment arm with a wheel of mass m moving about a vertical axis with speed v. If the wheel doesn't spin then the situation is as above. Now, suppose that the wheel spins with angular velocity **Ω**, where the axis of spin points in the radial direction. If we examine the picture, we have a real gravitomagnetic field due to the mass moving with speed **v** and two equivalent gravitomagnetic fields, namely that due to the angular velocity of the wheel **Ω** and that due to the angular velocity of precession ω. The situation is quite complicated and the alternative view provided by picturing the gravitomagnetic fields is not helpful.

Let us make use of part of the gravitational analogue of the Lorentz force, namely $\mathbf{F} = m(\mathbf{v} \times \mathbf{b})$.

> The mass m is subjected to a combined equivalent induced gravitomagnetic field
> $\mathbf{b} = -2\omega\hat{\mathbf{z}} - 2\Omega\hat{\mathbf{r}}$
> So, the Lorentz force splits into two parts.
> $$\mathbf{F} = -\frac{1}{2}m(v\hat{\boldsymbol{\theta}} \times 2\omega\hat{\mathbf{z}}) - \frac{1}{2}m(v\hat{\boldsymbol{\theta}} \times 2\Omega\hat{\mathbf{r}})$$
> $$= -m\frac{v^2}{r}\hat{\mathbf{r}} - mv\Omega\hat{\mathbf{z}}$$

The first term is the centripetal force. The second term is a force in the vertical direction.

The second term makes the wheel appear to be lighter (as demonstrated by Professor Laithwaite) or heavier. Although considering gravitomagnetic fields provides an alternative view of the flywheel motion, gyro results are more simply understood using Newtonian mechanics.

A purely gravitational means of thrust would be available if we had negative mass. We are only aware of ordinary (positive) mass. But does negative mass exist somewhere in the Universe? In 1957, Professor Sir Hermann Bondi, of King's College London, pointed out that there is nothing in physics that says negative mass can't exist. However, some academics disagree and claim that negative mass is forbidden in Einstein's general relativity.

Theoretically, positive and negative point masses obey Newton's inverse square law of gravitation. This shows that like masses attract and unlike masses repel. Note that the free mass, whether positive or negative, is in free fall. It seems that the interactions between the masses are obvious, but they are not. When the directions of the accelerations of the free masses are examined, the predicted reactions are as given below. You may find them hard to believe.

- A free positive mass accelerates towards a fixed positive mass.
- A free negative mass accelerates towards a fixed positive mass.
- A free positive mass accelerates away from a fixed negative mass.
- A free negative mass accelerates away from a fixed negative mass.

In Earth's gravity field, to slow down a falling positive (ordinary) mass you must push upwards on it, or tug it upwards. We can use a spring balance to weigh a positive mass. The spring pulls the mass up and stops it from falling. From the spring extension, we can determine the weight of the positive mass. In theory, to slow down a negative mass falling in Earth's gravity field you must push down on it, or tug it downwards. To weigh a negative mass, we must push down on it with a pressure meter until the mass stops falling. From the pressure reading, we can determine the weight of the negative mass. It's bizarre.

The reactions indicate that positive masses are likely to form large conglomerates, such as stars and planets, and that smaller lumps of positive mass will be attracted by larger bodies and fall as meteorites. However, the reactions indicate that negative masses do not form conglomerates. If negative mass does exist, it is likely to be in the form of subatomic particles or single atoms.

Based on the inverse square law for point masses, there are three cases to consider where positive and negative masses interact:

1. The magnitudes of the positive and negative masses are equal.

This leads to the mass dipole and the idea of gravitational thrust.

2. The magnitude of the negative mass is greater than that of the positive mass. This case is unlikely, as negative matter, if it exists, appears in a fundamental form. However, if the situation does arise, the distance between the two particles will increase, leading to particle separation.

3. The magnitude of the positive mass is greater than the magnitude of the negative mass. This is the most likely case. If negative matter does exist in a fundamental form, it will attach itself to a positive mass.

A particle of matter and its anti-matter particle have opposite electric charge but are both expected to have positive (ordinary) mass m. An example is an electron with charge -e and a positron with charge +e, both with positive mass $m_e$. And a proton with a charge +e and an antiproton (observed in cosmic rays) with a charge –e, both with mass $m_p$. If a particle of matter and its anti-matter particle come into contact, they will annihilate each other, and their combined mass will be converted into energy $E = 2mc^2$, which is radiated away.

The nucleus of an anti-hydrogen atom is an antiproton with charge –e, which is orbited by a positron with charge +e. In 2020, we are awaiting the results of the ALPHA experiment at CERN, where anti-hydrogen molecules will be allowed to free fall in Earth's g-field. If the anti-hydrogen molecules fall down, we cannot say whether they have positive or negative mass. If they fall up, physics will have some explaining to do.

There seems to be no reason why an electron and a positron each with a negative mass can't exist. In this case, a cloud-chamber experiment would not be able to distinguish a negative mass electron from a positron. In the same way, a negative mass positron could not be distinguished from an electron. The same is true for the proton and the antiproton. It seems to me that we can't rule out the possibility that negative mass atoms are sometimes created during the explosion of a star. They would have a nucleus containing an antiproton with negative mass and a neutron with negative mass surrounded by orbiting positrons with negative mass. The negative atom would still be held together by electromagnetic force.

Negative mass may exist somewhere in the Universe. But what happens when negative matter comes into direct contact with positive matter? Negative matter is not anti-matter, so we should not expect a violent explosion when it comes into contact with positive matter. Mass neutralisation may occur, or the two types of matter might co-exist. Without experimental evidence, we can't say what happens.

Suppose positive and negative matter do co-exist. We can imagine positive matter dust and negative matter atoms settling on the airless Moon's surface. The negative matter atoms would attach themselves to the positive matter dust. If this is so, astronauts would find the moon dust tacky, due to its negative matter content.

In the case of the Earth, which has an atmosphere, any negative matter atoms attracted to Earth's surface would undergo a number of interactions with air molecules, water droplets, positive matter dust, etc., before reaching the Earth's surface. It may be that specks of negative mass exist on the Earth combined with ordinary matter in such a way that we haven't recognised its presence. It may be that the presence of negative matter in a sample of positive matter changes the density of the sample. Some thought needs to be given as to how to conduct a search for specks of negative mass hidden in Earth's positive mass surface. The difficulty is how to detect their presence and separate them from their positive matter host, and how to store negative matter.

We now consider the case of the mass dipole. There exists the possibility of natural microscopic mass dipoles made of fundamental positive and negative mass particles. If we can store fundamental negative masses to form a large negative mass, then the possibility exists of forming a macroscopic mass dipole. The positive mass, if free, would accelerate away from the negative mass, while the negative mass, if free, would accelerate with the same acceleration towards the positive mass. The combination forms an accelerating mass dipole with the positive mass at the leading edge and the negative mass at the trailing edge. Such a macroscopic gravitational propulsion system was first considered in the early 1960s by the US scientist, the late Dr Robert Forward, at that time a senior scientist at the Hughes Aircraft Research Center in Malibu, California. The mass dipole may be a science-fiction concept, but it doesn't disobey any physical principles. Although it has no inertia, it satisfies the conservation of energy and the conservation of linear momentum as it accelerates. Some academics assume that the mass dipole must accelerate in a runaway mode to infinite speed and, for this reason, rule it out. However, Einstein's mass velocity formula applies to positive and negative masses, so the terminal speed of the mass dipole is the speed of light.

Overall, a balanced mass dipole has zero mass. Slightly unbalanced micro-mass dipoles may exist, with a tiny mass signature. A microscopic example in nature may be the neutrino. According to particle physicists, there are three different types of neutrino, all moving at the speed of light. There are billions of neutrinos passing through the Earth every second. These are formed and ejected from the centres of stars and are a product of the weak nuclear force.

Perhaps radioactive materials are associated with positive and negative mass. One of the types of particles escaping from a radioactive source is an electron-neutrino, along with α particles (Helium nuclei) and β particles (Electrons or positrons) and electromagnetic γ (gamma) rays. It is very difficult to detect neutrinos, as they have a tiny mass signature. When a neutrino escapes from radioactive matter, it accelerates away to the speed of light. If a neutrino is a slightly unbalanced mass dipole, how might we catch it and then separate the negative mass from the ordinary mass and store it?

In 2017, Professor Peter Engels, at Washington State University in the USA, used a Bose-Einstein condensate to create the effect of negative mass. Although not real negative mass, perhaps their experiment has given them a clue as to how negative mass might be stored.

In the quantum world of nanotechnology, investigations are underway to see whether the Casimir force can be used to create banks of negative energy micro-cavities with macroscopic negative mass characteristics.

The EM Drive, developed by Roger Shawyer, is a device using a resonating microwave signal in a cone-shaped metal cavity which creates thrust in one direction. The cone cavity is truncated with the narrower region loaded with dielectric. The signal fed into the cavity has a frequency of 5 GHz, with a bandwidth of ±20 MHz. Perhaps the device has moved beyond the speculative stage, because tests in the UK and the USA have shown that the device develops a thrust of 20 mN in the direction from the wider end to the narrower end of the cavity. Shawyer has used special relativity to explain why the thrust occurs. Others disagree with his explanation and suggest that the thrust is due to the way the device is mounted and tested.

I have speculated that, due to the cone shape of the EM Drive, cut-off of some frequencies will occur as the photons travel towards the narrower end. This is massive red shift to extinction for some frequencies and means that energy is transferred from these photons to the quantum vacuum. Thus, an energy gradient is formed in the cavity, leading to a thrust. In other words, I suspect that the EM Drive thrust is due to the formation of a mass dipole.

You may think that the idea of a gravitational dipole propelling itself is too strange to be true. In 2013, an optical version of a gravitational dipole was demonstrated by Professor Ulf Peschel at the University of Erlangen-Nuremberg, in Germany.

It is still very early days in the development of a space-drive device based on gravitational propulsion, but scientists and engineers are gradually moving forward in that direction. Arthur C. Clarke, in his book, *Profiles of the Future*, guessed that we would have a space-drive by 2070. Control of gravity, as well as playing a part in space-drives, will also have many other applications.

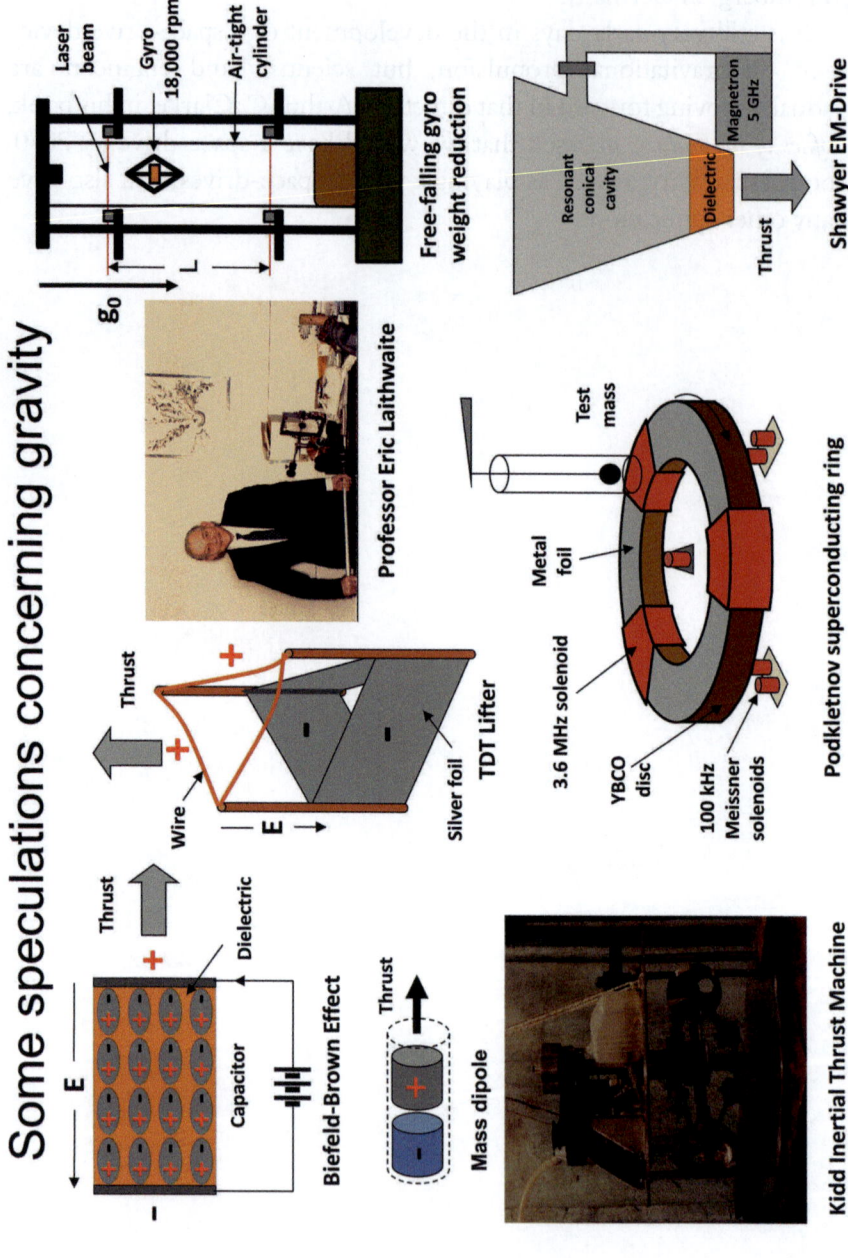

CHAPTER 38

# PROJECT GREENGLOW

Within the Advanced Technology Group of British Aerospace's Military Aircraft Division, at Warton Airfield in Lancashire, I had been running a part-time activity keeping an eye on developments in gravity research since late 1986. The two concerns were 'seeing with gravity', as a counter-stealth measure, and 'gravity field control', as a disruptive futuristic means of propulsion. Stealth research work is generally classified, but work on gravitational propulsion, being futuristic, is not at the moment, so I was able to write about it. In March 1990, we sponsored a round-table meeting with a number of UK universities to learn about the current state of gravity research. A copy of my unclassified report on the meeting had been sent to NASA in exchange for NASA reports. The late Brian Young, then the Technical Director, took a close interest in our activity. He arranged for a small study to be carried out by Advanced Project engineers to explore the benefits of using a gravity engine to propel an aerospace vehicle. A fictitious gravity engine was proposed, allowing several designs to be investigated. The subsequent artwork based on the designs provided some nervous fizz to the work. With the information gained, Brian Young gave several lectures on the subject with the provocative title, 'Anti-Gravity! The End of Aerodynamics?' Naturally, this alerted the outside world to our gravity study and stimulated the interest of the media. Brian Young was interviewed on BBC radio and quizzed about British Aerospace's gravity programme. The newspapers added their comments with a mixture of light support and 'we know better' amused sarcasm.

During the 1950s, most US aircraft companies had had some involvement in classified gravity research programmes, possibly linked with the UFO scare at the time and the onset of the Cold War. Had the Soviet Russian scientists developed a quiet gravitational propulsion system for flying saucers? By the mid-1970s, all the US aerospace firms had ended their gravity research work, with no hint of any breakthrough having been made. Or had the technology gone black? In 1989, Bob Lazar claimed that flying saucers using gravitational propulsion were being flight-tested at the highly secret Groom Lake facility (Area 51) in the Nevada Desert, where US stealth aircraft were tested. We had no idea what to believe.

In July 1995, the Interstellar Propulsion Society (IPS) was started in the USA. The first article in their quarterly newsletter was by Marc Millis, from the Space Propulsion Technology Division, NASA Lewis Research Center, Cleveland, Ohio. It was called, *Emerging Possibilities for Space Propulsion Breakthroughs*. The Board of Advisers was mostly made up of scientists and engineers from NASA, USAF, Boeing and Martin-Marietta.

In June 1996, there was an international conference on 'New Ideas in Natural Science' held in St Petersburg, Russia. The conference was hosted by the St Petersburg Physics Society, the Russian Geological Society, the Russian Geographical Society and the St Petersburg High School Teachers' Association. An adventurous programme of topics was presented, with a very strong emphasis on gravity-related research.

Dr Tom Shelley wrote an editorial in the *Eureka* Magazine (November 1996) about the Russian conference. He highlighted the fact that engineers from Panasonic, one of the world's largest electronic companies, had presented a joint paper at the conference, with academics from the Faculty of Engineering at Tohoku University, Japan, on the apparent anti-gravity property of spinning gyros in free fall. An earlier paper on the subject had appeared in *Physical Review Letters* (Vol. 63, No. 25, pp.2701–2704) in 1989, and many scientists had dismissed the results as being flawed. Although the gyro weight change was very tiny, Dr Shelley suggested that perhaps we should stop laughing and start to take the subject more seriously. He wrote, *The science is completely obscure, but there is no reason why anti-gravity might not be possible. Unlike perpetual motion machines, anti-gravity defies no fundamental laws of physics. Many accepted phenomena defy sensible explanation and many accepted explanations of well-known effects turn out to be wrong. The sensible attitude to take to any idea is: can you demonstrate it? Can you make it work? And, most importantly, how much is it likely to cost?*

In July 1996, the NASA Breakthrough in Propulsion Physics (BPP) Program was started, with a steering group containing scientists and engineers from NASA, DoD, DoE and USAF. The NASA BPP Program was led by Marc Millis and had the following visionary goals:

1. To propel a vehicle without propellant mass.
2. To attain the maximum transit speed possible.
3. To create a new energy production method to support the above.

The research programme envisaged a field propulsion system, with spacecraft moving at a significant fraction of the speed of light, to make interstellar, not just interplanetary, exploration a possibility. As was made clear, the goals listed were currently beyond existing scientific knowledge, with no guarantee that they were achievable. But without them, interstellar travel was ruled out for humans, due to the relatively short lifespan of individuals. What was being sought was a breakthrough, a huge step change, in propulsion physics and engineering.

The NASA BPP program was aimed at seeking a breakthrough in spaceship propulsion. It has long been recognised that, in terms of using conventional rockets, exploring the Solar System is going to be extremely limited because of the very long journey times. For example, a trip to Mars currently takes 6 months, while a trip to an outer planet takes years. This has almost ruled out manned exploration for the foreseeable future.

Back in the 18th century, explorations of the Pacific Ocean, made by the likes of Captain James Cook and Captain William Bligh, both of the Royal Navy, took years to accomplish. Nowadays, holidaymakers can fly from the UK to Australia, or New Zealand, in about a day. Presumably, something similar will happen with planetary exploration, but the comparison with the seafaring explorations does, unfortunately, suggest that we may have to wait another century or so before we have the technology needed to zip around the Solar System.

If a gravity propulsion system could be realised, just operating at an acceleration of one-g (equivalent to Earth's surface gravity), then trips to Mars could be made in just over 48 hours. (Theoretically, the speed of light imposes an ultimate limit on the speed.) One-g flight would also remove the muscle-wasting health problems associated with long periods of weightlessness from which astronauts currently suffer. However, continuously accelerating at one-g implies providing fuel from somewhere, since only a limited amount can be carried on board. We need some form of fuel/energy scoop, but of what?

Perhaps we should also bear in mind that within a few centuries we are likely to have used up much of the land-based mineral wealth of the Earth and be stretching energy demands to near breaking point. So, unless we can mine minerals from deep under the oceans and find new sources of fuel/energy, we are likely to enter a period of political instability as powerfully armed countries attempt to secure what they can of dwindling resources for their own populations. Global wars over such limited resources cannot be ruled out and, indeed, have probably already begun. Eventually, such wars may see the end of planet Earth, in terms of a haven for humans. So, perhaps we shouldn't think of a new propulsion system and a new source of energy just to allow humans to explore interplanetary space, but think of it as a necessity for the long-term survival of mankind.

During the 1990s, Dr Andrew May, an astrophysicist from Cambridge University, was the desk officer responsible for managing MoD research programmes in Aerodynamics, Propulsion and Guidance within the Science (Air) Directorate (TG3). In 1996, I was required to report progress to him on an MoD-funded British Aerospace research programme. It was not to do with gravity but, in passing, I spoke of my interest in gravity and mentioned that we were carrying out a small study. Dr May showed an interest so, later, I sent him a British Aerospace report describing our gravity study. In a letter, commenting on the study report, Dr May wrote:

*I found it very entertaining and thought provoking, with the promise of a high payoff if the ideas amount to anything. As you say, the real value of such work is in its ability to stimulate speculation and discussion in completely new areas. For myself, it's made me realize just how incomplete current physics is (i.e. gaps and inconsistencies), and how much scope there is for new phenomena without violating accepted theories or experiments.*

Although Dr May's interest and support never resulted in any MoD funding, it did provide the British Aerospace study with a level of credibility within the two organisations. This was quite important, as gravity field propulsion might otherwise have been dismissed as anti-gravity nonsense. Dr Gari Owen, from MoD Technical Intelligence, also took an interest in our study, as it provided him with some clues of what to look for in foreign scientific developments in this area.

In March 1997, the late Professor John Allen gave a lecture to engineers from British Aerospace's Future Concepts Group on 'The long-term future of aerospace'. In 1970, John Allen had edited a book entitled, *The Future of Aeronautics*, published by Hutchinson. In one chapter, he had mentioned revolutionary changes that occurred from time to time, suggesting that auto- (not anti-) gravity levitation might be a possibility in the future. John Allen had been the Chief Engineer Hawk at the time of the aircraft's inception at British Aerospace Kingston, the home of the Harrier. Later, he became the Chief Future Project Engineer at Kingston. In 1983, he took early retirement and in 1997, he was the Visiting Professor of Aerospace Design at the College of Aeronautics at the Cranfield Institute of Technology. We asked John Allen to carry out a more in-depth gravity study on the basis of 'What if we had a gravity engine – How would it change aerospace vehicle design?' This speculative approach is used by a number of large engineering corporations, particularly in the automobile industry. The study eventually culminated in a technical paper in the journal, *Progress in Aerospace Sciences* (Vol. 39, pp. 1–60, 2003).

During the summer of 1997, Walter Johnston, from British Aerospace's Operational Analysis Department, and I visited Lancaster University to meet Professor Robin Tucker, Dr Charles Wang and Dr Jonathan Gratus, all experts in general relativity. We discussed the idea of setting up a small UK version of the NASA BPP programme. They agreed that they were interested and that the funding would be on a shared basis. I then had to get our share of funding from British Aerospace to support the venture. In November 1997, I submitted a 1-page proposal entitled *Advanced Propulsion Study*, applying for funding from the Warton Technologists' research budget. The proposal had the subtitle, *Project Greenglow*; the name Greenglow coming from Brian Young's earlier tongue-in-cheek comment that anti-gravity might emit a green glow. Although we would be responsible for the programme, the work was to be carried out primarily by university academics. Professor Tucker was designated as the programme Academic Adviser and, later, Professor Allen accepted the role of Technical Design Consultant. The application was successful.

The Greenglow programme started in late spring 1998. Several down-to-earth topics were proposed, to get the programme underway. Firstly, a review of global gravity research was to be carried out. Secondly, an investigation was to be made of the linearisation of Einstein's gravity equation leading to the gravito-electro-magnetism (GEM) equations, resulting in a paper in the journal *Classical and Quantum Gravity* (Vol. 17, pp. 4125–4157, 2000). Thirdly, an examination of the 'gravi-craft' concept; whereby an orbiting spacecraft can change its altitude by altering its angular velocity. This led to a paper being published in *Acta Astonautica* (Vol. 53, pp. 161–172, 2003). The concept, originally proposed in 1968 by the Russian Soviet scientists, Beletsky and Givertz, was not what we had in mind for gravity propulsion when we started out. But that's what happens when you begin to investigate a subject; other routes open up that you hadn't thought of. Fourthly, the dynamics of slender orbiting structures was to be analysed, linked with detecting gravitational waves, and this resulted in a PhD Thesis in 2002. A fifth idea, to look at using tethered satellites to extract gravitational energy from the Earth's field, was abandoned due to tether instability problems.

We had put feelers out searching for any other UK companies that might like to participate with us on Project Greenglow. Only Rolls-Royce responded, and Dr Eddie Williams, the Chief of University Research Liaison, and Dr Michael Provost, a senior project engineer, were invited to join Project Greenglow's progress meetings.

A website was set up for Project Greenglow, voluntarily managed initially by Colin Brown and then, later, by Paul Baker, both from the Warton Electromagnetics Test Group. We stated that Greenglow was the beginning of a new adventure in the realm of aerospace. We hoped that other enthusiastic scientists from academia, government and industry might like to join us, particularly those who believed that the gravitational field was not restricted to passivity. The world-wide response by the public was overwhelming, with people wanting more information and offering us their own ideas for future propulsion systems. We were totally swamped by the interest shown. We hadn't realised the size of the public's interest in Project Greenglow. We tried our best to reply to all email enquiries, but it was extremely time-consuming. One has to realise that Project Greenglow was a part-time activity for those British Aerospace engineers involved with it.

To try to give the Greenglow programme a European dimension, we originally planned to invite academics from Europe to talk about their gravity research work, with particular emphasis being placed on gravitational propulsion studies. The first of these was a joint lecture held in September 1998 at Lancaster University. Dr Costas Kyritsis, from the National Technical University of Athens, spoke about 'A unified derivation of non-linear electromagnetism and gravitation – implications in electromagnetic propulsion'. This was followed by Stavros Dimitriou, from the Technological Institute in Athens, who spoke about the Biefeld-Brown phenomenon and 'Thrust from time-derivatives of the electric charge'.

To make a breakthrough, one needs to be creative and think outside the current physics dogma, which both Modern Greek philosophers tried to do. But, in exposing their ideas, they were subjected to a rigorous grilling by the audience, which consisted of academics and engineers from various institutions. We could have been back in ancient Greece, where scientific debate was more adversarial! I can't say that I was wholly comfortable about the process, but those trying to wrestle Nature's secrets from her must expect a tough time.

On reflection, I felt that for speculative topics, a round-table meeting, or a brainstorming session, would be better, as there was then an opportunity for all to participate in proposing, challenging and defending new ideas, rather than one person defending a way-out idea against the rest. Public lectures on very speculative topics, I decided, were best avoided, as their controversial nature often elicited a great deal of scepticism, which is rather negative and can be demoralising for those trying to come up with new ideas.

So, there was a change of approach; not so much trying to generate new ideas but instead learning more about current ideas of interest. Among the academics that agreed to talk to Greenglow audiences, the better known included Professor Sir Roger Penrose, from Oxford University, who spoke about current gaps in physics; Professor John Barrow, from Cambridge University, who spoke about patterns in nature; and Professor Ed Hinds, from Surrey University, who spoke about quantum physics and the Casimir force.

Dr May, at the MoD, was staggered at British Aerospace's boldness in setting up Project Greenglow and going public with it. He said that he had tacit approval from the Director of Science (Air) for him to continue his interest. However, because there was a fear within the MoD that certain elements of the mass media would be likely to twist the facts if it openly supported Greenglow, say, with a headline like, *Government Sponsors UFO Programme*, the MoD would have to keep its distance. Nevertheless, he remained keen to help and suggested that we should read an unclassified USAF report written by Dr Robert Forward, entitled *Mass Modification Experiment Definition Study (Journal of Scientific Exploration*, Vol. 10, No. 3, pp. 325–354, 1996). The underlying theme of the report was about the quantum fluctuations of the vacuum and the possible link with inertia. Dr May said that since the USAF were willing to sponsor such speculative studies, it was not inconceivable that the MoD might fund work along these lines, too, perhaps from the MoD Technology Watch Budget.

By the end of 1999, we had added three more small research activities to Project Greenglow. The first was a variation of the concept of the 'photon' rocket, being investigated by Dr Paul Smith at Dundee University in collaboration with Professor Sergei Vinogradov, visiting from the University of Kharkov, in the Ukraine. The basis of the idea was the microwave illumination of an open spherical conducting cavity which, at resonance, experienced an enhanced, although small, thrust above straightforward radiation pressure. Scientists in the USSR were among the first to investigate microwave cavity thrust. Similar studies have been carried out in the USA.

When the open microwave cavity thrust study came to a premature, but inconclusive, end in 2002, we tried to re-allocate the Greenglow funding to support the study of a closed cavity microwave thrust device. This was the EM Drive, which achieved thrust without expelling any obvious exhaust. The brainchild of the device was Roger Shawyer, a former British Aerospace employee. However, when BAE Systems sold Astrium, its satellite business, to Airbus (EADS) in 2002, it signed a caveat agreeing not to work on anything connected with space for a limited period. Consequently, we were prevented by Head Office from supporting research on the EM Drive.

The second study was a small experimental programme led by Dr Clive Woods at Sheffield University, to investigate certain aspects of the claim that a YBCO superconducting disc could shield a test mass from Earth's gravity. The claim had originally been made in 1992 by Dr Evgeny Podkletnov, a Russian materials scientist.

Following Robert Matthews' article in the *New Scientist* magazine (21st September 1996) and his mention that gravitomagnetism might be linked with gravity shielding, I wrote to Dr Podkletnov in September 1996 and asked him for more information. He replied a month later, saying that he was an expert in materials science and he had reported a strange gravity effect that he had noticed while carrying out Type II YBCO superconducting experiments. He had never worked on 'anti-gravity' research.

And there the matter rested, for a while. Then, in the March 1998 issue of the science magazine *WIRED*, Charles Platt wrote an article about Dr Podkletnov's gravity shielding claim. In the article, he wrote that Dr Podkletnov had been in contact with students at Sheffield University who had been investigating his reported gravity effect. I made some enquiries and found that Dr Clive Woods, a senior lecturer in the Electronic and Electrical Engineering Department at Sheffield University, was responsible for overseeing the final-year student projects which had included some investigating of Podkletnov's experiment. I contacted Dr Woods and asked him what he thought of Podkletnov's claim. He said that he had no idea whether the claim was true, but that several students were interested in examining some aspects of Podkletnov's experiment for their final-year degree projects. Although the topic was very controversial, no one in the department had seen any reason to dissuade them from trying. It is a valid scientific endeavour to attempt to reproduce the claims of others that have been published in the correct scientific manner. The role of the staff was to ensure that the students followed a proper research procedure in their experiments.

In May 1998, I visited Sheffield University to learn more about the results of the students' experiments. I was quite keen to get British Aerospace to fund some further studies, but was very aware of the scepticism and hostility of many UK scientists to Dr Podkletnov's claim. The turning point was an article by Charles Seife in the *New Scientist* (6th February 1999), which disclosed that NASA had initiated Project Delta-G, with funding of $600,000, to investigate Dr Podkletnov's claim.

By April 1999, Dr Woods had provided us with the details of a small research programme, to be carried out by Dr Steve Cooke. British Aerospace funding for the programme was agreed by the Warton Technologists' research board in July and the work started at Sheffield University in November 1999. The cost of the programme included funding a visit to the UK by Dr Podkletnov. He arrived in February 2000 and gave a lecture on his superconducting research work to a group of invited guests from academia, industry and the media. He stressed that he was a materials scientist, not a gravitational physicist.

Limited funding meant that Podkletnov's method of using Meissner levitation and interacting radio frequency (rf) fields to generate rotation of the YBCO disc was not attempted. Instead, a small YBCO disc, of 50-mm diameter, formed the lid of a cryostat mounted on the vertical axle of a motor. While the cryostat rotated and the YBCO disc was in its superconducting state, the weight of a test mass suspended just above the disc was observed. No weight loss was detected. In a modified experiment, a superconducting YBCO disc was excited with a 13.56 MHz signal but, again, no weight loss of the test mass suspended overhead was noticed. So, the limited Sheffield experimental results provided no support for Dr Podkletnov's claim of local gravity modification.

The third activity was a small study looking at possible macroscopic applications of the Casimir force, conducted by Peter Laurie, a private researcher. This was our tentative venture into the realm of the quantum vacuum.

These activities, together with Professor Robin Tucker's gravitational research programme and Professor John Allen's design study, formed the basis for Project Greenglow.

During the late 1990s, the British Aerospace Board was exploring the idea of a merger with DASA, the German aerospace company. Then, suddenly, in 1999, it was announced that British Aerospace had agreed to purchase Marconi Electronic Systems from GEC and intended merging the two defence companies to form BAE Systems in the millennium year of 2000. It came as a complete surprise to most of us, who had assumed that we were teaming up with the Germans.

There followed a protracted period during which time the executive positions of British Aerospace and Marconi Electronic Systems were briefly amalgamated, after which all executives had to apply for the reduced number of new posts in BAE Systems. Of course, there were redundancies, but fortunately for me, being on the bottom rung of the executive ladder, I was retained.

During this transitional period, funding for Project Greenglow was withheld, while responsibilities were sorted out. After the merger, it was decided that the Military Aircraft Division at Warton should no longer fund speculative research work as this was now the responsibility of the new company's Advanced Technology Centres (ATC). The post of engineering director for the ATC was filled by Dr Carl Loller, based at what had been the British Aerospace Sowerby Research Centre at Filton. The chief technologist of the ATC was Dr Brian Wardrop, based at what had been Marconi's research centre at Great Baddow, near Chelmsford.

It was agreed that the ATC would fund Project Greenglow for a further 3 years, as part of the ATC speculative research programme called OUTlook! As head of the OUTlook! programme, Dr Bill Martin took Project Greenglow under his wing. The OUTlook! budget manager was Dr Vaughan Stanger, who had responsibility for funding Project Greenglow. I continued my part-time role as the technical coordinator of Project Greenglow.

To put Project Greenglow on a more formal research footing, the ATC decided that a goals and metric study was needed. This was undertaken by Professor Colin McInnes, Professor Matthew Cartmell and Dr Spencer Ziegler of Glasgow University and was jointly funded by BAE Systems (OUTlook!) and Rolls-Royce. The report recommended a slight shift in emphasis away from the classical physics of general relativity towards quantum mechanics and the zero point energy field of the vacuum of space. In 1994, Hal Puthoff, Bernhard Haisch and Alfonso Rueda had investigated the idea that inertia might be a drag force experienced by a mass as it accelerated through the zero point energy field. The 'mass modification' report by Dr Forward was an added reason for making a slight shift in the direction of research towards investigating quantum mechanical effects linked with gravity and acceleration.

In January 2001, I was invited to attend the inaugural meeting of the IoP (Institute of Physics) Gravitational Physics Group, by the chairman, Professor Mike Cruise of Birmingham University. The specialist group of IoP members were closely involved with gravity research. The meeting was well attended. From the presentations given, it was clear that most interest was in the possible detection of gravitational waves. There were no presentations on fundamental gravity research looking for links with the other natural forces. My presentation on Greenglow activities was received with reserved silence. At the inaugural meeting, a committee was formed to pursue the gravitational physics groups' interests in research topics and funding sources. Professor Robin Tucker, the Greenglow academic adviser, became a member, and Dr Walter Johnston, a BAE Systems engineer from Warton, accepted the role of industrial liaison officer.

At the end of January 2001, I attended a 3-day international conference at Sussex University with the focus on the search for new ideas for space vehicle propulsion, with particular emphasis on force-field control. I was initially doubtful about attending the conference as it seemed, to me, to be a UFO convention in disguise. But my misgivings were wrong. As I learnt later, the conference was sponsored by the British National Space Centre (BNSC), through the offices of Mike Geer, then the technology manager.

The Field Propulsion Conference was co-chaired by Graham Ennis and Dr Anders Hansson. To mention just a few of the presenters, there was: Dr Alan Holt of NASA; Dr Hal Puthoff of Advanced Studies at Houston; Professor John Allen of Kingston University; Stavros Dimitriou of the Technical Institute in Athens; Professor Jean-Pierre Vigier of the Laboratoire de Gravitation et Cosmologie Relativistes; Dr Anders Hansson of the 3rd Commission of the International Academy of Astronautics (IAA), Paris; Tony Cuthbert, a UK inventor; Dr Claudio Maccone of ALENIA; Richard Obousy of Qinetiq; Dr Jean-Paul Petit, the Director of the CNKS Laboratoire at Marseille; and many others, including me. And to mention some of the attendees, there was: Nick Pope, formerly the MoD desk officer responsible for assessing UFO sightings in the UK; several aviation journalists, including Nick Cook of *Jane's Defence Weekly*, Malcolm English of *Air International* and Alexandre Szames, who writes for *Air & Cosmos*; and several science reporters, including Ian Sample who writes for *New Scientist* and Jonathan Leake of *The Sunday Times*. Tony Edwards, a TV producer of science programmes, was there. And there were a number of authors of popular science books present, too.

The conference was well run and I enjoyed it. However, it did have the trappings of a media extravaganza. On the first day, the conference organiser, Graham Ennis, was interviewed by John Humphrys on the BBC Radio 4 *Today* programme. On the last day, a full-page article about the conference appeared in the *Guardian* newspaper, written by James Meek. A comprehensive write-up of the topics presented at the conference appeared in the March/April 2001 issue of *UFO Magazine*, prepared by the editor, the late Graham Birdsall. Mark Pilkington also reported on the conference in his article entitled, *Fields of Dreams*, in the February 2001 issue of the *Fortean Times*. There were no bombshells! But, like many conferences, it was a good place to meet people and discuss ideas.

With a view to making contact with Marc Millis, the team leader of the NASA BPP Program, Dr May advised us to contact Major Jerry Sellers PhD at the USAF EOARD (European Office of Aerospace Research and Development) in London. Major Sellers was very helpful. We learnt that the NASA BPP team was scheduled to visit Europe during September 2000, to drum up support for their programme. Arrangements were made for the NASA team to visit BAE Systems at Warton, but first they visited the European Space Agency's Research and Technology Centre (ESA-ESTEC) in Holland. There, among other scientists, they met Dr Martin Tajmar and Dr Clovis de Matos.

The NASA meeting at BAE Systems Warton was hosted by Dr Bill Martin. Mike Geer, the technical manager of the British National Space Programme (BNSC) was present, representing the official UK link with NASA. The Greenglow team of academics present included Professor Robin Tucker (Lancaster University), Professor Colin McInnes and Professor Matthew Cartmell (Glasgow University), Dr Clive Woods (Sheffield University), Dr Paul Smith (Dundee University) and Professor John Allen (Kingston University). Other attendees, linked with Greenglow, included Dr Eddie Williams and Dr Mike Provost from Rolls-Royce. Also present were Dr Anders Hansson (space consultant) and Tony Cuthbert (inventor), along with engineers from BAE Systems.

The exchange was very successful. Marc Millis extended an invitation to some of the Greenglow academics to present papers at the forthcoming 37th AIAA/SME/SAE/ASEE Joint Propulsion Conference at Salt Lake City, Utah, during 8th to 11th July 2001. The USAF EOARD team offered to provide funding to enable this to happen; an offer that was accepted. Why couldn't the UK find the funds to do this? Following their attendance, several papers were published in the AIAA journal's coverage of the 2001 Joint Propulsion Conference by Greenglow academics.

During 2002, a further study was begun, in line with the Greenglow goals and metric study, investigating the Casimir force in a bit more depth. The modern view of the ether is as a quantum phenomenon, where teeming virtual photons of all possible frequencies flit briefly into and out of existence. Experimental support for this view stemmed from the 'Lamb frequency shift': a spectral observation of the shift in the Zeeman lines in hot hydrogen gas. In 1948, the Dutch physicist Hendrick Casimir predicted that a tiny conducting cavity, of micron dimensions, would experience a tiny force, of order $10^{-3} N/m^2$, due to its buffeting by the virtual photons of the quantum vacuum (a new name for the ether) as the dimensions of the cavity would exclude some virtual photons from being able to pop into existence inside the cavity. Thus, the zero point energy within the cavity was reduced below the natural zero point energy of the vacuum. The energy gradient across the cavity wall would give rise to a force on the wall. Or, the resulting imbalance in buffeting would lead to the cavity experiencing a tiny force. Evidence of this effect was first obtained in 1958. Nowadays, the use of atomic microscopes has made measurement of the curious Casimir effect much easier. Most research is now aimed at understanding micro-cavity designs to exploit the Casimir force. The Casimir force research study was led by Dr Clive Speake, the Head of Experimental Gravity and Space Research at Birmingham University, with support from Dr Giles Hammond.

Annual Greenglow progress meetings were held at various universities to allow researchers to tell of their work. All the studies were of an unclassified nature. The meetings were attended by academics, industrial scientists and engineers, and government scientists.

The NASA BPP Program ended in 2002. At the end of 2004, the ATC management decided not to continue funding Project Greenglow. We had carried out a successful programme of work but had made no dramatic breakthroughs in understanding how to control gravity.

Following the British Aerospace-Marconi merger to form BAE Systems in 2000, there were still internal changes taking place for several years afterwards. I retired in February 2005. The Sowerby ATC Research Centre at Filton was closed in 2014. The airfield at Filton was closed earlier in 2012.

Today, the Space Systems Department at Dresden University in Germany is the European centre for breakthrough space propulsion and gravity research studies, led by Professor Martin Tajmar. In the US, Marc Millis heads the privately funded Tau Zero foundation, begun in 2008, which is focussed on research on interstellar flight. Also in the US, the 100-Year Starship Program is a small funded initiative by DARPA (US Defense Advanced Research Project Agency).

# DRAWING & PHOTOGRAPHS

Centre, then from 12 o'clock, clockwise.

1. BAE Systems drawing EAG-14322 of a VTOL Combat Aircraft Concept, produced by Martin Kennedy of Advanced Studies. The futuristic propulsion system was assumed to be a hybrid anti-gravity and conventional jet engine combination.
2. The futuristic anti-gravity aircraft was painted by Alan Groves of Graphic Support. The green anti-gravity rays, added for effect, led to the project title of Greenglow. Other designs were also prepared. All pictures were used as a focus for discussion.
3. Members of the Future Concepts Group at Warton. In the centre is Professor John E Allen (Greenglow Technical Design Consultant). On his right is Peter Liddell, Head of Advanced Studies. On his left is Dr Ron Evans, Leader of Project Greenglow.
4. The NASA meeting at BAE Systems (Warton). Marc Millis, the NASA BPP Leader talking to Dr Bill Martin, from the BAE Systems Sowerby Research Centre, Filton.
5. Lecture by Dr Evgeny Podkletnov, at Sheffield University, to describe his superconductivity research and the possible link with gravity. Dr Podkletnov is talking to Nick Cook, an aviation journalist from *Jane's Defence Weekly* and Sqn Ldr Bruce Holley from RAF Cranwell.
6. Meeting at Birmingham University. In the centre is Dr Clive Speake.
7. Lecture by Professor John D Barrow at BAE Systems (Warton). From the left, Ron Evans, Derek Reeh (BAE Systems Chief Test Pilot), Professor John D Barrow (Cambridge University), Professor Robin Tucker (Lancaster University) and Professor Phil Bissell (UCLAN).
8. Lecture by Professor Sir Roger Penrose at BAE Systems HQ, London. On the left, Professor Penrose (Oxford University) and Professor Robin Tucker (Lancaster University).
9. Meeting at Lancaster University. From the left, Professor Robin Tucker, Dr Jonathan Gratus and Dr Clive Woods (Sheffield University).

# Project Greenglow

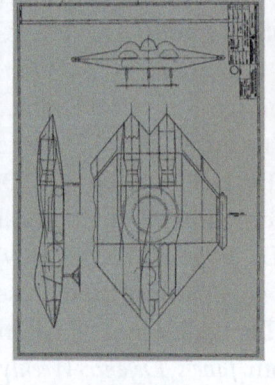

CHAPTER 39

# THE BOOK AND THE BBC *HORIZON* PROGRAMME

Brian Young, the British Aerospace Director of Strategic Projects (formerly the Technical Director), was interviewed by Alun Lewis on the BBC Radio 4 programme *Science Now* in May 1992 and gave some early details of our small gravity study. Further details appeared in Sandy Kidd's book, *Beyond 2001:The Laws of Physics Revolutionised* and in Nick Cook's book, *The Hunt for Zero Point*, but we had done much more since then.

After I retired from BAE Systems in February 2005, I spent some considerable time writing down what had happened during the Greenglow years while the events were still fresh in my memory. Although we had made no breakthroughs in understanding how to control gravity, I thought that we had carried out a good programme of work. And I wanted to leave a record of what we had done and try to inspire others to follow on. Furthermore, I wanted to correct the impression that we were a bunch of idiotic engineers working on an anti-gravity programme, as some elements of the press would have you believe. And, for the record, Einstein never ruled out the possibility of anti-gravity. Project Geenglow was a very speculative study, with strictly limited funding, delving into the possibility of controlling the force of gravity. It had some very bright university minds working on it. It was a daring thing to do, rather like studying aeronautics during the 19$^{th}$ century. So, it was with these thoughts that I worked away at producing a book describing British Aerospace/BAE Systems' gravity programme.

Finally, when I'd finished the book, which had forty-two chapters, I didn't know what to do with it. However, other things happened in my life, including working as a consultant for Wales Aerospace for a while, so I had to put the book to one side. But, eventually, things returned to an even keel and I looked at the book again. I produced a trimmed-down version of it and in 2010 started looking for a UK agent to help me get it published. I was not successful. The book fell between two stools. It was not a wholly academic book, nor was it a popular science book. It was a semi-technical book. I had no experience in the writing field and no one to help me. I wrote to UK publishers to see whether they had any interest. By and large, they didn't. Finally, after more than two years of trying to get help, I decided to self-publish. And there I had some good luck. I contacted Jeremy Thompson, the managing director of Troubador Publishing Ltd. After viewing my manuscript, he said his company would be willing to publish it. I had a very good experience with the company and have no hesitation in recommending Troubador to other would-be science authors.

The book, *Greenglow & The Search for Gravity Control*, was published by Matador in April 2015. Apart from a press release by Matador, there was very little publicity. However, in May 2015, Dr Andrew May wrote a supportive review in *Popular Science*. At the end of the review, he wrote,

> *So if the book is not aimed at the typical pop-sci reader, the typical textbook reader or the typical alternative science reader, who is going to enjoy reading it? The answer is all of the above! The writing, if you're prepared to skip over the (generally unnecessary) equations, is as lucid and well-structured as the best popular physics book I've read.*

The book appeared on the Amazon website and attracted the wrath of the arch- sceptic Dr David J. Fisher, who wrote a blistering review of the book, calling it an utterly disgraceful and very worrying book. Ever since we began our interest in gravity research at British Aerospace Warton, Dr Fisher has asked me for information about what we were doing, and I sent him reports and details. I knew he was a sceptic and I have no problem with that. If we are to make scientific progress, we can't just accept what Aristotle, Galileo, Newton and Einstein tell us. We need to check for ourselves. If an apparent breakthrough is made, we must remain sceptical until it is confirmed, or otherwise. And we know that the reactions of scientists to claims of breakthroughs are very aggressive.

Dr Fisher's hostile review elicited a number of favourable responses in support of the book. One reviewer suggested that Dr Fisher was championing current gravitational physics orthodoxy, thereby ruling out any possibility of further advances being made in this area of study. This comment led Dr Fisher to write that the positive reviewers were members of the lunatic fringe. I prepared my own reply to Dr Fisher's flamboyant criticisms but was advised by Matador not to send it. I wanted to tell Dr Fisher that the Greenglow studies were undertaken by university researchers with the highest 5★ UK government ratings and that they had published papers on their work in respectable academic journals, but I didn't.

In July 2015, Sarah Taylor, the marketing manager at Troubador Publishing Ltd, passed on an email enquiry to me from Steve Crabtree, the editor of the BBC's prestigious science programme *Horizon*. As Sarah added to her email, *Great News!* Indeed it was. Steve asked whether I would be interested in making the book into a TV documentary. I wrote back to him saying that I would. In fact, while working for British Aerospace/BAE Systems, I had had two previous enquiries from earlier editors of the *Horizon* programme, asking whether I would participate in their TV programme, but I had declined, in order to avoid any adverse publicity for the company and not to upset senior management. But now I was a free agent and I thought that the publicity from a TV documentary would be good for my book sales, although, as it transpired, my book was not mentioned in the documentary.

Steve arranged for me to meet his producer, Nic Young, and the programme researcher, Annie Mackinder. We had an initial meeting in Blackpool, and then a meeting at BAE System Warton. I was keen to get Warton backing for the proposed TV documentary, and Warton were keen to take part, too. But then, BAE Systems' Head Office management stepped in and placed an embargo on any further contact with the BBC *Horizon* programme. I wrote to Nigel Whitehead, the Group Managing Director of Programmes and Support, in the hope of getting the embargo lifted. For a short while during the 1980s, Nigel and I had both worked in the Aerodynamics Office at Warton. However, I only managed to get a reply from Nigel's executive assistant confirming the decision that BAE Systems would not participate in the making of the programme. I was disappointed, as were a number of working and retired BAE Systems engineers. But making the TV documentary went ahead, regardless.

Nic Young had a well-thumbed copy of my book and from it he had produced a storyboard for the TV documentary. He sent me a copy and asked for my comments. I spent some time going over his proposed programme, which looked very interesting, and sent him some comments back. The programme story followed my book fairly closely, with some added material about things that had happened since Greenglow had ended a decade before. Nic said that the programme must also be entertaining so, at the beginning, the impact of gravity was looked at from the point of view of people's experiences on rides at Blackpool's fun fair, where gravity plays an important part in the thrill of the rides. Filming started in October 2015.

Nic was keen to stage a re-enactment of Professor Laithwaite's gyro experiment at the Royal Institution in 1974. I had mentioned it in my book, but I had reservations about repeating it for the TV documentary, knowing the difficulties that it caused to Professor Laithwaite. But to be in the lecture theatre of the Royal Institution was exciting. Dr Adam Wojcik, an engineer from University College London, performed the gyro re-enactment. As another part of Nic's programme, I was asked to read aloud some of Michael Faraday's notes from his famous laboratory diary. I am not an actor and I fumbled over the words, which I knew so well. I was disappointed with my effort. Unsurprisingly, this did not appear in the *Horizon* programme.

Nic and Annie went off to interview Dr Evgeny Podkletnov in Moscow about his claim that superconductors could create gravity fields. In the Greenglow programme, we had provided Sheffield University with limited funding to enable them to replicate some aspects of Podkletnov's original experiment but, in their subsequent investigation, they didn't detect any gravity effect. Dr Podkletnov also claimed that using a superconductor, he had built a gravity beam device.

Professor John Ellis, a particle physicist from King's College London, was filmed talking about gravity. In his opinion, anti-gravity was an impossible dream. He was dismissive of Podkletnov's claim that a superconductor could be used to create a gravity field. He was not alone; other scientists had taken a similarly strong view. Later, I was filmed at the hotel in London where I'd secretly met Dr Podkletnov on a second occasion to discuss his gravity beam device. I was asked what I thought about it. My answer was that I didn't know. We don't understand gravity. I was there to listen to what Dr Podkletnov had to say.

Nic and Annie went to the USA and interviewed Marc Millis about the NASA Breakthrough in Propulsion Physics Program that began in 1996 and ended in 2002. Dr Millis said that they were looking for a revolution in space propulsion technology. Rockets were not the answer for space vehicle propulsion to explore the Solar System and beyond. He was asked about Dr Podkletnov's work. He said that although NASA had fully copied Dr Podkletnov's gravity beam experiment, with Dr Podkletnov in attendance, nothing was detected.

The voice-over told us that since the Greenglow and NASA programmes had both ended, the baton for gravity research and the investigation of new means of propulsion had been passed to Professor Martin Tajmar at Dresden University, in Germany. Nic visited Professor Tajmar's laboratory, where they had investigated Dr Podkletnov's gravity beam experiment, again without any success.

Professor Tajmar's view was that for control of gravity we need negative mass. Coupling positive and negative mass together, to form a gravity dipole, would lead to a propulsion system, or a warp drive. From theory, the movement of the gravity dipole is in the direction from the negative to the positive mass. Professor Tajmar said that an optical version of such a device was already in existence.

Professor Clifford Johnson, a theoretical physicist from the University of Southern California, said that compared with electromagnetism, gravity is tiny. The gravity of the whole Earth could not pull down a fridge magnet stuck to the metal door of a car. The force of electromagnetism is 10 to the power 40 times greater than the force of gravity. It was only after a repeat viewing of the Horizon documentary, *Project Greenglow: The Quest for Gravity Control*, that I realised that there was one very important word not mentioned in the programme – gravitomagnetism. It is the missing element in gravity control. Gravitomagnetism is the dual force of gravity, like magnetism is the dual of electricity. If electromagnetism is linked with gravity, as many scientists suspect, then the coupling between the two might be between magnetism and gravitomagnetism. If so, the $10^{40}$ factor might work in reverse. A tiny change in magnetism would give rise to a large change in gravitomagnetism and, via duality, with gravity. Professor Johnson went on to say that negative mass is impossible, because if it existed, it would warp space-time into a hill. If negative mass was coupled with ordinary (positive) mass, causing a dip in space-time, the two masses would create a gravitational dipole, or warp-drive; a runaway phenomenon for which, Professor Johnson claimed, there was no supporting evidence in the Universe. When the interviewer mentioned that gravitational propulsion, in the form of a warp drive, was just the device that engineers were looking for, Professor Johnson said it was hilarious.

Nic interviewed Professor Tamara Davis, an Australian astrophysicist, on a mountain in Switzerland. She seemed to contradict Professor Johnson's view, saying that anti-gravity does appear to exist in the far reaches of the cosmos, where it is responsible for pushing the galaxies apart and causing their outward acceleration at the boundary of the Universe.

The programme then turned to particle physics. Dr Dragon Hajdukovic, at CERN, is investigating whether anti-matter, as well as possessing opposite electric charge, possesses opposite, or negative, mass. At the Alpha experiment laboratory, they will let anti-hydrogen particles drop in the Earth's gravity field to see whether they fall up or down. As I wrote in my book, most scientists believe that anti-matter has positive mass, but this has yet to be confirmed. Also, as I explained in my book, negative mass cannot be used for anti-gravity purposes on its own, but must be coupled with positive mass to form a propulsive gravity dipole. This was alluded to by Professor Tajmar. Dr Hajdukovic suggested that dark matter may be linked with anti-matter and with anti-gravity and may be responsible for the galaxies pushing apart.

Colonel Dr 'Coyote' Smith, at one time the Chief of Future Concepts at the Pentagon's Space Office, mentioned the EM Drive; a device which appears to be able to propel itself using microwaves bouncing around in a closed cone-shaped conducting cavity. The EM Drive has been developed by Roger Shawyer, a UK aerospace engineer. As Colonel Smith noted, the fact that theoretical physicists said it couldn't work, hadn't stopped engineers from testing it to see whether it did work. Initial tests seem to show that it produced a tiny, but useable, thrust similar to that of an ion drive. However, the results of tests done on a German version of the EM Drive by Professor Tajmar, at Dresden University, had been inconclusive. We must await the outcome of further tests, said Colonel Smith. If they confirm that the EM Drive works then it would be a revolution in propulsion systems. Although not mentioned in the TV documentary, in the Greenglow programme, we had funded Dundee University to investigate a similar device using microwaves and an open spherical cavity. In my book, I suggested a possible reason for why the Shawyer EM Drive might experience a thrust was because it created a gravity dipole.

The *Horizon* documentary ended with a visit to the UK MoD Research Centre at Porton Down, near Salisbury, where I met Dr Neil Stansfield. Neil is responsible for MoD programmes investigating emerging and disruptive technologies. British Aerospace's original interest in gravity was to determine whether it was possible to use gravity as a counter-stealth measure. Thirty years ago, it was not possible.

Nowadays, however, quantum engineering is making new technologies possible and gravity detection of hidden, or camouflaged, objects is just one idea being considered. We were shown the MoD's quantum gravity accelerometer, which uses a set of lasers to trap 1 million rubidium atoms in a small, extremely cold, vacuum chamber where they form a Bose-Einstein condensate which is extremely sensitive to changes in gravity fields or acceleration. Today, the detection range of a small mass is of the order of a few metres but, with further development, it is hoped to extend the detection distance considerably. Asked whether quantum engineering might play a part in gravitational propulsion, Neil replied, "It could do." Since the end of filming, it has been reported that researchers at Washington State University have used a Bose-Einstein condensate to create a fluid with negative mass properties, the exotic part of a warp drive!

Scott Chasserot took a wonderful publicity photo of me, with a green ball floating just above my outstretched hand. The *Horizon* TV programme was first transmitted on 23 March 2016. The voice-over for the documentary was done by Peter Capaldi, the twelfth *Doctor Who*, with just the right sense of gravity. Project Greenglow, he said, was an incredible scientific adventure, investigating the mysterious force of gravity: the force which holds us on the planet Earth and binds us to the Universe. The programme contained an interesting collection of scientists discussing a subject which is at the forefront of today's physics. I felt privileged to have taken part in it.

# The book and the BBC *Horizon* documentary

Filming at the Royal Institution

BBC Copyright / Scott Chasserot

Publisher Matador (2015)
ISBN 9781784620233
eISBN 9781784627065

**BBC *HORIZON* Programme Advert**

**Reach for the Stars.** (23 March 2016)
Gravitation is what holds our feet to the ground, so if we could defeat it, or even just disturb it, then it could send us to the stars.

Reported by Peter Capaldi, the 12th Doctor Who.

CHAPTER 40

# THE SEARCH FOR GRAVITOMAGNETISM

The search is begun on the basis that gravitomagnetism holds the key to gravity control. Gravitomagnetism occurs because changes in the gravity fields of moving masses take time to travel. We can think of gravitomagnetism as a special relativity effect which occurs between relatively moving masses. When the gravitomagnetic field **h** of a moving mass sweeps over another mass, it induces an effect in it which is modified by the gravitomagnetic permeability $\eta$ of the medium. We write the induced effect as $\mathbf{b} = \eta\,\mathbf{h}$. Assuming that gravitational changes in space take place at the speed of light, then $\eta = 1/\gamma c^2 = 4\pi \times 0.74 \times 10^{-27}$ m/kg. Consequently, **h** must be huge to induce a **b**-effect across space on another mass. Thus, in the NASA Gravity Probe-B experiment, which first detected gravitomagnetism (and space-time warping), one of the moving masses was the Earth. For masses, on the human scale of things, it seems that we can dismiss any interaction between moving masses across space via the gravitomagnetic field.

It is generally assumed that the gravitational term Big G (equal to $1/4\pi\gamma$, where $\gamma$ is the gravitational permittivity) remains constant inside matter and within the space of the Solar System. However, we guess that the gravitomagnetic permeability $\eta$ can change within matter. Inside some bodies, the speed of light remains high, for example, glass. Therefore, we might expect $\eta_{glass} = 1/(\gamma\,c^2)$ to be very low. For bodies which are impenetrable to light, such as iron, we might expect $\eta_{iron}$ to be very high. But, unless we can think of some way of experimentally testing materials to determine their $\eta$ value, we remain in the dark.

There is an important difference with the way that moving electric charges affect other charges, compared with the way that moving masses affect other masses. In the moving charge case, generally, the moving electric field is neutralised and only the magnetic field plays a part, while in the gravity case, the gravity field of the moving masses is always present (apart from the free-fall case), along with the gravitomagnetic field.

From the Bohr model of the atom, we arrive at the conclusion that the electric force is $10^{42}$ times more powerful than the gravity force. The same ratio holds for the magnetic force and the gravitomagnetic force. The huge size of the electric and magnetic forces explains why scientists of the past were able to explore their properties. By comparison, gravitational and gravitomagnetic forces are very much weaker in space, making exploring their properties and connections much more difficult.

At the moment, gravitomagnetism hardly seems to play any part in the physics of the natural forces. When we have a better understanding of gravitomagnetism, this view will probably change. As an aid to investigation, we need to be able to carry out experiments involving gravitomagnetism in a laboratory on Earth. Not only will that keep the costs low, but it allows probing during experimentation. This is not really possible in space, where extensive planning is required, since changes to experiments, once in orbit, are very difficult. At first sight, Earth-based experiments would seem to be out of the question, given the smallness of $\eta$ in space. So we need to be imaginative and think of some new ideas where gravitomagnetism might play a part before our very eyes that we haven't realised before. We need to be daring, too, and risk considering ideas that many scientists might think are crazy, or absurd. If they are not, such ideas are unlikely to lead anywhere. We must be willing to examine and question the reasoning behind these strange ideas and not just dismiss them as being unlikely. The aim is to devise Earth-based laboratory experiments to detect gravitomagnetism and to confirm (or otherwise) the Maxwell-type linearisation model linking gravity with gravitomagnetism.

Most scientists are convinced that electromagnetism and gravity are linked, somehow. Discovering the link would surely lead to the control of gravity. One of the first questions we might try to address is, 'Is the magnetic field **H** linked with the gravitomagnetic field **h**?' An electron has a negative point charge e, given by

$$e = -1.602 \times 10^{-19} \, C$$

And it has a radial electric field **E**. But an electron also has a point mass $m_e$, given by

$$m_e = 9.109 \times 10^{-31} \text{ kg}$$

Thus, an electron must have a radial gravitational field **g**, too. **E** and **g** can be thought of as twinned fields. We have speculated that the following relationship might exist:

$$\mathbf{g} = -\left(\frac{m_e}{e}\right)\left(\frac{\varepsilon}{\gamma}\right)\mathbf{E}$$

This indicates that the ratio of the intensity **g** of the gravity field to the intensity **E** of the electric field is about $10^{-31}$. Since force equals intensity times source, this is consistent with the view that the ratio of the gravitational force to the electric force is $10^{42}$.

It seems reasonable to suppose that a magnetic field **H**, created by a moving electric charge, is twinned with a gravitomagnetic field **h**, created by a moving electron mass. We have speculated that for a spinning electron:

$$\mathbf{h} = \left(\frac{m_e}{e}\right)\mathbf{H}$$

We might speculate that this relationship applies in general. So, in the electromagnetic case, where **E** and **H** fields are present, we expect that **g** and **h** fields will be present, too. This is in line with the idea of twinned fields. For vibrating matter, at the atomic and molecular level, we associate the vibrations with oscillating **g** and **h** fields. For high-frequency vibrations matter radiates heat and, possibly, light, which means the presence of **E** and **H** fields. At low temperatures, as an approximation, we assume that for stationary matter, only a **g** field is present and that for moving matter, an **h** field is also present. For these two cases, we assume that the **g** and **h** fields do not harbour noticeable **E** and **H** fields. More generally, using some form of screening, it might be possible to separate a **g** field existing within an **E** field and separate an **h** field existing within an **H** field.

The assumption that **H** and **h** are twinned fields suggests that methods for detecting **H** may be similarly used to detect **h**. For example, in the NASA Gravity Probe-B experiment, the interaction between rotating masses, the analogue of magnetically interacting current-carrying coils, was used to detect the Earth's gravitomagnetic field (and space-warping).

From our idea that **H** and **h** are twinned fields, we expect that the effect of Faraday rotation will apply to gravitomagnetism, too, although this has not been demonstrated, as yet. In fact, one wonders whether Faraday rotation only occurs for a magnetic field because it harbours a gravitomagnetic field. If it is possible to separate the two fields, it would be interesting to see whether Faraday rotation still occurs in a magnetic field free of gravitomagnetism.

At the molecular level, there is a direct link between magnetism and the spin (angular velocity) of ferro-magnetic molecules, which possess magnetic dipoles. When a ferro-magnetic body is placed in an external magnetic field, greater molecular dipole alignment occurs and the **B** field increases. Internally, the dipoles precess and the body's angular velocity increases. This phenomenon was investigated theoretically by Professor Sir Joseph Larmor and Professor Owen Richardson at Cambridge University in the early 20$^{th}$ century. At the macroscopic level, or visible scale, we must multiply the effect by the number of molecules involved. Experimental studies of rotating a bar magnet with an angular velocity about its longitudinal axis, confirmed the increase in induced magnetic field **B**. The work was done by the American physicist Professor Samuel Barnett, in 1915.

Also in 1915, Albert Einstein and the Dutch physicist Wander de Haas showed that the effect works in reverse, too. An iron bar is placed axially inside a solenoid, or long coil. The magnetic permeability $\mu$ of iron is much greater than the free-space value, so when the current in the coil is switched on, a powerful magnetic field is induced in the iron bar. What Einstein and de Haas did was to suspend an iron bar by a thread, attached to one end, so that the bar dangled inside the coil. The bar was left to settle until it was stationary and then the coil current was switched off. The induced magnetic field **B** in the bar, generated by the coil, vanished, but the bar rotated very slightly. The induced magnetism **B** lost by the bar appears as rotation. This result, it is claimed, demonstrates the conservation of microscopic quantum angular momentum at the macroscopic level.

A variation of the Einstein-de Haas experiment might be tried with a cylindrical superconductor (Type I) suspended on a thread above the pole of a permanent magnet. The advent or loss of superconductivity, and the angular momentum associated with the supercurrent, may cause the cylinder to rotate.

The effect of the decay of the **b** field and the **B** field in the Einstein–de

Haas experiment suggests that the **h** field can be separated from the **H** field. Is there a better way of separating the **h** and **H** fields?

Methods exist to screen against electric **E** fields and magnetic **H** fields.

1. Faraday Cage. Used to screen against electric field **E**, but not **g**.
2. Mu-metal shield. Used to screen against magnetic field **H**, but not **h**.
3. Plasma sheet. Used to screen against electromagnetic field.

In the absence of negative mass, it is not possible to screen against a gravity field **g**. Likewise, it is assumed that it is not possible to screen against a gravitomagnetic field **h**.

In the NASA Gravity Probe-B experiment, the gyros were screened from electromagnetic effects (particularly Earth's magnetic field), while leaving them exposed to the Earth's gravitomagnetic field, the Earth's gravity field being cancelled through free fall.

We can use a coil and superconductor to create a very strong vertical magnetic field **H**. To detect the presence of the gravitomagnetic field **h** just above the pole of the magnetic superconductor, we might try using a pendulum with a non-magnetic bob. The path taken by the swinging bob ought to be slightly curved by the gravitomagnetic field, thereby increasing the periodicity of the swing. This is a variation of the Einstein–de Haas experiment.

As an extension of the above experiment, we could introduce a horizontal mu-metal shield, or flat plate, just above the magnetic pole. The mu-metal plate should deflect the magnetic field horizontally through the plate. So, above the plate, the magnetic field should be removed, or at least greatly reduced, leaving the unscreened gravitomagnetic field **h**. Again, to try to detect the presence of the gravitomagnetic field, we might use a pendulum with a non-magnetic bob.

Imagine we have a thick copper wire. When a current i flows along the wire, a magnetic field **H** arises which circles around the wire. We can detect the magnetic field using a compass held near to the wire. This was first noticed by Hans Oersted in 1820. Another method of detecting the magnetic field is to use Faraday rotation, an effect discovered by Michael Faraday in 1845. A linearly polarised ray of light co-linear with a magnetic field has its plane of polarisation rotated as it propagates. The angle of rotation θ is dependent on the magnetic field strength H and the path length $\ell$ of the light ray parallel to the magnetic field.

θ = VHℓ (see Chapter 33 for units)

We now consider the special case where a fibre optic is wound around a straight length of thick copper wire to form a fibre optic coil. Linearly polarised light is input at one end of the fibre coil and detected at the other end. If a current i is sent through the copper wire, a circular magnetic field **H** will be created around the wire which is coincident with the fibre optic coil. If the length of the unwound coil is ℓ then the plane of polarisation at the output end will undergo a rotation θ, which can be detected and measured. From the above formula, knowing the Verdet constant V for the fibre optic, the strength of the magnetic field H can be calculated. The current strength i can then be determined from the formula derived by the French scientists Jean Biot and Felix Savart.

$$H = \frac{i}{2\pi a} \text{ A/m} \quad \text{where a is the radius of the wire.}$$

This method, proposed by Boeing, was successfully used during lightning strike tests on a metal structure at BAE Systems Warton.

Suppose a long mu-metal sleeve, or cylinder, is placed around a length of the thick copper wire carrying a current. A simple compass should indicate the absence of the magnetic field just outside the mu-metal sleeve. If the fibre optic coil is wrapped around the mu-metal sleeve, we would like to know whether Faraday rotation still occurs. This has not been tried. Does Faraday rotation stay the same, suggesting that the effect is due to the **h**-field? Is it reduced? Or, is it non-existent?

The gravitomagnetic field separated from a magnetic field seems, to me, to be what some Russian scientists call a torsion field. The Russian method of generating a torsion field employs an electromagnet. Although the Russian experiments in this area have been largely dismissed as pseudoscience, perhaps their method of detecting torsion fields should be examined further. They claim that a torsion field is the spin-polarisation of the quantum vacuum and that it can reduce the electrical resistance of a conductor. More pertinently, the Russian scientists working on torsion fields claim that it is possible to screen against torsion fields using particular forms of polyethylene films, which is contrary to the assumption made by most scientists (particularly NASA Gravity Probe-B scientists) that it is not possible to screen against the gravitomagnetic field.

To detect an internal gravitomagnetic field **h**, we might consider wrapping a fibre optic coil around a pipe with circular cross-section, of radius r = a, containing a uniform flow of fluid (opaque or clear) and shining linearly polarised light in at one end of the fibre and looking for any rotation of the plane of polarisation at the other end. Although there is no relative motion between the detecting system (the fibre optic coil) and the pipe wall, it is assumed that the gravitomagnetic field associated with the fluid current is not screened by the pipe wall. Bearing in mind the Russian view – that polyethylene may act as a screen against gravitomagnetism – a pipe with a thin metal wall should be used. Although it seems unlikely that we would see an effect, who knows? Fluids with different densities and opacities may have different gravitomagnetic permeabilities $\eta$.

A variation of the piped fluid experiment would be to place an axially aligned fibre optic coil (radius r < a) in the uniform fluid flow, trying not to disturb the flow. In fact, the circular gravitomagnetic **h** field may be the natural source of turbulence in a fluid. We could use the magnetic analogue to predict the strength of the gravitomagnetic field in the fluid within the pipe. The gravitomagnetic analogue of the Biot–Savart formula, for a pipe of radius a, gives the magnitude of the internal **h**-field at a radius r as

$$h_r = \frac{Ir}{2\pi a^2}$$

where I is the mass current.

If the analogue holds true, then the rotation $\theta$ of the plane of polarisation, in a coil with n turns, would be given by

$$\theta = V_{GM} h_r \ell_r \quad \text{where } \ell_r = 2\pi r n \text{ is the length of the fibre optic coil.}$$

The constant $V_{GM}$ is unknown for gravitomagnetism. Liquids with different densities and opacities would need to be experimented with. Detecting a change in $\theta$ is the important point at this stage, as it would signal the presence of the **h**-field.

In a 1961 paper (Proc. IRE, page 1442) Dr Robert Forward, of the Hughes Aircraft Company, pointed out that anti-gravity had been shown to exist theoretically since general relativity was formulated in 1918. To give an example of anti-gravity, Forward considered a toroid (a solenoid formed into a circle). When an electric current i flows around the toroid the accompanying magnetic field is contained within the loops of the toroid. However, when the current i changes the magnetic field **H** changes and theory predicts that an external transient electric dipole-field

**E** is created, which is aligned with the axis of the toroid. This result can be demonstrated experimentally. Assuming that the toroid contains N loops of wire of radius a and that the radius of the toroid is R, Forward gave a rough approximation (R > a) for the magnitude of the transient E field.

$$E = -\mu N \left(\frac{\partial i}{\partial t}\right) \frac{a^2}{R^2}$$

In his 1988 book *Future Magic*, Dr Forward pointed out, tongue-in-cheek, that if the toroid was made of a tubular winding and that an extremely dense fluid was pumped around the tube with increasing speed (an increasing mass current I), then an axial transient gravity dipole-field **g** would be created. Based on analogy, the magnitude of the transient gravity dipole-field would be

$$g = \eta N \left(\frac{\partial I}{\partial t}\right) \frac{a^2}{R^2}$$

However, Dr Forward ruled out such a possibility in reality because η, the gravitomagnetic permeability of space, is so small.

From twinning we expect that any change in the magnetic field **H** will be accompanied by a change in the gravitomagnetic field **h**, giving rise to a tiny dipole-field **g**, co-existing with the dipole-field **E**. So, in the case of the toroid it seems that we can create a gravity field, which is an important conclusion. However, even if electric screening is used, it is highly unlikely that the **g**-field created in this manner would be measurable. Nevertheless, before dismissing the idea, note that the creation of transient **E** and **g** fields is dependent on the permeability parameters μ and η. If the toroidal wire is embedded in a material with a very high η, then the **g** field might predominate, creating a transient gravitational dipole and giving the material a thrust. If the embedded toroid contained three separate windings and the three currents were phased then a constant **g** field might be created causing the toroid to accelerate. It may seem unlikely, but only an experiment can say what does happen. We need to know more about the value of η within materials.

I have not ruled out the idea that superconductivity is linked with gravity and gravitomagnetism. Dr Podkletnov claimed that a horizontal superconducting (Type II) annular disc rotating within a current-fed toroid created a steady gravitational effect above it. The experiment was screened to prevent any electrical fields leaking out. Gravity fields cannot be screened. The set-up has some similarity with Dr Forward's idea described above.

Several groups of scientists have repeated Dr Podkletnov's experiment but failed to substantiate his result, so his claim has been dismissed. If a gravitational effect does arise with the Podkletnov set-up, it may be due to the formation of circular Abrikosov magnetic vortices within the Type II annulus. A change of the toroid current will cause a change in the magnetic vortices and, from twinning, a change in the circular gravitomagnetic field within the superconducting annulus. From Faraday's law of induction, we know that a transitional electric field is created when a changing magnetic field occurs. The gravitational analogue indicates that a transitional gravity field is created when a gravitomagnetic field changes. The direction of the gravity field is vertical, in line with the axis of rotation. So, if a gravitational effect does occur with the above superconducting experiment, I feel that it is likely to be fleeting and probably tiny, making its detection doubly difficult.

I have also wondered whether the gravitational red shift above the letter box-shaped 3.6 MHz solenoids (see illustration in Chapter 37), with a superconducting core, of Podkletnov's apparatus plays a part in the weight change of the test mass.

Let us now consider some phenomena where gravitomagnetism might be playing a part in the gravity phenomenon which hasn't been realised before. The more absurd the idea seems the less chance that it has been considered before.

Let us start with the idea that gravitomagnetism is responsible for inertia. When a mass is accelerated, there is a resisting force that we call inertia. Think of trying to push a loaded railway truck. It's very hard to get the truck moving, but once it is moving, it keeps going without having to apply any further force. But then, try to stop it moving! If you are running, you cannot just stop on the spot; you are forced to take a few steps further during the period of deceleration. Why?

Einstein was fond of thought experiments. He is quoted as saying that one of the happiest thoughts of his life was when he realised that a man who had fallen from a roof was free of gravity while he was falling. It was this thought that provided Einstein with the method to tackle accelerating frames of reference, leading to his general theory of relativity. Choosing any point in 4-D space-time to be the origin of an accelerating system, he could replace acceleration with gravity and then use the theory of special relativity. The general interpretation of Einstein's thought was that the falling man was weightless and, by extension, that this was due to the equivalence between gravity and acceleration. An equally important observation is that, while accelerating downwards, the falling man had no inertia.

And there is more. An accelerating mass radiates away energy. But, a mass accelerating (falling) in a gravity field does not. So, we can see that inertia is linked with energy radiation.

According to Maxwell's extended Newtonian model of gravity, when a moving mass is observed from a stationary viewpoint, it is accompanied by a gravitomagnetic field **h**, both inside and outside the mass. When the mass accelerates, the surrounding gravitomagnetic field will change with time. From the gravitational analogue of Faraday's law of electric field induction, we have

$$\nabla \times \mathbf{g}_i = \eta \frac{\partial \mathbf{h}}{\partial t}$$

In theory, the changing gravitomagnetic field $\partial \mathbf{h}/\partial t$, multiplied by the factor $\eta$, induces a gravity field $\mathbf{g}_i$. But in the space outside the accelerating mass, since the gravitomagnetic permeability $\eta$ is so small, the induced gravity field $\mathbf{g}_i$, there, will be incredibly weak and can be ignored. But inside the mass $\eta$ may be different. We have speculated that $\eta_{glass} << \eta_{iron}$. But, whatever the internal value of $\eta$, the change of induced gravitomagnetic field with time $\partial \mathbf{b}/\partial t$ is the same for an accelerating mass, so that the internal gravity field $\mathbf{g}_i$ that is created (induced) is the same. So, for example,

$$\eta_{iron} \frac{\partial \mathbf{h}}{\partial t}\bigg|_{iron} = \eta_{glass} \frac{\partial \mathbf{h}}{\partial t}\bigg|_{glass} \quad \text{which implies that} \quad \frac{\partial \mathbf{h}}{\partial t}\bigg|_{iron} << \frac{\partial \mathbf{h}}{\partial t}\bigg|_{glass}$$

We speculate that the induced internal gravity field $\mathbf{g}_i$ is responsible for the resisting force during periods of mass acceleration.

Newton made use of data about the Moon free falling in Earth's gravity field to confirm his inverse square law for the force of gravity. Faraday, in his first gravity experiment, used a cylindrical test mass falling in Earth's gravity field and looked for an effect associated with the accelerating mass. Faraday's goal was the unification of the known (at that time) forces of nature. He was searching for a link between gravity and electricity. He wound a copper wire around the cylindrical mass and attached the ends of the wire to a galvanometer. With the axis of the cylinder vertical, he let the mass fall and looked for an induced (circumferential) electrical current as the mass accelerated downwards. But, there was no induced current.

Now is the time to repeat Faraday's gravity experiments of 1849 and 1859, with some modifications.

We begin by considering Faraday's first gravity experiment, with his falling cylindrical mass surrounded by a conducting wire coil. Faraday searched for a current induced in the coil as it accelerated downwards, but found nothing. Our Maxwell model of gravity suggests that we should look, instead, for a changing gravitomagnetic field $\partial \mathbf{h}/\partial t$ surrounding the falling cylindrical mass. A glass cylinder might be best. To detect the changing gravitomagnetic field around the falling mass cylinder, it is proposed to use a fibre optic coil wound around the cylinder and look for changing Faraday rotation along the optical fibre. Based on our twinned field idea (between **H** and **h**), this assumes that Faraday rotation also occurs when a polarised light beam is aligned with a gravitomagnetic field, although this is unproven at present.

The experiment to detect a changing gravitomagnetic field is difficult.

1. Faraday rotation for gravitomagnetism is unproven at present.
2. The period of free fall is very limited, being only a second or so.
3. During free fall, the predicted **h** field increases uniformly with time.
4. Consequently, the plane of polarisation rotates uniformly with time.
5. The change in plane of polarisation angle θ is accompanied by a change of light intensity at the fibre optic output, which must be interpreted electronically to determine the rate of rotation.

If the experiment is successful, it would show just how prescient Michael Faraday was in his thinking. It would be the first detection of gravitomagnetism in an Earth-based laboratory. A variation of the experiment might be to remove the cylindrical test mass and just drop the fibre optic coil, since the coil has its own mass.

Another experimental set-up would be to use a fibre optic coil in the form of a toroid. In this case, the detection method relies on the Sagnac effect, as used in the fibre optic gyro (FOG). The test mass, if used, would form a doughnut shape within the toroid. As with a conventional FOG, a light source would be split and fed clockwise and counter-clockwise around the toroidal coil. For stationary conditions, the output light signals would emerge in-phase.

For a conventional FOG, with a fibre optic coil in the form of a solenoid, for constant rotation (angular velocity) aligned with the axis of the coil, there is a constant phase shift between the two emerging light signals. If the rotation increases, then the phase shift increases.

For free fall, our Maxwell gravity model predicts that the doughnut mass at the centre of the fibre optic toroid will contain a changing gravitomagnetic field which, from equivalence, will be an increasing angular velocity, or rotation. Thus, we expect the emerging light signals to exhibit an increasing phase shift between them, indicating a uniform increase in rotation. There is a close resemblance to Dr Forward's toroidal gravity dipole.

We now come to Faraday's third gravity experiment, with two parts. The first part of the experiment involved the raising and lowering of a mass of mercury to observe whether it resulted in a change in temperature. Faraday's thinking was along the right lines, fitting in with Joule's conservation of energy concept. However, the experimental method was flawed. Nevertheless, the idea has been demonstrated by others and the principle proved, so there is no need to redo the experiment.

So, this leaves us to consider the second part of Faraday's third gravity experiment, namely that charge may appear on a mass as it changes its position in the Earth's gravity field. Also, whether a current might be enabled to flow between two masses fixed at different heights in Earth's gravity field, connected by wire. These static experiments must be conducted in the open air, not inside a building. Near the Earth's surface, the Earth's electric field $E = 100$ v/m, which suggests that if any effects are detected, they will not be due to gravity.

Also of interest is the dynamic gravity experiment that Faraday vaguely suggested in 1859, but didn't do. Suppose we have a conducting cylindrical test mass to which a wire is attached to the top and another wire is attached to the bottom, with the ends connected to a galvanometer to complete the circuit. Faraday wondered whether an induced current might flow around the circuit when the uncharged test mass fell in Earth's **g**-field; he was not aware of the Earth's **E**-field.

We have assumed that if a natural **H**-field exists then it contains an **h**-field, but that the reverse is not true. So, in the case of free fall, the appearance of $\partial \mathbf{h}/\partial t$ should not mean the appearance of $\partial \mathbf{H}/\partial t$. So, in free fall, there should be no **E**-field induced; so no current should flow around the circuit. But, since Faraday considered the idea and since the result of experiment is the final arbiter, perhaps the experiment should be tried.

There are a couple of other test variations to be considered:

1. Suppose the test mass is charged ±Q. In what direction does the current flow when the charged mass falls in Earth's **g**-field and **E**-field?
2. When the current flows, is the mass acceleration different from $g_0$?

Some thought needs to be given as to whether the outcomes are solely due to electromagnetic effects, or whether there is a link with the **g**-field.

Newton's pendulum experiment, described in Book I of his *Principia*, used a pendulum with a hollow bob which could be filled with different substances. Regardless of the substance in the bob, Newton found that for small angles, the period τ of oscillation was constant. This showed the equivalence between gravitational mass (associated with the weight of mass) and inertial mass (associated with accelerating mass).

Between 1914 and 1926, the US scientist Dr Charles Brush repeated Newton's pendulum experiment as part of his investigation into his kinetic theory of gravity. Brush recorded that for some substances he measured a slight change in the oscillation period τ of the pendulum, in disagreement with Newton's findings. Subsequently, Brush carried out a series of free-fall experiments with sealed hollow tubes containing different substances. Brush tested Galileo's hypothesis that masses, of whatever substance, all fall at the same rate. But Brush reported that his measurements showed, again, contrary to expectation, that some masses accelerated downwards with an acceleration very slightly less than $g_0$. Brush noted that for such masses, some substances exhibited a slight heating effect during the very short time of fall. Brush's results do not fit in with conventional Newtonian gravitational theory, so little attention has been paid to his work,

We have speculated that the magnetic field **H** and the gravitomagnetic field **h** are twinned. Ferromagnetic substances are subject to the Curie temperature effect, above which they lose their strong magnetic property. What happens is that the magnetic permeability $\mu$ of the substance is greatly reduced through heating. Under static conditions, we are not aware of any substances that possess natural gravitomagnetic fields, analogous to natural magnets. However, our Maxwell model of extended Newtonian gravity predicts that under dynamic conditions, a moving mass is associated with a gravitomagnetic field **h**. The gravitomagnetic field is a manifestation of a mass viewed in a framework where it is moving. The field $\mathbf{h}$ arises from special relativity and cannot be destroyed. External to the moving mass, the gravitomagnetic permeability $\eta$ of space is virtually zero, so the external induced gravitomagnetic field $\mathbf{b} = \eta\mathbf{h}$ will be virtually zero, too (unless the moving mass in question is planetary-sized, as in the NASA GP-B experiment). Inside a moving mass, however, $\eta_i$ may be large and the induced gravitomagnetic field $\mathbf{b} = \eta_i\mathbf{h}$ large, too. Twinning suggests that the gravitomagnetic permeability $\eta_i$, perceived inside a moving mass, might be reduced for some substances due to a Curie-type temperature effect. If so, this would affect the internal induced gravitomagnetic field $\mathbf{b}$. For an accelerating mass, this would mean a reduction in the size of $\partial\mathbf{b}/\partial t$, which would affect the inertia of the accelerating mass.

This suggests a new version of Newton's pendulum experiment should be undertaken, with a lagged (so no change in bob surface temperature occurs) hollow bob filled with substances that can be internally heated to see whether the period of oscillation is affected by temperature change.

During the oscillatory motion of the pendulum bob, the Maxwell gravity model predicts there to be a cyclic change in the gravitomagnetic field within the bob. So, the pendulum experiment should be carried out for a range of substances over a range of internal bob temperatures, and a change in the period $\tau$ looked for. There is no need to restrict the oscillations to small angles for these experiments.

Any change in period $\tau$, or rapid decay in swing, with change in temperature, suggests that the gravitomagnetic permeability $\eta_i$ within the mass of the bob has changed, resulting in a weakened changing induced gravitomagnetic field $\partial\mathbf{b}/\partial t$ within the bob. If so, this would support the model of gravitomagnetism. Note that the bob's changing gravitomagnetic field $\partial\mathbf{h}/\partial t$ is not affected. If there is a gravitomagnetic Curie temperature for a particular substance, then we have to hope that it is fairly low (For iron, the magnetic Curie temperature is 1043 K); otherwise, we are unlikely to notice it. Let's hope that serendipity smiles on our efforts. If heat can be applied to alter the internal induced gravitomagnetic field asymmetrically, then the plane of swing may alter. If the gravitomagnetic field is susceptible to disturbance by heat, then it offers the possibility of gravity control.

Another phenomenon which I feel needs to be investigated is friction. During the early 1950s, the brilliant mathematician John Nash, while a young graduate student at Princeton, considered the idea that gravity was linked with friction. At that time, Einstein was at Princeton, so Nash took the opportunity to discuss his idea with him. But, Einstein told Nash that his idea was absurd and that he needed to study more physics. Nash's main interest was in game theory and it was for work in this field that he was awarded the Nobel Prize in Economic Sciences in 1994. The Oscar-winning film, *A Beautiful Mind*, describes the ups and downs of Nash's life. Nash was not deterred by Einstein's comments and he continued to work on general relativity theory over many years. In 2003, he gave a lecture at Penn State University on his 'interesting' extension to Einstein's work. A few days before he died in a car crash, in 2015, he told the French mathematician Cédric Villani that he had discovered a new equation for general relativity. But whether he had or hadn't, my view is that the highly complex theory describing general relativity is unlikely to lead to an understanding of how to control gravity. The way forward is more likely to be through a series of Earth-based experiments.

I suspect that friction and gravitomagnetism are very closely related. Following Einstein's fondness for thought experiments to develop his understanding of a problem, I will try the same approach. Let's start with the electromagnetic analogue of two conducting rings placed one above the other. When a steady current passes around the lower ring, a steady axial magnetic field is created. Although the constant magnetic field of the lower ring passes through the upper ring, the upper ring remains unresponsive. When the current around the lower ring changes, the axial magnetic field passing through both rings changes. The response of the upper ring, to the changing magnetic field passing through it, is to create a current around the upper ring. This is described by Faraday's law of induction.

Now let us consider the gravitational analogue. We replace the two conducting rings with two large flat washers placed one above the other. If we start the lower washer rotating with a steady mass current, there is no effect on the upper washer. However, if we place the upper washer on the lower washer then friction will cause the upper washer to turn. Once both washers are moving with the same speed, friction disappears.

Now let us interpret the thought experiment in terms of gravitomagnetism. If the washers aren't touching, we wouldn't expect any effect because the gravitomagnetic permeability of free space $\eta$ is virtually zero. This thought experiment is an Earth-based variation of the NASA Gravity Probe-B space experiment. When the two washers are in contact, when we start the lower one rotating, there is a change in the gravitomagnetic field of the lower washer which induces a mass current in the upper washer. Once both washers are rotating with the same speed, the gravitomagnetic field between the washers disappears.

In my earlier book, I suggested a friction-type experiment to try to detect the gravitomagnetic field arising between two flat blocks of glass moving relative to one another.

For static friction, when a horizontal force is applied to a flat block resting on a flat surface, the inter-atomic bonds between the two contact surfaces are strong enough to prevent any movement. This situation continues as the horizontal force is increased until the coefficient of static friction $\mu_s$ reaches a maximum value. Once the horizontal force produces sliding, the value for the coefficient of kinetic friction $\mu_k$ reduces, indicating a reduction in the atomic bonding between the surfaces. For example, for dry contact between two blocks of glass at room temperature, $\mu_s$ = 0.94 and $\mu_k$ = 0.4. In the classical approach of fields, as the glass block in contact with the flat glass surface moves, we expect a gravitomagnetic field to arise. From the perspective of the stationary surface, it is speculated that the moving block creates an induced gravitomagnetic field **b** which at the interface is perpendicular to the direction of motion. We can liken the induced gravitomagnetic field at the interface to a sheet of parallel rotating cylinders. Remember that an induced gravitomagnetic field line **b** is equivalent to an angular velocity in the opposite direction. By shining a linearly polarised laser beam through the moving glass block, near to the contact surface and parallel to the predicted gravitomagnetic field, it may be that Faraday rotation can be used to detect the presence of the gravitomagnetic field **h**. Have we really been oblivious to the existence of gravitomagnetism in the classroom friction experiment all this time?

If the block accelerates freely across the surface when we apply a horizontal force to it then we only have inertia to consider. We have speculated that an induced gravity field **g** is created which opposes the acceleration. Taking friction into account, too, means that the induced gravity field **g** due to inertia is modified. Assuming that the block reaches a steady speed, then the **h** and **g** induced fields are both constant. The induced fields **g** and **h** are perpendicular, and their vector cross product **g** × **h** (analogous to the Poynting vector in electromagnetism) gives rise to a flow of energy across the interface into the moving block. This is the frictional heat that you feel when you push down on a copper coin and slide it across the desktop.

$$\mathbf{g} \times \mathbf{h} = JkT = \text{rate of energy flowing across unit contact surface area.}$$

Initially, there will be an orderly flow of energy into the moving glass block and a temperature gradient field **T** forms. But, due to the finite dimensions of the block (and depending on its thermal conductivity k), it will quickly heat up, resulting in chaotic thermal vibrations associated with high temperature. Heat can destroy magnetic fields, so the analogue suggests that heat can destroy gravitomagnetic fields, too. Consequently, the detection of Faraday rotation needs to be attempted quickly.

With regard to frictional heating of moving mass, it is interesting to 'read-across' to the electrical analogue of charge moving in a conducting wire due to an electric field **E**. During a very brief initial acceleration phase (analogous to free fall), the charges move without resistance (no electrical inertia), and the lost charge electrical potential energy is converted into magnetic energy, resulting in the formation of a magnetic field **H**. But once the charges reach a steady (average drift) velocity, no further loss of charge potential energy can be transferred to the magnetic field. The further continued loss of energy appears as heat and the wire gets hot. The rate of outward energy flow per unit surface area of the wire is given by **E** × **H**. This result, known as the Poynting vector, was given by the English scientist John Poynting in 1884.

The analogue also applies to mass falling under gravity in a resisting (viscous) medium which reaches a terminal velocity. The mass gets frictionally heated (some of which is transmitted to the liquid). The classroom friction experiment may be treated as a special case of extreme viscosity.

Reduction in temperature of the materials in sliding contact may reduce friction. In the analogue with electricity, at very low temperatures, some materials become superconducting and electric charges move without resistance, creating large magnetic fields in the process. Can this idea be 'read-across' to friction? Could we create large gravitomagnetic fields using frictionless rubbing, or is the idea too absurd?

Let us now look at an acoustic analogue of the EM Drive. Physics students will be familiar with the resonating solid bar, of length $\ell$ and uniform cross-section. Theory assumes that longitudinal compression waves exist in the bar. Their interference gives rise to a standing wave, where the resonant wave-length is $\lambda = 2\ell$. The resonant standing-wave frequency is

$$f_{RES} = \frac{c_S}{2\ell}$$

where $c_s$ is the speed of sound in the bar.

In Chapter 29, we looked at acoustics and suggested (following Maxwell) that the longitudinal acoustic compression wave was a simplification of a transverse T-R wave, analogous to an electromagnetic E-H wave. For the acoustic analogy of the EM Drive, we have a solid truncated cylinder, with a vacuum chamber at the narrow end and an acoustic source, of narrow frequency band, tuned for resonance at the wide end. We speculate that some of the lower frequency content of the acoustic signal will suffer 'cut-off' (not a feature of a compression wave) at the narrower end of the solid truncated cone. The internal decaying frequencies appear as external radial acoustic radiation from the narrow end. We might expect the narrow end to get hot. Internally, there is less acoustic energy contained in the narrow end than there is at the wide end. Since force is the negative gradient of energy ($F = -\nabla E$), we might expect the truncated cone to experience a force in the direction from the wide end towards the narrow end, as though a gravitational dipole exists within the cone. Until an experiment is carried out, the analogue remains just an unproven thought experiment. Whether Roger Shawyer's EM Drive forms a gravitational dipole, so that the EM Drive is actually a gravitational propulsion device, remains an open question at the moment.

The medium for electromagnetism is the ether of flat space-time. The classical theory of electromagnetic interactions is described by Maxwell's set of partial differential equations satisfied by **E** and **H** fields. The theory is linear, given that the addition of any two solutions for the differential equations is also a solution. In other words, the vector properties of **E** and **H** abide by the parallelogram rule of addition. The changes in energy during interactions occur in a smooth, or continuous, way. The adjoint to Maxwell's model is quantum electrodynamics, where energy changes during electromagnetic interactions occur in discrete, or quantum, amounts. In the absence of electromagnetic fields, Maxwell's luminiferous ether is empty and contains no energy. By contrast, the quantum view of the ether, in the absence of electromagnetic fields, is a bubbling cauldron of discrete energy particles, called photons in electromagnetic terms, which fleetingly appear and disappear. The quantum ether has an inherent energy level, called zero point energy. Both forms of ether support electromagnetic waves. In classical terms we recognise a wave as a sinusoidal variation in amplitude of **E** and **H** fields with distance, while in quantum terms we treat the propagating wave as a stream of photons.

To see how this 'reads-across' to the gravitational analogue, we must adopt the linearised version of Einstein's equations for general relativity.

These are the GEM equations, or the Maxwell-type differential equations linking gravity and gravitomagnetism. The classical ether supports **g** and **h** fields. In gravitational terms the bubbling cauldron of discrete energy particles of the quantum ether are called gravitons.

In trying to form a quantum theory of gravity, the analogue suggests that a good place to start is with the linearised form of the equations for general relativity. We could then try to exploit the pattern with electromagnetism to search for any discrete properties associated gravity and gravitomagnetism. Although we are not aware of any discrete properties of gravity, there are hints that gravitomagnetism does have some. What is the relationship between **h** and Planck's constant h? Is spin, or quantum angular velocity, related to **b**? Looked at in this way, it seems that quantum mechanics and quantum gravity must merge together.

The SQUID is a superconducting quantum interference device used to measure tiny changes in a magnetic field. In very simple terms, the device is a superconducting ring containing a Josephson junction. When the ring is placed in a magnetic field, it only allows a discrete amount of magnetic field (called the magnetic flux) to pass through the ring. The quantum of magnetic flux is $\phi_0$ = h/2e, where h is Planck's constant and e is the charge of an electron.

The twinned field relationship between **H** and **h** suggests that the analogue of the quantum of gravitomagnetic flux is $\Phi_0$ = h/2$m_e$, where $m_e$ is the mass of an electron. So, inroads into a quantum theory of gravity might be gained via the backdoor, starting with quantising gravitomagnetism. The rate of change of gravitomagnetism would then imply that inertia is quantised, which fits in with the fact that inertia is linked with change in energy, which is known to be quantised.

Recent gravity experiments on Earth have largely focussed on detecting gravitational waves resulting from collisions of massive astronomical bodies in space. Such waves are incredibly weak but were detected by US scientists in 2015. In the NASA GP-B space experiment, scientists detected space-time curvature caused by the Earth in 2007 and the gravitomagnetic field due to the Earth's rotation in 2011. Although supporting the non-linear Einstein model of gravity, they have not taken us any further forward in our understanding of the gravity phenomenon. On the other hand, in terms of a linearised theory of gravity, inertia can be viewed as an interaction between gravity and changing gravitomagnetism within matter. Moreover, the size of the fields involved suggests that Earth-based laboratory experiments are possible. If so, it would allow scientists to examine gravitomagnetism in more detail. Learning how to generate and control a gravitomagnetic **h**-field within matter is the key to gravity control. A glittering prize awaits the first successful team of experimenters able to do this, perhaps even a Nobel Prize.

We have come to the end of my book and you may be asking yourself, *So what?* It's certainly a question that I have asked myself. My question arose out of frustration. What I really wanted to know was *How can we move on from here?* I am a mathematical engineer, not an experimenter. Somehow or other, I have to persuade someone to carry out the experiments for me. So, to answer the *So what?* question, this book is aimed at publicising the above experiments in the hope that one, or two, will catch the eye of someone capable and interested in carrying them out.

Two hundred years ago, Hans Oersted discovered, quite by chance, the link between magnetism and electricity. Both subjects had been known about for at least 2,000 years, but no one had realised that they were coupled together. Oersted's discovery set off a flurry of activity by scientists looking for further effects. It led to Joseph Henry's and Michael Faraday's near synchronous experimental discovery of electromagnetic induction. And it led to James Maxwell's mathematical model of electromagnetism which predicted the existence of electromagnetic waves. There was no way that Oersted could have foreseen what his discovery would lead to. Today, we have all sorts of electromagnetic devices, including heating and lighting in buildings, radio and television and other uses of the spectrum of electromagnetic waves, computers, mobile telephones, electric cars, and so on. Until Oersted's time, people's lives were not vastly different from those of the ancient Greeks or the Romans. Following Oersted's discovery, changes began to occur with increasing rapidity. The control of electromagnetism has totally changed our lives.

I hope that some of the experimental ideas that I have described above will reveal that a gravitational analogue of Oersted's discovery does exist. If so, it will demonstrate the link between moving mass and gravitomagnetism. It is only the first step, but look at what Oersted's first step in linking moving electric charge (current) with magnetism led to. Being able to probe gravitomagnetism in laboratory experiments will lead to ways of manipulating gravitomagnetism, which is what is needed for gravity control. I suspect that manipulation of gravitomagnetism will be achieved by electromagnetic means (possibly laser light), leading to the much sought-after link between gravity and electromagnetism. Once we can control gravity, we can be sure that within a few years more miraculous discoveries concerning gravity will be made, but from our current understanding, we only have a tiny inkling of what they might be. As electromagnetism has changed the way that we live, we can be absolutely certain that the changes to our lifestyle brought about by gravity control will be mind-blowing.

# The search for gravitomagnetism

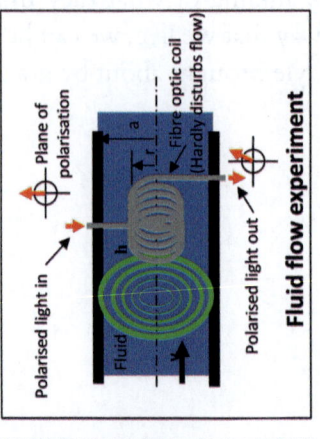

**Separation of H and h fields**

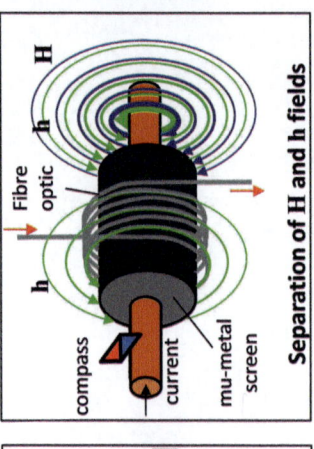

**Separation of H and h fields**

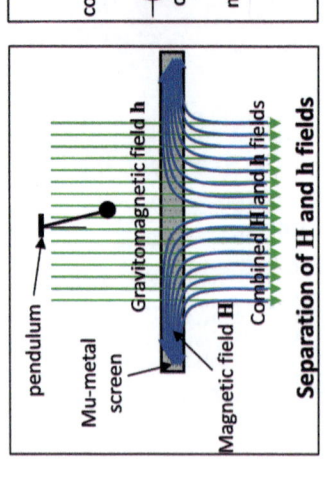

**Fluid flow experiment**

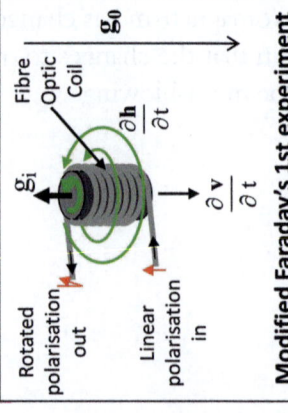

**Changing current in toroid**

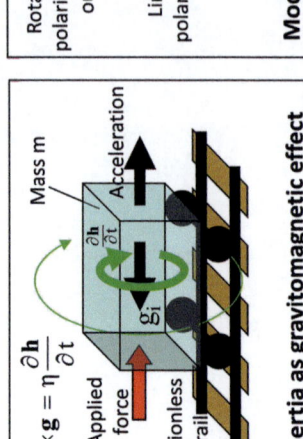

**Inertia as gravitomagnetic effect**

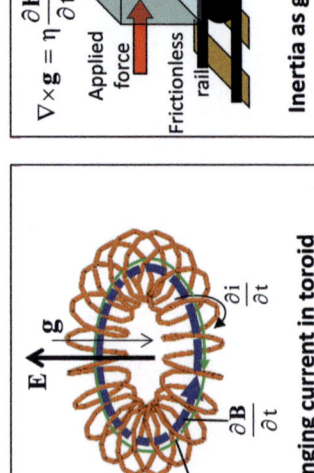

**Modified Faraday's 1st experiment**

# The search for Gravitomagnetism

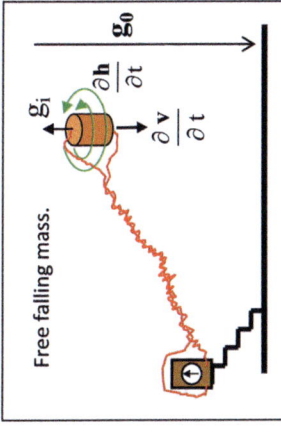

**Fibre optic toroid in free fall**

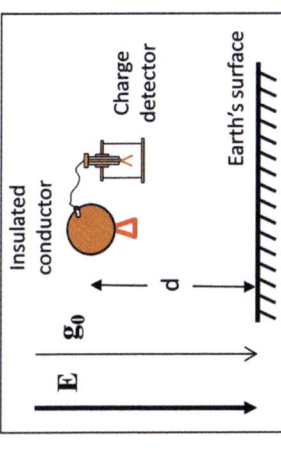

**Repeat Faraday's 3rd experiment**

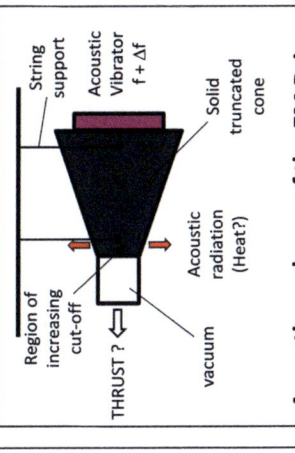

**Faraday's Dynamic 3rd experiment**

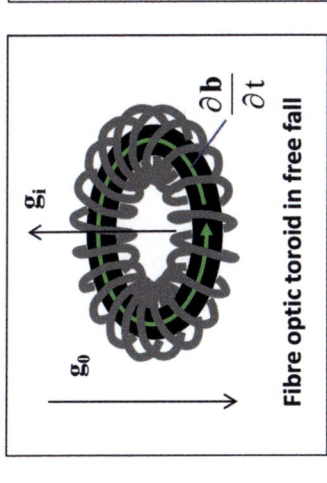

**New pendulum experiment**

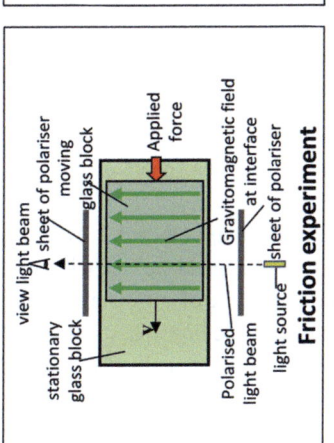

**Friction experiment**

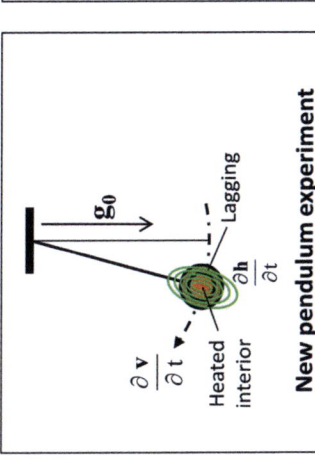

**Acoustic analogue of the EM Drive**

# APPENDIX
# EXTENDED GRAVITY

### A.1 Gravitational field of a mass shell

We need to use Gauss' theorem, which makes use of solid angles.
We deal with the concept of a solid angle first. Suppose we have a point P surrounded by a sphere of unit radius. The surface area of a unit sphere is $4\pi$.

The solid angle subtended by a partial surface dS at P is determined by forming a conical surface with the perimeter of the partial–surface dS being the base of the cone and the point P as the apex. This cone cuts the surface of the unit sphere. The surface area of the unit sphere cut off by the cone is the solid angle of dS viewed from P.

If P is totally surrounded by a surface S then the solid angle = $4\pi$ steradians.
If P is completely outside of a surface S then the solid angle = 0.

Gauss' theorem is as follows:
Given a spherical surface S of fixed radius r and a mass m then

$\iint_S \mathbf{g} \bullet d\mathbf{S} = -\dfrac{m}{\gamma}$ if the mass is inside S.

$\quad\quad\quad\quad = 0$ if the mass is outside S.

Suppose the mass m is a spherical shell of radius a.
If the mass shell is inside S (a < r) then outside the shell we have

$\iint_S \mathbf{g} \bullet d\mathbf{S} = g_{Ext} \iint_S dS = g_{Ext} 4\pi r^2 = -\dfrac{m}{\gamma}$

Therefore, $\mathbf{g}_{Ext} = -\dfrac{m}{\gamma 4\pi r^2}\hat{\mathbf{r}}$ and potential $\phi_{Ext} = -\dfrac{m}{\gamma 4\pi r}$

So, outside the shell, the shell acts like a point mass at the centre of the shell.

If the mass shell is outside S (a > r) then inside the shell we have
$$\iint_S \mathbf{g} \cdot d\mathbf{S} = g_{Int} \iint_S dS = g_{Int} 4\pi r^2 = 0$$
Therefore, $g_{Int} = 0$ and potential $\phi_{Int}$ = constant.

### A.2  Gravitational field of a solid sphere of mass M and radius a

Suppose at radius r (r < a) we have a very thin annular cavity.
Then we have an outer mass shell and an inner sphere of mass $M_{Int} = M\dfrac{r^3}{a^3}$.
Suppose P is a point in the thin cavity.
The gravity field inside a mass shell is zero. Therefore, the mass of the shell makes no contribution to the gravity field at P. The gravity field at P is just due to the inner spherical mass $M_{Int}$.

$$\mathbf{g}_{Int} = -\frac{M_{Int}}{\gamma 4\pi r^2}\hat{\mathbf{r}} = -\frac{Mr}{\gamma 4\pi a^3}\hat{\mathbf{r}} \quad \text{and} \quad \phi_{Int} = \frac{M}{\gamma 4\pi r}(r^2 - 3a^2)$$

Externally (r > a), the gravity field is the same as that for a point mass M.

$$\mathbf{g}_{Ext} = -\frac{M}{\gamma 4\pi r^2}\hat{\mathbf{r}} \quad \text{and} \quad \phi_{Ext} = -\frac{M}{\gamma 4\pi r}$$

### A.3  Gravitational field of an infinitely long line mass

Use cylindrical polars $(R, \theta, z)$.
Assume a thin line mass lies along the z-axis, where length $\ell$ has mass M.
Element of mass dm is at A on the line. R = OP is the perpendicular distance of the observation point P from the line. The angle OPA = $\alpha$.
The mass element $dm = \dfrac{M}{\ell}dz$ where $z = OA = R\tan\alpha$. $dz = R\sec^2\alpha \, d\alpha$

$$d\mathbf{g} = -\frac{dm}{\gamma 4\pi (AP)^2}\cos\alpha \, \hat{\mathbf{R}} \quad \text{where } AP = \frac{R}{\cos\alpha} = R\sec\alpha$$

$$\mathbf{g}_{Ext} = -\frac{M}{\gamma 4\pi \ell}\int_{-\frac{\pi}{2}}^{\frac{\pi}{2}} \frac{R\sec^2\alpha}{(R\sec\alpha)^2}\cos\alpha \, d\alpha \, \hat{\mathbf{R}} = -\frac{M}{\gamma 2\pi R \ell}\int_{-\frac{\pi}{2}}^{\frac{\pi}{2}}\cos\alpha \, d\alpha \, \hat{\mathbf{R}}$$

$$= -\frac{M}{\gamma 4\pi R \ell}[\sin\alpha]_{-\frac{\pi}{2}}^{\frac{\pi}{2}} \hat{\mathbf{R}} = -\frac{M}{\gamma 2\pi R \ell}\hat{\mathbf{R}} \qquad \phi_{Ext} = \frac{M}{\gamma 2\pi \ell}\log R$$

Suppose the line mass has a cross-sectional area $\pi a^2$.
Then internally for $R \leq a$ the internal mass giving rise to $g_{Int}$ is $M_{Int} = M\left(\dfrac{R^2}{a^2}\right)$.
Therefore,

$$\mathbf{g}_{Int} = -\frac{M\dfrac{R^2}{a^2}}{\gamma 2\pi R \ell}\hat{\mathbf{R}} = -\frac{MR}{\gamma 2\pi a^2 \ell}\hat{\mathbf{R}} \qquad \phi_{Int} = \frac{M}{\gamma 4\pi \ell}\frac{R^2}{a^2} + \text{constant}$$

## A.4  1–D gravitational field of a mass m with a face area A

Use rectangular coordinates (x,y,z), where A lies in the xy-plane.

$$g = -\frac{m}{\gamma A}\hat{z} \qquad \phi = \frac{m}{\gamma A}z$$

## A.5  Work done equals the change in energy

When a mass m is moved a distance dz in a 1-D gravity field g work dW is done by the field and gravitational energy is expended. The change in energy is equal and opposite to the work done.

$$dE = -dW = -F.dz = -mg.dz = -m(-\nabla\phi).dz = m\frac{d\phi}{dz}dz = md\phi$$

Integrating $E = m\phi$.

## A.6  Self-energy of a point mass m

The self-energy of a point mass m is $\tfrac{1}{2}m\phi$. The $\tfrac{1}{2}$ arises because the contribution to energy E is counted twice. Once as the mass m in its own potential $\phi$ and secondly as the potential $\phi$ created by its own mass m.

Since a point mass has no interior all its energy is external.

$$E_{Ext} = \frac{1}{2}\{m\phi\} = -\frac{1}{2}\lim_{r\to 0}\left\{\frac{-m^2}{\gamma 4\pi r}\right\} = -\frac{m}{\gamma}\cdot\lim_{r\to 0}\left\{\frac{m}{8\pi r}\right\}$$

Speculate that $\lim_{r\to 0}\left\{\dfrac{8\pi r}{m}\right\} \to \eta \neq 0$ such that $\dfrac{1}{\gamma\eta} = c^2$.

If so, then $E_{Ext} = \lim_{r\to 0}\left\{\dfrac{m\phi}{2}\right\} = -mc^2$

The speed of a gravitational disturbance is c. If $\eta = 0$ then $c = \infty$.

## A.7  Change of energy when mass $m_2$ is placed in the gravity field of $m_1$

Treat the Earth as a 1-D mass $m_1$ with surface area A. When a mass $m_2$ is introduced at a height z above the Earth's surface, the change of energy is

$$E = m_2\phi_1 = m_2\left\{-\frac{m_1}{\gamma A}z\right\} = m_2 g_1 z = m_1\left\{-\frac{m_2}{\gamma A}z\right\} = m_1 g_2 z.$$

Both masses experience the same force $F = m_2 g_1 = m_1 g_2$.

For a volume $V = Az$, we can write the energy density as

$$\frac{E}{V} = \frac{m_1 g_2 z}{Az} = -\gamma\left(\frac{-m_1}{\gamma A}\right)g_2 = -\gamma g_1 g_2.$$

## A.8 Energy of a continuous distribution of mass

The total energy E of a distribution of mass contained in a volume V is given by

$$E = \frac{1}{2}\iiint_V \rho\phi dV \quad \text{where } \rho = \text{density of mass in volume V.}$$

The $\frac{1}{2}$ arises because the contribution to the energy sum E is counted twice for each mass element.

Using a vector identity, the divergence theorem and gravitational boundary conditions across the surface of the volume V, it can be shown that

$$E = -\frac{\gamma}{2}\iiint_V g^2 dV = -\frac{\gamma}{2}\iiint_{V_{Int}} g_{Int}^2 dV - \frac{\gamma}{2}\iiint_{V_{Ext}} g_{Ext}^2 dV$$

The gravitational energy density for a distributed mass $= \frac{E}{V} = \frac{1}{2}\gamma g^2$.

## A.9 Gravitational energy of a length $\ell$ of an infinite line mass

$$E = -\frac{1}{2}\gamma\iiint_V g^2 dV \quad \text{Where } dV = (dR)(R\,d\theta)(dz) \text{ in cylindrical polars.}$$

$$E = -\frac{1}{2}\gamma\iiint_{V_{Int}}\left(\frac{-M}{\gamma 2\pi\ell}\frac{R}{a^2}\right)^2 (dR)(R\,d\theta)(dz) - \frac{1}{2}\gamma\iiint_{V_{Ext}}\left(\frac{-M}{\gamma 2\pi R\ell}\right)^2 (dR)(R\,d\theta)(dz)$$

$$= -\frac{1}{2}\gamma\left(\frac{-M}{\gamma 2\pi\ell a^2}\right)^2 \int_0^a R^3 dR \int_0^{2\pi} d\theta \int_0^\ell dz - \frac{1}{2}\gamma\left(\frac{-M}{\gamma 2\pi\ell}\right)^2 \int_a^\infty \frac{dR}{R} \int_0^{2\pi} d\theta \int_0^\ell dz$$

$$= -\frac{1}{2}\gamma\left(\frac{-M}{\gamma 2\pi\ell a^2}\right)^2 \left[\frac{R^4}{4}\right]_0^a [\theta]_0^{2\pi} [z]_0^\ell - \frac{1}{2}\gamma\left(\frac{-M}{\gamma 2\pi\ell}\right)^2 [\log R]_a^\infty [\theta]_0^{2\pi} [z]_0^\ell$$

$$= -\frac{1}{2\gamma}\left(\frac{M}{2\pi\ell a^2}\right)^2 \frac{a^4}{4}(2\pi)\ell - \frac{1}{2\gamma}\left(\frac{M}{2\pi\ell}\right)^2 (\log\infty - \log a)(2\pi)\ell$$

$$= -\frac{1}{\gamma}\frac{M^2}{16\pi\ell} - \infty$$

$$= -\frac{1}{2}Mc^2 - \infty \quad \text{if } \eta = \frac{8\pi\ell}{M}$$

## A.10 Gravitational energy of a solid sphere of mass M

$$E = -\frac{1}{2}\gamma \iiint_V g^2 dV \quad \text{Where } dV = (dr)(rd\theta)(r\sin\theta d\varphi) \text{ in spherical polars.}$$

$$E = -\frac{1}{2}\gamma \iiint_{V_{Int}} \left(\frac{-Mr}{\gamma 4\pi a^3}\right)^2 dV - \frac{1}{2}\gamma \iiint_{V_{Ext}} \left(\frac{-M}{\gamma 4\pi r^2}\right)^2 dV$$

$$= -\frac{1}{2}\gamma \int_0^a \left(\frac{-Mr}{\gamma 4\pi a^3}\right)^2 r^2 dr \int_0^\pi \sin\theta d\theta \int_0^{2\pi} d\varphi - \frac{1}{2}\gamma \int_a^\infty \left(\frac{-M}{\gamma 4\pi r^2}\right)^2 r^2 dr \int_0^\pi \sin\theta d\theta \int_0^{2\pi} d\varphi$$

$$= -\frac{1}{2}\gamma \left(\frac{-M}{\gamma 4\pi a^3}\right)^2 \int_0^a r^4 dr \int_0^\pi \sin\theta d\theta \int_0^{2\pi} d\varphi - \frac{1}{2}\gamma \left(\frac{-M}{\gamma 4\pi}\right)^2 \int_a^\infty \frac{dr}{r^2} \int_0^\pi \sin\theta d\theta \int_0^{2\pi} d\varphi$$

$$= -\frac{1}{2}\gamma \left(\frac{-M}{\gamma 4\pi a^3}\right)^2 \left[\frac{r^5}{5}\right]_0^a [-\cos\theta]_0^\pi [\varphi]_0^{2\pi} - \frac{1}{2\gamma}\left(\frac{M}{4\pi}\right)^2 \left[\frac{-1}{r}\right]_a^\infty [-\cos\theta]_0^\pi [\varphi]_0^{2\pi}$$

$$= -\frac{1}{2\gamma} \frac{M^2}{(4\pi a^3)^2}\left(\frac{a^5}{5}\right)(2)(2\pi) - \frac{1}{2\gamma}\frac{M^2}{(4\pi)^2}\left(0+\frac{1}{a}\right)(2)(2\pi)$$

$$= -\frac{1}{\gamma}\frac{M^2}{40\pi a} - \frac{1}{\gamma}\frac{M^2}{8\pi a}$$

$$= -\frac{1}{\gamma}\frac{3M^2}{20\pi a}$$

For a point mass, $a \to 0$ and externally $E \to -mc^2$.

## A.11 Angular momentum

Suppose O is the fixed point about which a mass M rotates. If the mass is at P, the length OP = R is the radius of rotation, which can change.

Angular momentum $\mathbf{H} = M(\mathbf{R} \times \mathbf{v}) = MRv\sin\alpha \,\hat{\boldsymbol{\theta}}$, where $\alpha$ is the angle between the unit vector $\hat{\mathbf{R}}$ in the radial direction and the velocity vector $\mathbf{v}$. The unit vector $\hat{\boldsymbol{\theta}}$ is perpendicular to the plane containing $\mathbf{R}$ and $\mathbf{v}$.

For constant angular momentum, when the radius R decreases the velocity v increases and vice-versa.

## A.12 Mass M moving with velocity v along the z-axis

In cylindrical polars $(R, \theta, z)$.

$$\mathbf{h} = -\frac{1}{2}\frac{MRv}{V}\hat{\boldsymbol{\theta}} = -\frac{1}{2}\rho R v \hat{\boldsymbol{\theta}} \quad \text{Where } \rho = \frac{M}{V}.$$

$$\nabla \times \mathbf{h} = \frac{1}{R}\begin{vmatrix} \hat{\mathbf{R}} & R\hat{\boldsymbol{\theta}} & \hat{\mathbf{z}} \\ \frac{\partial}{\partial R} & \frac{\partial}{\partial \theta} & \frac{\partial}{\partial z} \\ 0 & \left(-\frac{1}{2}\rho R v\right)R & 0 \end{vmatrix}$$

$$= \frac{1}{R}\frac{\partial}{\partial R}\left(-\frac{1}{2}\rho R^2 v\right)\hat{\mathbf{z}}$$

$$= -\frac{1}{R}\left\{\frac{\partial}{\partial R}\left(\frac{\rho R^2}{2}\right)\right\}v\hat{\mathbf{z}} - \frac{1}{R}\left\{\frac{\partial}{\partial R}\left(\frac{\rho R^2}{2}\right)\right\}v\hat{\mathbf{z}}$$

$$= -\rho v\hat{\mathbf{z}} - \frac{1}{R}\eta\frac{\partial v}{\partial R}\hat{\mathbf{z}} \quad \text{where } \eta = \frac{\rho R^2}{2}$$

$$= -\rho v\hat{\mathbf{z}} - \eta\frac{1}{ct}\frac{1}{c}\frac{\partial v}{\partial t}\hat{\mathbf{z}} \quad \text{where } R = ct$$

$$= -\rho v\hat{\mathbf{z}} - \gamma\frac{g}{t}\hat{\mathbf{z}} \quad \text{where } c^2 = \frac{1}{\gamma\eta}$$

## A.13 Kinetic energy is equal to gravitomagnetic energy

A mass moving with speed v develops a gravitomagnetic field **h**. At radius R

$$E = -\frac{1}{2}mv^2 = -\frac{1}{2}m(R\Omega)^2 = -\frac{1}{2}(\rho R^2)\Omega^2 V = -\frac{1}{\eta}(\eta h)^2 V = -\eta h^2 V$$

## A.14 Gravitomagnetic energy of a line mass section moving with speed v

Assume that the line mass has cross-sectional area $\pi a^2$.
The gravitomagnetic energy for a distributed mass is

$$E = -\frac{1}{2}\iiint_{V_{Int}}\eta h_{Int}^2 \, dV - \frac{1}{2}\iiint_{V_{Ext}}\eta h_{Ext}^2 \, dV$$

$$= -\frac{1}{2}\iiint_{V_{Int}}\left(\frac{2}{\rho R^2}\right)\left(\frac{IR}{2\pi a^2}\right)^2 dV - \iiint_{V_{Ext}}\left(\frac{2}{\rho R^2}\right)\left(\frac{I}{2\pi R}\right)^2 dV$$

$$= -\frac{1}{2}\left(\frac{2}{\rho}\right)\left(\frac{I}{2\pi a^2}\right)^2 \int_0^a R\,dR \int_0^{2\pi}d\theta\int_0^\ell dz - \frac{1}{2}\left(\frac{2}{\rho}\right)\left(\frac{I}{2\pi}\right)^2 \int_a^\infty \frac{dR}{R^3}\int_0^{2\pi}d\theta\int_0^\ell dz$$

$$= -\frac{1}{2}\left(\frac{2}{\rho}\right)\left(\frac{I}{2\pi a^2}\right)^2 \left[\frac{R^2}{2}\right]_0^a 2\pi\ell - \frac{1}{2}\left(\frac{2}{\rho}\right)\left(\frac{I}{2\pi}\right)^2 \left[\frac{-1}{2R^2}\right]_a^\infty 2\pi\ell$$

$$= -\frac{1}{2}\left(\frac{2}{\rho}\right)\left(\frac{I}{2\pi a^2}\right)^2 \left(\frac{a^2}{2}\right) 2\pi\ell - \frac{1}{2}\left(\frac{2}{\rho}\right)\left(\frac{I}{2\pi}\right)^2 \left(\frac{+1}{2a^2}\right) 2\pi\ell$$

$$= -\frac{1}{2}\left(\frac{2}{\rho}\right)\frac{1}{(2\pi a^2)^2}\frac{M^2 v^2}{\ell^2}\left(\frac{a^2}{2}\right) 2\pi\ell - \frac{1}{2}\left(\frac{2}{\rho}\right)\frac{1}{(2\pi)^2}\frac{M^2 v^2}{\ell^2}\left(\frac{1}{2a^2}\right) 2\pi\ell$$

$$= -\frac{1}{2}\left(\frac{2}{\rho}\right)\frac{1}{(2\pi)^2}\frac{M^2 v^2}{\ell^2}\left(\frac{1}{2a^2}\right) 2\pi\ell - \frac{1}{2}\left(\frac{2}{\rho}\right)\frac{1}{(2\pi)^2}\frac{M^2 v^2}{\ell^2}\left(\frac{1}{2a^2}\right) 2\pi\ell$$

$$= -\frac{1}{4}\frac{1}{(\pi a^2 \ell)}\frac{M^2 v^2}{\rho} - \frac{1}{4}\frac{1}{(\pi a^2 \ell)}\frac{M^2 v^2}{\rho}$$

$$= -\frac{1}{4}M v^2 - \frac{1}{4}M v^2$$

$$= -\frac{1}{2}M v^2$$

## A.15 Kinetic energy without gravitomagnetism

Suppose the mass m falls in a gravity field with speed v, so that $dz = v\,dt$.
From equivalence $g = \dfrac{dv}{dt}$.
The mass m is weightless and has no gravitational energy.

$$dE = -mg\,dz = -m\frac{dv}{dt}v\,dt = -mv\,dv$$

Integrating

$E = -\dfrac{1}{2}mv^2$  The gravitational energy expended is converted to kinetic energy.

## A.16 The gravitomagnetic vector potential A
For vector operators the divergence of curl is always zero. That is $\nabla \cdot \nabla \times = 0$.
Since $\nabla \cdot \mathbf{b} = 0$ we can set $\mathbf{b} = \nabla \times \mathbf{A}$.
Dimensional analysis shows that the vector $\mathbf{A}$ is a velocity.

## A.17 Mass falling in a static gravity field
Assume 1 – D static gravity field acts downwards. $\mathbf{g}_{Static} = -g\hat{\mathbf{z}} = \nabla\phi$.

$$\nabla \times \mathbf{g}_{Total} = \frac{\partial \mathbf{b}}{\partial t} = \frac{\partial (\nabla \times \mathbf{A})}{\partial t} = \nabla \times \left(\frac{\partial \mathbf{A}}{\partial t}\right)$$

Therefore $\nabla \times \left\{ \mathbf{g}_{Total} - \frac{\partial \mathbf{A}}{\partial t} \right\} = 0$

For vector operators the curl of gradient is always zero. That is $\nabla \times \nabla = 0$.
Therefore we can introduce a gradient of potential $\phi$ such that

$$\nabla \phi = \mathbf{g}_{Total} - \frac{\partial \mathbf{A}}{\partial t}$$

Total gravity field $\mathbf{g}_{Total} = \nabla \phi + \frac{\partial \mathbf{A}}{\partial t} = \mathbf{g}_{Static} + \mathbf{g}_{Dynamic}$

In the static case $\frac{\partial \mathbf{A}}{\partial t} = 0$ and $\mathbf{g}_{Total} = \mathbf{g}_{Static} = \nabla \phi = -g\hat{\mathbf{z}}$.

This means $\nabla \times \mathbf{g}_{Static} = 0$. This is the condition for the gravity field to be conservative. That is, a mass can move around the static $\mathbf{g}$-field, but on returning to its starting position there is no change in energy.

In the dynamic case of free fall, $\mathbf{A} = \mathbf{v}$. An upward gravity field $\frac{\partial \mathbf{v}}{\partial t}$ is created, which cancels out with the downward static gravity field. Although accelerating, the falling mass is force-free. This is the root of equivalence.

## A.18 The induced gravitomagnetic field b is an angular velocity

$$\mathbf{b} = \nabla \times \mathbf{A} = \frac{1}{R}\begin{vmatrix} \hat{\mathbf{R}} & R\hat{\boldsymbol{\theta}} & \hat{\mathbf{z}} \\ \frac{\partial}{\partial R} & \frac{\partial}{\partial \theta} & \frac{\partial}{\partial z} \\ 0 & 0 & v \end{vmatrix} = -\frac{1}{R}\frac{\partial v}{\partial R}R\hat{\boldsymbol{\theta}} = -\frac{\partial (R\Omega)}{\partial R}\hat{\boldsymbol{\theta}} = -\Omega\hat{\boldsymbol{\theta}} = -\boldsymbol{\Omega}.$$

## A.19 Force on a mass rotating with angular velocity $\Omega$
Using cylindrical polar coordinates $(R, \theta, z)$. Suppose the rotation is about the vertical $z$-axis. Newton's 2nd law gives

$\mathbf{F} = m\frac{d\mathbf{v}}{dt} = m\frac{d(R\Omega\hat{\boldsymbol{\theta}})}{dt}$ The unit vector in the tangential direction is $\hat{\boldsymbol{\theta}}$.

$= mR\Omega\frac{d\hat{\boldsymbol{\theta}}}{dt} = mR\Omega(-\Omega\hat{\mathbf{R}})$ The unit vector in the radial direction is $\hat{\mathbf{R}}$.

$= -mR\Omega^2 \hat{\mathbf{R}}$

## A.20 The simple pendulum

Consider a pendulum of length $\ell$ with a bob of mass m suspended in a gravity field g. At the start, the angle between the vertical and $\ell$ is $\theta_0$. When released, the bob swings to and fro. The velocity v of the bob perpendicular to $\ell$ is $v = \ell\dot{\theta}$ where $\dot{\theta} = \dfrac{d\theta}{dt}$ is the angular velocity of swing.

From Newton's $2^{nd}$ law, resolving the forces on the mass m perpendicular to $\ell$.

$$-m\dfrac{dv}{dt} = mg\sin\theta$$

For small angles $\sin\theta \approx \theta$. Substituting for v gives the differential equation.

$$\ell\dfrac{d^2\theta}{dt^2} + g\theta = 0$$

This is simple harmonic motion.

The period of swing $\tau = 2\pi\sqrt{\dfrac{\ell}{g}}$ s and the resonant frequency $f_{Res} = \dfrac{1}{\tau}$ Hz.

## A.21 The electromagnetic analogue of the simple pendulum

The moving bob of mass m creates a mass current I. At angle $\theta$

$$I = \dfrac{dm}{dt} = \dfrac{m}{\ell\theta_0}\ell\dot{\theta} \text{ kg/s}$$

The angular velocity $\dot{\theta}$ is equivalent to an induced gravitomagnetic field, through which the mass m passes through. Based on the magnetic analogue (see chapter 25), the maximum gravitomagnetic energy of the simple pendulum is given by

$$E_{Max} = \dfrac{1}{2}LI_{Max}^2$$ where L is the gravitomagnetic self–inductance.

The maximum mass current $I_{Max}$ occurs when angle $\theta = 0$ and the angular velocity $\dot{\theta}$ is a maximum. For this condition

$$E_{Max} = \dfrac{1}{2}mv_{Max}^2 = \dfrac{1}{2}m(\ell\dot{\theta}_{Max})^2.$$

Hence, we can determine the self–inductance.

$$L = \dfrac{(\ell\theta_0)^2}{m} \text{ m}^2/\text{kg}.$$

The bob stores potential energy by rising in the gravity field g. In doing so, the gravity field is stressed, like an electric field is stressed when charge is stored in a capacitor. Based on the electrical analogue, the capacitance C of the simple pendulum is given by

$$C = \dfrac{m}{g\ell\theta_0^2} \text{ kg}^2/\text{J}$$

From the electromagnetic analogue of a tuning circuit, the resonant frequency of the simple pendulum is

$$f_{Res} = \dfrac{1}{2\pi}\dfrac{1}{\sqrt{LC}} = \dfrac{1}{2\pi}\sqrt{\dfrac{g}{\ell}} \text{ Hz.}$$

# NAMES INDEX

## A
Akimov, Anatoli, 307
Alcubierre, Miguel, 298
Allen, John, xi, 322, 329
Allen, Ken, x
Amontons, Guillaume, 206
Arago, Dominique Francois, 157
Archimedes, 5, 14, 20, 22
Aristarchus, 32
Aristotle, 9
Avogadro, Count Amadeo, 81

## B
Babbage, Charles, 166
Bacon, Francis, 183, 215
Baker, Paul, 324
Banks, Joseph, 156
Barish, Barry, 293
Barnett, Samuel, 346
Barrow, Isaac, 19, 20
Barrow, John, 325, 333
Bekenstein, Jacob, 228
Bell, Alexander Graham, 47
Bernard, Thomas, 204
Bernoulli, Daniel, 120
Biefeld, Paul, 304
Biot, Jean-Baptiste, 348
Birdsall, Graham, 330

Bissell, Phil, 333
Black, Joseph, 187
Brown, Thomas Townsend, 304
Bohr, Niels, 186
Boltzmann, Ludwig, 183
Bonaparte, Napoléon, 156
Bondi, Hermann, 313
Boulton, Matthew, 148, 196
Boyle, Robert, 182
Brahe, Tycho, 13
Branly, Edouard, 179
Brown, Colin, 324
Brown, Robert, 80
Bruno, Giordano, 273
Brush, Charles, 263, 301, 355
Bunsen, Robert, 156

## C
Cameron, Jeff, 304
Capaldi, Peter, 341
Carnot, Lazare, 198
Carnot, Sadi, 198
Cartan, Elie, 105
Cartmell, Matthew, 328
Casimir, Hendrick, 331
Cavendish, Henry, 40, 204
Celsius, Anders, 182
Charles, Jacques, 182, 183

Chasserot, Scott, 341
Christoffel, Elwin, 97, 99, 105
Clarke, Arthur C., 2, 317
Clausius, Rudolf, 201, 208
Cook, Nick, 330, 333, 335
Cooke, Steve, 327
Coriolis, Gaspard de, 130
Coulomb, Charles, 148, 202
Crabtree, Steve, xii, 337
Crompton, Arthur, 80
Cruise, Mike, 329
Ctesibius, 181, 194
Cuthbert, Tony, x, 329

**D**

Daimler, Gottlieb, 6, 210
D'Alembert, Jean, 127
Darwin, Charles, 196
Darwin, Erasmus, 196
Da Vinci, Leonardo, 138, 202
Davis, Tamara, 339
Davy, Humphry, 156, 205
De Haas, Wander, 346
De Matos, Clovis, 330
Descartes, René, 20, 25, 71
Dimitriou, Stavros, 324, 329
Doppler, Christian, 58
Du Fay, Charles, 147

**E**

Eddington, Arthur, 64
Edwards, Tony, 330
Einstein, Albert, 23, 55, 64, 79, 274, 346
Ellis, John, 338
Engels, Peter, 316
English, Malcolm, 330
Ennis, Graham, 329
Eratosthenes, 32
Eötvos, Roland von, 24

Euclid, 70
Euler, Leonhard, 120, 275
Everett, Frances, 283

**F**

Fahrenheit, Gabriel, 182
Faraday, Michael, 26, 118, 152, 156, 161, 165, 232, 251, 353
Ferdinand II (Grand Duke of Tuscany), 182
Feynman, Richard, 239, 246
Fisher, David, 336
FitzGerald, George, 50, 176
Fizeau, Armand H., 36
Fleming, John Ambrose, 161
Forward, Robert, 291, 315, 325, 349
Foucault, Léon, 37, 271
Fourier, Jean-Baptiste, 158, 190
Franklin, Benjamin, 7, 148, 151, 196

**G**

Galilei, Galileo, 9, 13, 15, 23, 35, 44, 65, 67, 114, 182, 273
Galvani, Luigi, 155
Gauss, Karl, 94, 123
Geer, Mike, 329
George, Catalina, xii
Giaever, Ivar, 113
Gilbert, William, 146
Gratus, Jonathan, 323
Gray, Stephen, 147, 155
Grossmann, Marcel, 97, 101
Groves, Alan, 333
Guericke, Otto von, 147

**H**

Haisch, Bernhard, 329
Hajdukovic, Dragon, 340

Halley, Edmund, 23
Hallwachs, Wilhelm, 79, 185
Hammond, Giles, 331
Hansson, Anders, 329
Harrison. H. Ron, xi, 310
Hauksbee, Francis, 147
Hawking, Stephen, 67, 229
Hayasaka, Hideo, 306
Heaviside, Oliver, 55, 174, 253
Helmholtz, Hermann von, 46, 132, 136
Henry, Joseph, 167
Hero, 181, 194
Hertz, Heinrich, 79, 172, 176
Hilbert, David, 102
Hinds, Ed, 325
Hipparchus, 33
Holley, Bruce, 333
Holt, Alan, 329
Hooke, Robert, 21, 182
Horrocks, Jeremiah, 15, 19
Hough, James, 292
Hubble, Edwin, 60, 110, 295
Hulse, Russell, 288
Humphrys, John, 330
Huygens, Christiaan, 15, 34

## J

Jacobson, Ted, 230
Jefimenko, Oleg, 268
Jennison, Roger, 299
Johnson, Clifford, 339
Johnston, Walter, 323
Joule, James Prescott, 189, 205, 207

## K

Kauffman, Walter, 55
Kay, Andrea, xii
Kennedy, Martin, 333

Kepler, Johannes, 13, 14, 15
Khayyam, Omar, 20
Kidd, Alexander (Sandy), 315, 335
Kleist, Ewald von, 150
Knight, Gowin, 147
Kozyrev, Nikolai, 305
Krim, Jacqueline, 202
Kvasnik, Frank, x
Kyritsis, Costas, 324

## L

Laithwaite, Eric, 309, 310, 338
Laplace, Pierre, 214, 217, 227, 271
Larmor, Joseph, 267, 346
Laurie, Peter, 328
Lazar, Robert, 320
Leake, Jonathan, 330
Leeuwenhoek, Antony van, 80
Leibnitz, Gottfried, 27, 258
Lenard, Philipp, 79
Lense, Josef, 382
Lenz, Heinrich, 167
Le Verrier, Urbain, 108
Levi-Civita, Tullio, 99, 102
Lewis, Alun, 335
Li, Ning, 308
Liddell, Peter, 333
Lodge, Oliver, 176, 178, 179, 299
Logunov, Anatoly, 104
Loller, Carl, 328
London, Fritz, 284
Lorentz, Hendrik, 52
Lorrain, Paul, 268
Lucas, Henry, 19

## M

Maccone, Claudio, 329
Mach, Ernst, 222
Mackinder, Annie, 337

Magnus, Heinrich, 139
Marconi, Guglielmo, 179
Martin, Bill, 328
Matthews, Robert, 308
Maxwell, James Clerk, 30, 45, 114, 132, 172, 218, 252
May, Andrew, 322, 336
Maybach, Wilhelm, 6, 210
Mayer, Julius, 205
McInnes, Colin, 328
McMahon, Noah, xii
Meek, James, 330
Michell, John, 39, 226
Michelson, Albert, 37, 46, 49
Millis, Marc, 321, 330, 333
Minkowski, Hermann, 70
Montgolfier, Jacques, 183
Montgolfier, Joseph, 183
Morley, Edward, 49
Moss, Darren, xi
Murdock, William, 195
Musschenbroek, Pieter, 151

# N

Nash. John, 356
Navier, Claude, 134
Newcomen, Thomas, 195
Newton, Isaac, 18, 39, 114, 195, 249

# O

Obousy, Richard, 330
Oersted, Hans Christian, 157, 165, 362
Ohm, Georg, 158, 191
Oldenburg, Henry, 147
Owen, Gari, 322

# P

Papin, Dionysius, 195

Pascal, Blaise, 20
Penrose, Roger, 227, 325, 333
Perrin, Jean, 81
Peschel, Ulf, 317
Petit, Jean-Paul, 330
Philon, 181
Pilkington, Mark, 330
Pixii, Hippolyte, 169
Planck, Max, 56, 102, 185
Platt, Charles, 326
Podkletnov, Evgeny, 326, 333, 338, 351
Poggendorff, Johann, 165
Poincaré, Henri, 55, 72
Pope, Nick, 330
Pound, Robert, 86
Powell, Baden, 310
Poynting, John, 359
Priestly, Joseph, 148, 196
Prigogine, Ilya, 209
Provost, Michael, 324, 331
Ptolemy, 33
Puthoff, Hal, 329

# R

Rankine, William, 120
Rebka, Glen, 86
Reeh, Derek, 333
Ricci-Curbastro, Gregorio, 99
Richardson, Owen, 346
Riemann, Bernhard, 97, 99
Roach, Peter, ix
Robertson, Norna, 292
Rockall, Mike, xi
Römer, Ole, 35
Rubin, Vera, 299
Rueda, Alfonso, 329
Russell, Geoff, 388
Rutherford, Earnest, ix, 179, 186

## S

Sagnac, Georges, 275
Sample, Ian, 330
Savart, Felix, 348
Savery, Thomas, 195
Schiff, Leonard, 283
Schwarzschild, Karl, 107, 228
Schweigger, Johann, 157, 165
Searle, George, 191
Seebeck, Thomas, 158
Sellers, Jerry, 330
Seyfang, George, x
Shawyer, Roger, 316, 340
Shelley, Tom, 320
Smith, 'Coyote', 340
Smith, Paul, 326
Smithson, James, 167
Sommerfeld, Arnold, 101
Speake, Clive, 331, 333
Stanger, Vaughan, 328
Stansfield, Neil, 340
Stevin, Simon, 9
Stokes, George, 134, 247
Sturgeon, William, 157, 162
Szames, Alexandre, 330

## T

Tajmar, Martin, 332, 339
Takeuchi, Sakae, 306
Taylor, Joseph, 288
Taylor, Sarah, 337
Thales, 146
Thirring, Hans, 282
Thomson, John Joseph, 148, 264
Thomson, William (Lord Kelvin), 176, 183, 205, 208, 258
Thompson, Benjamin (Count Rumford), 167, 203, 204
Thompson, Jeremy, 336
Thorne, Kip, 293
Torr, Douglas, 308
Tucker, Robin, xi, 278, 323, 328, 333

## V

Verdet, Marcel Émile, 263, 348
Vigier, Jean-Pierre, 329
Vinogradov, Sergei, 326
Volta, Count Alessandro, 155

## W

Wall, William, 151
Wallace, Henry, 309
Wallis, John, 19
Wang, Charles, 323
Wardrop, Brian, 328
Watt, James, 195
Weber, Joseph, 213, 290
Wedgewood, Josiah, 196
Weiss, Rainer, 293
Wheatstone, Charles, 166
Whitehead, Nigel, 337
Wickens, Alan, xi
Wien, Wilhelm, 185
Williams, Eddie, 324, 331
Williams, Pharis, 230
Windett, Dave, xii
Wojcik, Adam, 338
Woods, Clive, 326, 330, 333
Wright, Orville and Wilbur, 143

## Y

Young, Brian, 319, 323, 335
Young, Nic, 337

## Z

Ziegler, Spencer, 328
Zwicky, Fritz, 299

# SUBJECT INDEX

### Ancient Greeks
Museum at Alexandria, 180;
Archimedes & buoyancy, 5;
Archimedes' early form of infinitesimal calculus, 14;
Calculating the radius $R_E$ of the spherical Earth, 32;
Geometry & trigonometry, 33;
Pythagoras' theorem, 70;
Euclid's The Elements, 70;
Hero's formula for area of triangle, 181;
Gears, levers & pulleys, 181;
Pneumatics, 181;
Ctesibius & the water pump, 194;
Piston & syringe, 181;
Steam pressure, 181;
Hero & the Aeolipile, 194.

### Astronomical measurement
Earth's radius $R_E$, 32, 34;
Earth's mass $M_E$, 41;
Moon's mass $M_m$, 41;
Sun's mass $M_S$, 41;
Earth–Mars distance $D_{EM}$, 34;
Earth–Moon distance $D_{Em}$, 33;
Earth–Sun distance $D_{ES}$, 34.

## BAE Systems
British Aerospace BAe, 319, 322;
Kingston, 323;
Warton, Lancashire, 144, 319;
Merger between BAe & Marconi (2000), 328;
Advanced Technology Centres, 215, 328, 332.

## BBC
Radio 4 *Science Now*, 335;
Radio 4 *Today*, 330;
TV *Horizon*, 337;
TV *Tomorrow's World*, 311.

## Black hole
Point singularity, 226-228;
Escape speed, 227;
Event horizon, 227;
Schwarzschild radius $R_s$, 228;
Entropy of black hole, 228-229;
Evaporation, 230;
Tearing apart quantum ether, 229;
Hawking radiation, 230.

## Curved paths in space
Roller coaster, 66;
Clothoid 2-D teardrop shape, 66;
Parabolic flight path, 66; ZERO G Corporation, 67;
Weightlessness, 67, 308, 321;
International Space Station (ISS), 67;
Moon's orbit, 15, 22.

## Doppler effect
Acoustics, 58;
Light, 59;
Red shift of galaxies, 60;
Hubble's expanding Universe, 60.

## Einstein's miraculous year
Photo-electric effect, 79-80, 185;

Brownian motion, 80;
Calculation of Avogadro's Number $N_A$, 81;
Size of molecules, 81;
Atomic weight, 81, 302;
Energy $E = mc^2$, 83, 104, 261;
Special relativity, 52, 55, 77, 79, 82, 259, 261, 268;
Gravitational red shift, 84, 86, 223, 35.

## Einstein's special relativity

Inertial frame, 45, 55, 63, 75, 82;
Galilean transform, 45, 52;
Michelson's light interferometer, 47, 49, 291;
Michelson-Morley experiment fails to detect ether, 50, 55;
FitzGerald contraction, 50, 57;
Lorentz transform, 52, 72, 101, 159, 259;
Lorentz factor $\gamma$, 53;
Einstein ignores ether, 55;
Time dilation, 54, 57;
NASA Gravity Probe-A, 86;
Twin paradox, 54.

## Einstein's general relativity

Curvature in 3-D Euclidean space, 93;
Gauss' radius of curvature K, 94;
Weak principle of equivalence between gravity & acceleration, 89;
Strong principle of equivalence, 90;
Minkowski's space-time, 74;
Lorentz transform in matrix form, 73;
Geodesic shortest distance, 98;
Space-time metric – square of geodesic, 74;
Invariance of metric, 76, 98;
Transform of metric – Minkowski's flat space-time matrix, 74;
Non-Euclidean geometry, 97;
Grossmann's tuition on tensor calculus, 98, 101;
Contravariant & covariant vector bases for 4-D space-time, 100;
Principle of covariance, 101, 104;
Einstein's curved space-time matrix, 91;
Tensor equation for curved space-time due to mass and energy, 103;
Schwarzschild's solution for space-time curvature due to the Sun, 107;

Rubber sheet analogy, 108;
Precession of Mercury's perihelion, 108;
Einstein's cosmological constant Λ to control expansion of Universe, 109, 110, 124, 295, 300;
Accelerating expansion of Universe – dark energy & anti-gravity effect, 110, 301.

## Electricity — static
Amber petrified resin, 146;
Frictional electric charge, 147;
Von Guericke's rotating sulphur ball, 147;
Coulomb's torsion balance, 148;
Inverse square law for charge, 148;
Electrical permittivity ε, 148;
Electric potential φ, 149;
Electric field $E = -\nabla\phi$, 149;
Voltage V as difference in potential φ, 156;
Earth's electric field, 239;
Ionosphere, 239, 245.

## Electricity — charge flow
Electrical conduction, 147, 155, 158, 239;
Vitreous & resinous electrical fluid, 148;
Electric current I taken as flow of positive charge, 173;
Galvani's experiment with frogs' legs, 155;
Volta's pile, 156, 244;
Battery, 157;
Thermoelectricity, 158, 188;
Oersted's discovery, 157, 161, 165, 172, 219, 347, 362;
Resistance R to current flow, 25, 114, 348;
Ohm's law $V = IR$, 191;
Conductivity σ, 158, 175;
Speed of charges in current flow, 159, 172;
Magnetism as a special relativity effect, 77, 159;
Thomson's electron current, 149;
Current ring & the solenoid, 266, 357, 167;
Electrolysis, 156;
Wall's view of lightning, 151;
Franklin's kite experiment, 151;

Lightning conductor, 152, 246;
BAe lightning trials, 152.

### Electricity — charge storage
Leyden jar, 151, 157, 173, 176, 178;
Parallel plate capacitor, 151;
Capacitance C, 151, 176;
Dielectrics, 149, 151, 173, 297, 303, 316;
Model of atom, 186, 265;
Charge displacement, 149, 173, 303;
Faraday's ice pail experiment, 152, 249;
Electrical screening – Faraday cage, 152, 249, 297, 347.

### Electricity — devices
Faraday's simple motor, 161;
Magnetic analogue of Magnus effect, 162;
Fleming's Left Hand Rule, 161;
Electromagnet, 157;
Commutator, 161;
Sturgeon's electric motor, 162;
Galvanometer, 165.

### Electricity — induction
Arago's disc, 166;
Faraday's discovery, 169;
Lenz's law, 167;
Fleming's Right Hand Rule, 169;
Dynamo, 169;
First commercial electric current generator, 169;
Self inductance L of coil, 168;
Magnetic energy of coil, 168.

### Electromagnetism
Maxwell's equations, 77, 174;
Maxwell's curl $\nabla \times$, 132, 157, 174;
Heaviside's vector form of Maxwell's equations, 174.

### Electromagnetic waves
Maxwell's idea of vibrations in the ether, 45, 77, 174;

Maxwell's equations for electromagnetic waves, 175;
Maxwell's equation of telegraphy, 175;
Helmholtz's interest in the idea of electromagnetic waves in the ether, 175;
Leyden jar spark discharge and Simple Harmonic Motion, 175;
Resonant frequency of oscillations, 178;
Sympathetic acoustic vibration between sand piles on Chladni plates, 176;
Hertz's discovery of electromagnetic waves, 178;
Standing waves, 213, 276;
Spark gap in conducting ring, 178;
Syntonic tuning, 178;
The coherer – Branly's electromagnetic wave detector, 179;
The electromagnetic wave spectrum, 179, 204, 362;
Marconi & wireless communication, 179.

## Ether
Ancient Greek view, 9, 25, 45;
Fluid type property, 25, 45, 124, 254;
Descartes' ether vortices in space, 20, 25, 255;
Planets move through ether with zero resistance, 25;
Newton's negative view on ether vortices in space, 25, 46;
Newton accepts ether as a medium to support gravity influence, 25, 96;
Maxwell views ether as a medium to support electromagnetic fields, 45, 77, 173;
Michelson–Morley experiment fails to detect ether, 47, 49, 58;
Lorentz transform must leave ether unchanged, 55;
Einstein's avoids taking a view on existence of ether, 55;
Modern view – the quantum ether, 26, 175, 229, 296, 331, 360.

## Faraday
Assistant to Humphry Davy at the RI, 156;
Helping Davy repeat Oersted's discovery, 157;
Laws of electrolysis, 156;
Iron filings reveal magnetic field, 158, 172;
Magnetic analogue of the Magnus effect, 162;
Simple electric motor, 161;
William Sturgeon's electric motor, 162;
Experimenting on acoustic vibration with Charles Wheatstone, 166, 176;
Discovery of electromagnetic induction, 167;
Meets Joseph Henry at the RI, 167;

View of James Maxwell's mathematics, 118;
Magneto-optics & the Faraday effect (FE), 237, 264;
The inverse Faraday effect (IFE), 264.

## Faraday's gravity experiments — in search of the dual force of gravity
$1^{st}$ Gravity experiment at the RI, 232;
$2^{nd}$ Gravity experiment at the RI, 237;
$3^{rd}$ Gravity experiment at the Shot Tower, 241.

## Field propulsion & inertial thrust machines (ITM)
Arthur C. Clarke's prediction of space drive, 2, 317;
Sandy Kidd's ITM, 311, 335;
Eric Laithwaite's demonstration of an ITM at the RI, 310;
Euler's equation & gyro theory, 309;
Mass dipole acceleration & the speed limit imposed by Einstein's mass-velocity equation, 55, 315;
Optical version of mass dipole, 317;
First Field Propulsion Conference (2001), 329;
Space-time warp drive, 298, 339;
Shawyer's EM Drive, 316, 326, 340.

## Fluids — ideal flow
Incompressibility (density $\rho$ = constant), 96, 120, 188;
Zero viscosity ($\eta = 0$), 120;
Source & sink of fluid, 120;
Inverse square law, 120;
Velocity potential $\phi$, 131;
Fluid velocity $\mathbf{v} = -\nabla\phi$, 131;
Analogue with gravity, 131;
Streamlines, 121, 126, 131;
Bernoulli's equation, 120-123, 139, 142, 207;
Hele–Shaw apparatus, 139;
D'Alembert's zero drag paradox, 127.

## Fluids — flow with vorticity
Rotational & irrotational flow, 129;
Circulation $\Gamma$, 131;
Coriolis force & hurricanes, 130;

Line vortex, 129;
Vorticity and the curl of velocity $\zeta = \nabla \times \mathbf{v}$, 132;
Ring vortex, 134;
Helmholtz's vortex laws, 132;
Duality between vortex cores and streamlines, 136;
Magnus effect, 139;
Wing lift & the bound vortex, 142;
Vortex control for flight, 142;
Foreplanes & tailplanes for stability, 143;
Propeller for powered flight, 6, 127, 134;
Actuator disc, 134.

## Flow – with viscosity
Navier–Stokes equation, 134, 144;
Boundary layer, 134, 138-140;
Vortex street, 138, 209;
Whistling of telephone wires, 139;
Wing lift (bound vortex & shed vorticity), 140;
Shock waves form in supersonic flow, 143;
Swept back wings for reduced drag in supersonic flight; 143.

## Friction
Empirical laws of friction, 202;
Coefficient of static friction $\mu_\sigma$ and of kinetic friction $\mu_\kappa$, 202–203, 358;
Classroom friction experiment, 202–203;
Frictional generation of heat, 200-203;
Benjamin Thompson & the boring of cannon guns, 204;
Atomic friction, 202;
James Joule's experiments, 205–206;
Conservation of energy, 205, 208, 241;
Mechanical equivalent of heat $J = 4.12$ J/cal, 206–207;
Joule's honeymoon waterfall experiment, 207.

## Galileo
Simple pendulum, 15, 114, 262, 273;
Masses of any size in free fall undergo the same acceleration, 10, 67;
Leaning tower of Pisa, 10;
Drop shafts, 64, 235;
Equivalence between gravity and acceleration, 10;

Inclined plane, 10, 65;
Parabolic curve, 65;
Correspondence with Kepler, 13;
Convinced that planetary orbits were circular, 14;
Dispute with Church about Sun-centred Solar System, 13, 274;
Thermoscope, 182:
Telescope and moons of Jupiter, 11, 35;
Need for experiments to confirm ideas, 11;
Broadcast results through publication of books, 11.

## Gas laws
Archimedes & buoyancy, 5;
Air pressure at ground level, 182, 214;
Von Guericke's vacuum pump, 182;
Boyle's law: pV = constant, 182;
Charles' law, 182;
Balloon flights by the Montgolfier brothers and Charles, 5, 183;
Kinetic theory of gases – Maxwell & Boltzmann, 183;
Kelvin's temperature scale and absolute zero, 183;
Zero point energy ZPE, 296–297.

## Gravity
Ancient Greeks assume masses fall to centre of universe and that heavier masses fall faster, 9;
Galileo demonstrates that whatever their weight all masses fall with the same acceleration, 10;
Scientist view mass as a source of gravitational attraction, 22;
Newton's inverse square law, 26, 249;
Newton's Big G, 27, 41, 249;
Gravitational permittivity $\gamma$, 27, 41;
Newton calculates $g_0$, 22-23;
Cavendish's torsion balance, 40;
Gravitational potential $\phi$, 29, 66, 96, 252, 288;
Gravitational field $\mathbf{g} = -\nabla\phi$, 29, 96;
Analogue of Bernoulli's equation for gravity field, 123, 301;
Reduction in $g_0$ due to Earth's rotation, 273;
Weightlessness (zero gravity) in International Space Station (ISS), 67;
Artificial gravity in rotating space station (MGM film *2001: A Space Odyssey*), 67.

SUBJECT INDEX

## Gravity's dual field — gravitomagnetism
Newton's linear momentum, 27;
Leibnitz's vis viva, 28, 258;
Kelvin's kinetic energy, 258;
Newtonian mechanics without gravitomagnetism, 29, 258, 261;
Maxwell's extension of Newtonian gravity, 249- 253;
The missing curls, 252;
Gravitomagnetic field **h**, 251, 254;
Heaviside's view of existence of gravitomagnetism, 253;
Gravitomagnetic permeability η, 251, 257, 262, 343, 352, 365;
Induced gravitomagnetic field **b** = η**h**, 356;
Possible link between **h** and Planck's constant h, 265;
Equivalence between **b** and angular velocity Ω, 258, 267, 271–272, 354;
Special relativity predicts gravitomagnetism for moving mass, 259, 268, 343, 356;
Gravitational analogue of the Lorentz force, 107, 259, 272, 277, 290, 310, 312;
Inertia, 24, 64, 84, 315, 361;
A field view of inertia, 234, 258, 261, 302, 325, 351–352, 361;
Gravitomagnetic field **h** of line mass current, 255–256;
The gravitational analogue of Oersted's discovery, 251, 362;
Mass current ring & gravitomagnetic dipole, 266;
Gravitomagnetic monopoles come in pairs, 251;
Gravitational analogue of Larmor precession, 267;
Earth as a gravitomagnetic dipole, 278;
Gravitomagnetic dipoles within atom, 266.

## Gravitomagnetic field detection
Moving mass analogue of a vortex core, 250, 344;
Gravitomagnetic field due to moving mass, 170, 253, 259, 355;
Weakness of gravitomagnetism, 115;
Equivalence between induced gravitomagnetism and angular velocity, 258, 267, 271;
Foucault's detection of Earth's rotation, 271–272;
Gyro axis direction appears fixed in space, 272;
Satellite experiment (NASA Gravity Probe-B) to measure space-time curvature due to Earth and to measure the gravitomagnetic field of the rotating Earth, 266, 279, 282–284, 343, 347;
Earth-based experiment (ULTRA G-1) to detect Earth's angular velocity $Ω_E$, 277;

Earth's rotation $\Omega_E$ is slowing, 278;
Chandler wobble – nutation of Earth's axis of rotation, 278;
ULTRA G-1 not sensitive enough to detect Earth's gravitomagnetic field, 278.

## Gravity waves
Possibility of dipole-type gravity waves, 287;
Probability of quadrupole-type gravity waves, 288;
Hint of existence – decaying orbit of body orbiting a neutron star, 288;
Planar wave-front in far field, 290;
Predicted frequencies of gravity waves, 288;
Bar antenna to detect gravity waves, 290-291;
Forward's gravity wave interferometer, 291;
Glasgow laser interferometer, 292;
German – UK 600 gravity wave antenna in Hanover, 292;
US Laser Interferometer Gravitational-wave Observatory LIGO, 292;
US scientists detect gravitational waves, 287, 293.

## Gyroscope
Foucault's gimballed flywheel, 271;
Gyro axle direction fixed in space, 272;
Euler's equations of motion, 275, 309–310;
Baden-Powell's demonstration at RI, 310;
Laithwaite's demonstration at the RI, 309;
Japanese gyro weight loss experiments, 306, 320;
Sagnac effect & experiment to detect rotation, 275;
Ring Laser Gyro (RLG), 275-277;
World's biggest (ULTRA G-1), 277;
Fibre Optic Gyro (FOG), 277;
Atomic gyro, 115.

## Heat – static
Bacon's & Hooke's view of heat phenomenon as the vibration between contiguous particles of matter, 183;
Thermoscope, 182;
Thermometers and temperature scales, 182,
Temperature θ as thermal potential, 188;
Black's discovery of latent heat, 187;
Amount of heat Q in mass M, 187;

Specific heat c, 187;
Heat capacity $C_H$, 187.

## Heat — flow of caloric
Thermodynamics, 183;
Black's heat experiments, 188;
Temperature θ as thermal potential, 188;
Temperature gradient forms thermal field $T = -\nabla\theta$, 190
Analogue with gravity field, 190;
Heat sources & sinks, 188;
Inverse square law, 189;
Thermal conductivity k, 190;
Fourier's equation, 191;
Searle's bar experiment, 191;
Heat diffusion equation, 217;
Maxwell's extension of thermodynamics, 218;
Analogue with equation of telegraphy, 220;
The unknown R field, 221;
Is an acoustic compression wave a transverse wave? 220–221.

## Heat — flow of energy
Frictional generation of heat, 200–203;
Benjamin Thompson & the boring of cannon guns, 204;
James Joule's heat experiments, 205–206;
Conservation of energy, 205, 208, 241;
Mechanical equivalent of heat $J = 4.12$ J/cal, 206–207;
Joule's honeymoon waterfall experiment, 207;
Modified units for heat model, 190, 207, 219;
Modified form of Fourier's equation, 207;
Analogue of Bernoulli equation for heat flow, 207;
Phonon as analogue of photon, 222;
High-frequency thermal phonons, 222–223;
Low-frequency acoustic phonons, 222–223.

## Heat - the steam engine
Papin's steam engine, 195;
Papin's pressure cooker, 195;
Savery's first successful steam engine, 195;
Demonstration of Savery's steam engine at the Royal Society, 195;

Newcomen's first practical steam engine, 195;
Boulton–Watt's steam engine, 195;
Watt's indicator diagram, 198;
Fixed & mobile steam engines. 195-196;
The Industrial Revolution, 196;
Lunar Society, 148, 196;
Carnot's theoretical ideal steam engine, 198-199;
The Carnot cycle – indicator diagram, 199-200;
Efficiency, 200;
Effect of friction during working cycle of heat engine, 200;
Entropy S & change in entropy $\Delta$S, 201-202.

## Heat – Thermodynamics
Conservation of energy & the 1$^{st}$ law of thermodynamics, 205, 208;
The 2$^{nd}$ law of thermodynamics, 208, 209;
Kelvin-Planck version of 2$^{nd}$ law, 208;
Open & closed systems, 209-210;
Refrigeration & Time machines, 210;
Entropy & disorder, 209;
Far from equilibrium systems not subject to 2$^{nd}$ law, 209;
Heat death of universe ruled out, 209;
Query about quantum thermodynamics, 208;
Link between gravity & thermodynamics, 87, 230.

## Kepler
Mathematical assistant to Tycho Brahe, 13;
Advocate of Copernicus' Sun-centred Solar System, 13;
Laws of planetary motion, 13-14;
Planetary orbits suggest structure of space, 13;
Correspondence with Galileo, 13-15;
Book on optics,13, 20;
Book employing Archimedes' infinitesimal calculus, 20.

## Magazines
*New Scientist*, 308, 326, 327, 330;
*WIRED*, 326;
*Eureka*, 320.

## Magnetism
Feng shui, 146;
Magnetite stones from Magnesia, 146;
Magnetic compass for navigation, 145, 270;
Magnetic compass for detecting magnetic fields, 157, 165-166;
Gilbert's book *De Magnete*, 146;
Earth as a magnetic dipole, 147, 270;
Aurora borealis, 270;
Magnetic monopoles occur in pairs, 149, 172, 251–252;
Like magnetic monopoles repel, 149;
Inverse square law for magnetic monopoles, 149;
Magnetic permeability m, 149, 298, 305, 346, 355;
Magnetic field of coil, or ring current, 157, 346;
Virtual magnetic poles of coil, 266;
Magnetic energy of coil, 168.

## Mass
Source of gravity, 113, 124;
Weight in gravity field, 10-11, 39, 42, 63, 182, 234;
Weightlessness, 66-67, 89, 308, 351;
Gravitational & inertial mass, 24, 89, 355;
Possibility of negative mass, 124, 250, 287, 313;
Interactions between masses, 313;
Mass dipole, 314-316, 260.

## Maxwell
Equations for electromagnetism, 77, 114, 174, 360;
Light is an electromagnetic wave, 175;
Predicts the speed of light c, 175;
Master of analogues, 116, 218;
Extension of Newton's gravity as an analogue of electromagnetism, 107, 251, 323, 361;
Kinetic theory of gases, 183;
By analogy with electromagnetism extends thermodynamics, 218;
Heat diffusion equation and the equation of telegraphy, 175, 220.

## Newton
At Trinity College Cambridge, 19, 27;
Interest in work by John Wallis, 19-20;

Extended Binomial series, 19-20;
Also read books by Galileo, Kepler and Descartes, 20-21;
At home to escape from the plague, 21;
Experiment with glass prism, 21;
Built compact reflecting telescope, 21;
Became Fellow of the Royal Society (1672), 21;
Fluxions, calculus & Leibnitz, 21, 27, 29;
Correspondence with Huygens & Hooke, 21-22;
Wrote *Principia*, 23, 65, 214, 249, 355;
Dispute with Hooke, 26;
Laws of motion, 23-24;
Centre of gravity, 22;
Inverse square law of gravity, 22-23, 26, 39–40, 65, 67, 249, 288, 298, 302, 313, 352;
Calculated $g_0$, 22–23;
Earth's tides, 27;
Measured the speed of sound, 24, 114, 214;
View of ether, 25, 96;
Dismissed Descartes' ether vortices, 20, 25, 46, 255;
President of Royal Society (1703), 147, 195.

## Project Greenglow & related studies

Professor Allen, auto-gravity & *The Future of Aeronautics* (1970), xi, 322–323;
Start of BAe Warton Gravity study (1986), 319;
BAe Gravity Report sent to NASA (1990), 319;
BAe Advanced Projects study of gravity propelled vehicle (1991), 319;
Professor Young's gravity lectures (1991), 319;
Interstellar Propulsion Society (IPS) founded (1995), 320;
St Petersburg Conference (1996), 320;
Start of NASA Breakthrough in Propulsion Physics (BPP) Program (1996), 320–321;
Professor Allen's 'What if?' Anti-gravity drive design study for BAe (1997), 323;
Professor Allen's paper in *Progress in Aerospace Sciences*, on gravity propulsion (2003), 323;
Start of BAe Project Greenglow (1998), 323;
Gravity studies at Lancaster University, 323;
Microwave thrust study at Dundee University, 326;

Study of Podkletnov's gravity effect at Sheffield University, 326–327;
Goals & metric study for Project Greenglow at Glasgow University, 328, 331;
Study of the Casimir force at Birmingham University, 331;
Support from Rolls-Royce, 324, 328;
Greenglow lectures, 324–325;
BAE Systems meeting with NASA BPP (2000), 330-331;
Closure of NASA BPP Program (2002), 331;
BAE Systems Project Greenglow ended (2005), 332;
Professor Allen's Moon Club, xi;
Tau Zero Foundation (2008), 332;
US DARPA 100-Year Starship Program, 332;
EU Breakthrough in Propulsion Physics Centre – Dresden University, 332.

## Quantum Mechanics
Black body radiation experiments, 184–185;
Planck's idea of discrete radiated amounts of energy, 185;
Planck's constant h, 185-186, 222, 265, 279, 361;
Energy quantum $E = hf$, 56, 64, 85;
Quantum renamed the photon, 80, 185;
Photon as a virtual particle transmitting electromagnetic radiation, 289;
Einstein's explanation of photo-electric effect, 79–80, 185–186;
Bohr's model of atom, 186, 265;
Acoustic & thermal analogue – the phonon, 222–223;
Gravitational analogue – the graviton, 290, 296;
Quantum gravity via gravitomagnetism, 230, 264, 341, 361.

## Rotational motion
Centrifugal force, 14–16, 22, 41, 67, 259, 273, 299;
Centripetal acceleration, 15, 22, 259–260, 312;
Angular momentum, 15, 56, 185-186, 222, 254, 259, 265–266, 274, 289, 296, 309;
Conservation of angular momentum, 346;
Rotating galaxies & dark matter, 299–300.

## Royal Institution (RI)
Formation (1800), 167, 204;
Experimental demonstrations, 204.

## Royal Society
The Invisible College (the fore-runner), 18;
Formation as a gentlemen's club (1660), 18;
Latin motto, 18;
Influence of Galileo & Bacon, 18;
Experimental demonstrations, 18;
Royal patronage (1662), 18;
Hooke curator of experiments, 182;
Newton becomes President (1703), 147, 195.

## Savile Club
Presentations on science topics, ix.

## Scientific discoveries
Chance, luck & serendipity, 114-118, 156, 157, 167, 304, 307–308, 351, 356, 362;
Knowing too much, 113;
Investigating experimental quirks, glitches & anomalies, 157, 311, 307;
Perseverance, 111;
Bacon on Nature's forms, 115-116;
Maxwell, the master of analogues, 116, 218;
Analogues, read-across and missing patterns, 115, 117, 129, 134, 136, 162, 192, 258, 359;
Synchronicity, 102, 117, 167.

## SI Units
Absolute temperature $\theta$ – Kelvin K, 25, 183, 229;
Capacitance C – Farad F, 151, 176;
Conductance (reciprocal of resistance) – Siemen S, 158;
Electric charge Q – Coulomb C, 148;
Electric potential difference $\Delta\phi$ – Volt V, 151, 156, 159, 191;
Electrical resistance R – Ohm $\Omega$, 158;
Electric current i – Amp A;
Energy E – Joule J, 56;
Force F – Newton N;
Inductance L – Henry H, 168;
Length $\ell$ – metre m;
Magnetic flux – Tesla T;
Magnetic monopole m – Weber W;

Mass M – Kilogram kg;
Mole M – Amount of matter, 81;
Power P – Watt W;
Thermal conductivity k – Cal/s.m.K ( Modified: Joules/s.m.K), 188, (207);
Time t, T – second s, 10, 48, 54, 82.

## Smithsonian
Foundation (1846), 167.

## Space-time
Cartesian 3-D space geometry & algebra, 45, 71, 100, 174;
Poincaré's 4-D space-time, 72–74;
Minkowski's 4-D space-time, 74–75;
Lorentz transforms in matrix form, 73–75;
An event and its light-cone, 75–76, 90, 104;
Simultaneity, 54, 74, 76, 82.

## Speculative means of propulsion
Electro-gravitics & the Biefeld–Brown effect, 298, 304-305;
TDT Lifter, 304;
Kidd's ITM, 311–312, 335;
Shawyer's EM Drive, 316, 340;
Acoustic analogue of the EM Drive, 359;
Mass dipole thrust, 260, 314–316;
Effective mass optical dipole thrust, 317;
Space-time warp, 298, 339.

## Speculative ideas related to gravity
Brush's kinetic theory of gravity & experimental results, 301–303, 355;
Professor Jennison's idea to investigate link between e and Big G, 299;
Can a change of γ (or Big G) explain dark matter?, 299–301, 305;
What is effect of twinning between **E** & **g** and between **B** & **b**?, 259, 350.

## Speculative experiments to detect gravitomagnetism
Use mu-metal magnetic screen to separate **H** & **h**, 347;
Modification of Faraday's 1$^{st}$ gravity experiment using fibre optic coil, 353;
Repeat of Faraday's two-part 3$^{rd}$ experiment in open air, 354;

Variation of Newton's pendulum experiment, 355–356;
Friction experiment using linearly polarised light, 358–359

## Speed of light c
Galileo's failed Earth-based attempt to measure c (1610), 35;
Romer's astronomical method to measure c (1676), 35;
Maxwell speculates that light is an electromagnetic wave and calculates c (1864), 46, 175;
Fizeau's Earth-based method to measure c (1849), 36–37;
Foucault's short- baseline method to measure c (1862), 37;
Michelson's version of Foucault's experiment (1924), 37.

## Speed of sound $c_s$
Newton's measurement in air, 24, 214;
In solid, liquid and gas, 214.

## Stars
Star formation, 225;
White dwarf, 226;
Neutron star, 226, 288;
Dark star or Black hole, 227–230, 293;
Supernova, 226, 288;
Binary stars, 288.

## Tensors
Scalar, 71, 99;
Vector, 26, 71, 99, 122;
Matrix, 72, 99.

## Vector Operators
Gradient operator $\nabla$ , 28-29, 84, 131, 149, 190–192, 207, 217, 297;
Divergence operator $\nabla \bullet$ , 123, 174, 217–219;
Curl operator $\nabla \times$ , 132, 157, 174, 252.

## Waves
Longitudinal waves, 212;
Transverse waves, 177, 214.

For exclusive discounts on Matador titles,
sign up to our occasional newsletter at
troubador.co.uk/bookshop